# MARX'S RADICAL CRITIQUE
# OF CAPITALIST SOCIETY

# Marx's Radical Critique of Capitalist Society

## A RECONSTRUCTION AND CRITICAL EVALUATION

N. SCOTT ARNOLD

OXFORD UNIVERSITY PRESS
New York   Oxford

Oxford University Press

Oxford   New York   Toronto
Delhi   Bombay   Calcutta   Madras   Karachi
Petaling Jaya   Singapore   Hong Kong   Tokyo
Nairobi   Dar es Salaam   Cape Town
Melbourne   Auckland

and associated companies in
Berlin   Ibadan

Library of Congress Cataloging-in-Publication Data
Arnold, N. Scott.
Marx's radical critique of capitalist society / N. Scott Arnold.
p. cm. Bibliography: p. Includes index.
ISBN 0-19-505879-8
1. Marx, Karl, 1818–1883. Kapital.   2. Capitalism.
3. Communism.   I. Title.
HB501.M5A76   1990
335.4'1—dc20   89-34746
ISBN 0-19-505879-8; ISBN 0-19-507264-2 (pbk.)

2 4 6 8 9 7 5 3 1

Printed in the United States of America
on acid free paper

*For Theresa*

# PREFACE

It has been widely remarked that Marx's theoretical writings pay less attention to ethical or normative issues than one might initially suppose. Those writings focus more on questions of economic, historical, and sociological theory. Nevertheless, it is obvious that Marx found many features of capitalist society deeply objectionable. What connects the negative assessments of capitalist society and the theoretical writings is the fact that Marx believed that the objectionable features of capitalism are rooted in the workings of its basic institutions, notably its economic system. In other words, Marx was a radical critic of capitalism. Radical criticism goes to the root of things, in this case, the major institutions of capitalist society. The primary theoretical task for Marx was to understand these institutions, and part of that understanding consists in explanations of how or why these basic institutions are responsible for social ills such as exploitation and alienation. Marx's perspective implies that these defects of capitalist society are not surface phenomena whose elimination requires only the correction of certain "abuses." It is the system that is "abusive," according to the radical Marx.

The main purpose of this study can be stated quite simply: It is to reconstruct and critically evaluate Marx's radical critique of capitalist society. The opening section of the first chapter investigates in more detail the requirements for a successful radical critique, whether Marx's or someone else's. Feminists, libertarians, and other malcontents, as they might be called, should find this discussion of some independent interest. One of these requirements is suggested above, namely, that systemic social ills must be identified and their existence explained in terms of the workings of society's basic institutions. For Marx, this requires explanations of the phenomena of exploitation and alienation in terms of the workings of the economic system. Chapters 2 through 5 address this requirement for a successful radical critique. (Chapter 3 consists of a foray into the history and philosophy of economics. The main issues are both technical and historical, and those unfamiliar with some of the debates in classical economics may find much of this hard to follow or of lesser interest. This chapter can be skipped; even though arguments in Chapter 4 depend on the main conclusions of Chapter 3, those conclusions are indicated in Chapter 4.)

Because the radical critic believes that society's problems are rooted in the workings of its basic institutions, he believes that these problems can be significantly addressed only by replacing existing institutions, that is, by revolutionary transformation of the latter. I argue in detail in the first chapter that this implies that a successful radical critique requires an account of the alternative institutions that the radical critic believes will replace or should replace those of the existing order; it also requires an argument to the effect that these institutions will not be responsible for the same systemic social ills found in existing society.

In Chapters 6 and 7 I give Marx the best case I can on these questions. The constructive element in my reconstruction of Marx's radical critique is most pronounced in these two chapters. Marx (and Engels) did not extensively discuss what would happen "after the revolution," though they had more to say than most people think. Fortunately, the materialist theory of history provides a useful guide for reconstructing an account of post-capitalist society. However one understands Marx's theory of history (and there has been considerable controversy on this score), this much is clear: If one knows how a society organizes production, that is, what its relations of production are, it is possible to infer many other things about that society. After all, this is the basis for Marx's explanations of various features of both pre-capitalist and capitalist societies. Specifically, what I do in these chapters is to assume post-capitalist relations of production (workers control of the means of production and the abolition of wage labor) as given and then ask what else can be inferred about post-capitalist society, given these assumptions.

Construed in this way, the theory of history provides some of the materials out of which to reconstruct Marx's account of post-capitalist society. Other materials come from what he says explicitly in various writings, notably in the *Critique of the Gotha Program*. However, the account that emerges from all this is at best incomplete vis-à-vis the demands for a successful radical critique identified in Chapter 1. This should come as no surprise to those familiar with Marx's writings, though I hope to provide a more detailed and systematic identification of the gaps in his account of post-capitalist society than has been done heretofore.

It would seem that this identification would provide a set of directives for Marxists who want to do the socially necessary labor required to complete the radical Marxist critique of capitalist society. However, before these Marxists grab their means of production and set to work, they would be advised to read Chapters 8 and 9. In these chapters I argue that Marx's radical critique of capitalist society is fundamentally and fatally flawed because his vision of post-capitalist society cannot be realized. In other words, his alternative to capitalism is historically impossible. It has often been said that Marxism, or communism, may work just fine in theory, but that it won't work in practice. My contention in these chapters is that it won't work in theory either.

In preparing to write this book, I read extensively in the secondary literature on Marx. I was disappointed to find (or, to be more honest, elated to find) that almost no one talked about Marx's views on post-capitalist society. One major exception was Allen Buchanan. His book, *Marx and Justice*, gave me the idea that the radical character of Marx's critique of capitalism made him highly vulnerable to a certain kind of burden of proof objection. More exactly, Buchanan suggested that Marx's indictment of capitalist society presupposes the possibility of a society not beset by capitalism's systemic social problems. Moreover, Marx seems to have simply assumed without argument that (his vision of) post-capitalist society would not be beset by these ills.

The question that occurred to me was, 'Given that Marx has failed to meet a crucial burden of proof, can anything more be said?'. Specifically, can it be shown that his conception of post-capitalist society cannot be realized and what follows if that is true? In attempting to answer these questions, I have been led to a set of questions and problems that I believe to be of more general philosophical interest.

How do we talk in an intelligent and informed way about societies that do not exist in reality? How could anyone know what life would be like in a society that does not as yet exist? On the face of it, this would appear to be very difficult. Indeed, most Marxists have demonstrated their modesty on these questions—perhaps unintentionally—by keeping silent about the details of post-capitalist society. The key to answering these questions is to be found in Marx's institutional perspective on existing societies. Both his theory of history and his economic theory conceptualize capitalism in terms of its institutions; Marx's theoretical writings attempt to describe and explain how those institutions function at a certain level of abstraction from their historical embodiments (though empirical evidence is brought in for illustrative and confirmatory purposes throughout his writings). Similarly, we can discuss societies that do not exist in reality by discussing their institutions at a level of abstraction that transcends whatever historical embodiment might come to exist. Though it is unreasonable to demand a detailed account of what daily life would be like in post-capitalist society, it is perfectly reasonable to demand an abstract description of that society's main social institutions—especially its economic system.

But what is the "right" level of abstraction to think about these institutions? Answering this analytical question is a necessary precondition for answering one of the central questions of social and political philosophy, namely, 'What are the institutions of the good society?'. I suggest an answer to the analytical question in Chapters 6 and 7, and implicit in Chapters 8, 9, and 10 are some suggestions for a framework for thinking about the central substantive question. I also have something to say in Chapter 10 about the relation between social ideals and social realities. In sum, I hope that this book raises some questions and addresses some issues that go beyond Marx and his radical critique of capitalist society.

Parts of Chapter 4 first appeared in "Capitalists and the Ethics of Contribution," *Canadian Journal of Philosophy* 15 (March 1985). Some of the material in Chapters 6 and 9 can be found in "Marx, Central Planning and Utopian Socialism," *Social Philosophy and Policy* 6 (Spring 1989), and parts of the first chapter were published as "Radical Social Criticism," in *Reason Papers* no. 4 (Spring 1989).

In addition to Allen Buchanan's book, I have benefited considerably from the writings of both G. A. Cohen and Peter Rutland. It was Cohen's *Marx's Theory of History* that first convinced me it was possible for an analytic philosopher to make a contribution to discussions of Marx. I have profited enormously from Cohen's clear and rigorous thinking about Marx, even when I have disagreed. Peter Rutland's *The Myth of the Plan* is an excellent discussion of central planning and the Soviet experience. It contains about the right mix of theory and hard empirical data. If economists could bring theory and data together in the way that Rutland does, economics would have much more to tell us about the real world than it currently does.

There are many people I have to thank for help on this project. First and foremost is Jeff Paul of the Social Philosophy and Policy Center at Bowling Green State University in Bowling Green, Ohio. It was on Jeff's initiative that I was brought to the Social Philosophy and Policy Center as a Visiting Scholar for the summers of 1987 and 1988 during which most of this book was written. I am grateful for the financial support provided by the individuals and foundations who support the work of the Policy Center. Without the time off from other responsibilities that my position there made possible, this book simply could not have been written. Moreover, the environment provided by the Center was ideal for the sustained writing I was able to accomplish during my two summers there. Thanks also are due to Fred Miller, the Executive Director, and Ellen Paul, the Deputy Director, as well as to the staff of the Policy Center for their many kindnesses.

Many people read and commented on drafts of various parts of the manuscript and/or preparatory material. They are: John Ahrens, Peter Cloyes, Allen Buchanan, John Gray, Harold Kincaid, Ellen Paul, James Rachels, and Lynn Stephens. My thanks to them for their efforts. Portions of Chapters 5 and 9 were discussed with the Austrian Economics Colloquium at New York University. Mario Rizzo and Israel Kirzner had penetrating questions and comments on that material. Thanks also to Cynthia Read, my editor at Oxford University Press, for her support and enthusiasm for the project. Responsibility for all errors remains mine.

The members of the Philosophy Department of the University of Alabama at Birmingham, especially its chair, George Graham, provided a congenial and supportive environment in which much of the preparatory work for this book was done. Special thanks go to Harold Kincaid, who has taught me most of what I know about the philosophy of the social sciences and who patiently read more of my writings on Marx and related topics over the past six years than I had a right to ask of him. Lynn Stephens's good humor and tolerance for some of my truly malevolent feelings toward Marx have enabled me to keep those sentiments out of my writing; his general philosophical wisdom is responsible for whatever skepticism I have managed to retain about my own views. Finally, this book owes more than I can say to my wife Theresa to whom it is dedicated.

*Birmingham, Alabama*                                                    N. S. A.
*February, 1989*

# CONTENTS

# A NOTE ON SOURCES

To keep down the number of endnotes, I have, where it is feasible, included references to the works of Marx and Engels in the body of the text. Citations will be given not only for direct quotations but also on occasion for paraphrases that the reader might want to follow up on. These citations will be given in parentheses immediately following these claims about what Marx said. When abbreviated, these references will be given according to the following schedule:

|  |  |
|---|---|
| *A-D* | *Anti-Dühring* |
| *EPM* | *Economic and Philosophic Manuscripts of 1844* |
| *CCPE* | *A Contribution to the Critique of Political Economy* |
| *CGP* | *Critique of the Gotha Program* |
| *CM* | *The Communist Manifesto* |
| *GI* | *The German Ideology* |
| *PP* | *The Poverty of Philosophy* |
| *MECW* | *Collected Works* of Marx and Engels |

Short titles will be used for infrequently cited works (e.g., *Eighteenth Brumaire*); full bibliographical information can be found in the bibliography. When reference is made to a work that appears in the *Collected Works* of Marx and Engels now being jointly published by Lawrence and Wishart, International Publishers, and Progress Publishers, this will be designated in the citation according to the above schedule, followed by the abbreviation 'MECW', the volume number, and the page number. For example, a quotation from *The German Ideology* would be cited as follows:

(*GI*, MECW, vol. 5, p. 47)

At present, most of the volumes of the *Collected Works* have been published with the major exception of *Capital* and all of the preparatory material for it, including *CCPE*.

This system is, I believe, the best way to direct the reader to the primary sources without unduly interrupting the discussion.

# MARX'S RADICAL CRITIQUE
# OF CAPITALIST SOCIETY

# 1

## The Very Idea
## of a Radical Critique

Nearly any reflective person has grounds for dissatisfaction with the social system in which he finds himself. Most of us are social critics of one sort or another, though some are more severe than others. A rough distinction can be drawn between the moderate or reformist critic and the radical critic: The former believes that the system is fundamentally sound, and/or that his society is basically a good society. Any society falls short of its ideals, and given that we are all sinners, it is not surprising that things don't go as well as they might. This critic believes that existing institutions can and should be modified or augmented in various ways to permit or encourage society to approach more closely the appropriate ideals. The fact that most reflective people are at least moderate critics is not surprising. They usually have enough imagination to conceive of ways in which society might be better. Few such people believe that, at the level of social institutions, this is the best of all possible worlds.

By contrast, radical critics believe that existing social institutions are fundamentally unjust or immoral. It is a philosophically interesting question to ask what sort of challenge the radical critic offers. To get a handle on this, there are some suggestive parallels in epistemology that are worth pursuing.

Most epistemologists believe that they and others really do know something about the world. One of the most fundamental questions in epistemology is whether or not this is true. Because this question is so fundamental and because (good) philosophers like a good fight, the skeptical challenge to all or most of our knowledge claims is sometimes regarded as the main problem in epistemology. Skeptical arguments, such as those found in Descartes's first two meditations, seek to call into question whole categories of belief. Comprehensive skeptical arguments are supposed to show that most of the things we think we know are not really known at all. All belief is mere opinion.

The radical social critic aims at a parallel result. He believes that, contrary

3

to popular opinion, the basic social institutions are unjust or immoral. Just as the skeptic challenges the ordinary knowledge claims that we make, the radical social critic challenges widely accepted pre-theoretical judgments about the justice or goodness of our basic social institutions.

Marx often conceived of himself primarily as a scientist whose chief aims were to explain the workings of the capitalist economy (his economic theory) and to explain the dynamics of class societies generally (his theory of history). However, it is reasonably clear that he also believed that capitalism was a bad thing in some sense, however historically necessary it might be. Furthermore, he obviously believed that the problems with capitalist society are deeply rooted in its main institutions. Since a radical critique is aimed at a society's main institutions, Marx can be conceived of as offering some sort of radical critique of capitalist society. The main purpose of this book is to explicate and evaluate this radical critique. My aim in this first chapter is twofold: First, I shall investigate some of the presuppositions of, or necessary conditions for, a successful radical critique of a society. That is, what must a radical critic do if he is to succeed? Second, I shall address some broadly based challenges to the project of explicating or reconstructing Marx's radical critique of capitalism. For example, someone might argue that the conception of radical critique developed here is inappropriate for Marx or inconsistent with his theory of history or his conception of morality. Addressing these challenges will permit a deeper understanding of the requirements of a radical critique and its relation to Marx's thought.

## The Very Idea of a Radical Critique

In epistemology, the skeptic's adversaries have often argued that the skeptic has set impossibly or unreasonably high standards for what counts as knowledge. Consequently, even if his arguments succeed, they only show that knowledge is unachievable in some nonstandard sense of 'knowledge'. Whether or not this objection is well taken, it points to an absolutely central question in the dispute between the skeptic and his opponents: 'What must the skeptic show in order for his position to be sustained?'. An answer to this question will in part define skepticism itself. It also makes clear that the skeptic bears a burden of proof. He cannot simply assert that everything we believe about the world might be false or not known to be true; arguments have to be produced to show that genuine knowledge is unachievable. A parallel question arises in the dispute between the radical social critic and his more moderate adversaries: 'What must the radical critic show for his critique to be successful?'. Put another way, 'What are the presuppositions of a (successful) radical critique of a society?'. The radical critic, like the skeptic, bears a burden of proof. What are those burdens?

The following are necessary conditions that a successful radical critique of a society must satisfy. All of them have a certain amount of intuitive appeal, but each will require some discussion and argumentation.

> i. A radical critique must identify widespread and pervasive social ills or injustices characteristic of the society in question (e.g., alienation and exploitation). It must be shown that these ills or injustices are both pervasive and rooted in the basic socioeconomic structure of the society. Failure to do this would leave the critic open to the moderate reformer's contention that these problems can be significantly ameliorated without changing the basic structure of the society. Forestalling this challenge requires the radical critic to have in hand a fairly substantial theory that identifies the basic structure of society and that explains how the relevant social problems arise from this structure. For example, as I shall argue in subsequent chapters, Marx maintains that alienation is rooted in the commodity nature of production under capitalism, and that exploitation is rooted in capitalist relations of production.

Let us call this the Critical Explanations requirement.

> ii. The radical critic needs a normative theory to explain, or an argument to justify, the negative judgments referred to in (i). For Marx, this requires answers to such questions as, 'What is wrong with exploitation?' and 'Why is alienation a bad thing?'. A full-scale ethical theory would be sufficient to meet this condition, but it is unclear that it is necessary as well. It may be that only part of a theory is needed to substantiate the relevant claims. (More on this shortly.)

Let us call this the Normative Theory requirement.

> iii. The radical critic needs to specify a set of alternative social institutions that he believes should and/or will replace the existing ones. This specification must meet the following conditions:
>
>> a. These institutions meet the conditions for a good or just society insofar as the latter are specified by the theory or partial theory called for in (ii). Or, more weakly, it must be shown that these alternative institutions at least do not have the problems that face existing institutions identified in (i).
>> b. A plausible description/explanation of how the institutions will function can be given.
>> c. These institutions can persist as stable social forms. Or, more weakly, there is some reason to believe that they are stable.

Let us call this the Alternative Institutions requirement.

iv. A plausible story can be told about how existing institutions can be or will be destroyed or set on a course of fundamental change. That is, the so-called "problem of the transition," as Marxists call it, must be shown to be solvable, at least in principle.

Let us call this the Transition requirement.

These conditions are motivated by the ultimate purposes of a radical critique: To know the truth about the defects of the existing order and to lay the intellectual foundations for radical social change. Meeting the first requirement constitutes a successful diagnosis of the social or moral illnesses afflicting the existing society. It is the etiology of the society's problems. Although it is possible to change the existing order for the better on the basis of a misunderstanding of the nature of systematic social problems, one is less likely to be successful than if one has correctly diagnosed those problems.

Are there any restrictions on the kind of explanations that a radical critic must offer? As noted above, they must appeal to the basic structure of the society, but there may be other requirements as well. In particular, some theorists have argued that explanations in the social sciences require elaborations at the level of individual actions if they are to be genuine explanations.[1] That is, the critical explanations in (i) must be given microfoundations. If microfoundations are required, then the radical critic must explain how the fundamental social institutions negatively affect individuals. In other words, how do these institutions shape and constrain the actions of individuals to produce the social evils to which the critic calls attention?

Whether or not genuine explanations in the social sciences require microfoundations, the provision of microfoundations for critical explanations is at least desirable. By doing this, the radical critic accomplishes two things: (1) He identifies more clearly objectionable patterns of behavior and their consequences; and (2) He identifies the institutional incentives and constraints that tend to produce the systematic injustices or social ills. This gives the radical critique a certain definiteness it might otherwise lack; it also provides implicit negative advice about how to structure the new institutions that are intended to be the result of radical social change.

The second condition (the Normative Theory requirement) is the most distinctively philosophical one. A successful theoretical inquiry enables the radical critic to substantiate claims of injustice against the existing order and/ or to identify and justify the ultimate (positive) values that have been systematically frustrated. Secondly, some sort of a theory can serve as a guide in evaluating the proposed alternative institutions that the critic hopes will constitute the new order. Without some theoretical support, the claim that the desired changes are in fact desirable remains unsubstantiated. From a philosophical point of view, these two purposes are the most important ones served by the Normative Theory requirement. Finally, such a theory may tell us something about what actions ought (and, perhaps more importantly, ought not) to be taken to bring about and sustain radical social change.

A normative theory need not be an ethical theory. The former is broader than the latter. This point is of some importance in light of some interpretations of Marx's general position on ethics or morality. What I mean by 'normative theory' is, roughly, any systematic attempt to identify fundamental values, behavioral dispositions, ("virtues") and/or action-guiding principles. How to distinguish moral from nonmoral values, virtues, and imperatives is a controversial question.[2] However, it is clear that social institutions and individual actions can be evaluated along a number of different dimensions, and some of these evaluations may issue in imperatives that agents believe are in conflict with, and even override, the demands of morality.

Two examples illustrate. A statesman may embark on a certain course of action that he believes is in the national interest, and yet he may also believe that this course of action is immoral. (Only the Americans believe in a pre-established harmony between the national interest and the demands of morality.) He decides to subordinate the demands of morality to those of the national interest—and not from weakness of the will. To take another example, an Old Bolshevik may believe that cooperation in his own show trial is immoral, that is, he has a moral obligation not to participate, and yet he might think that, all things considered, he ought to cooperate because the ultimate success of the revolution requires it. Bukharin and Radek may well have come to believe this during their famous Moscow show trials in the Thirties.

Kant would maintain that these attitudes are inconsistent: The demands of morality necessarily override all other considerations; moral theory issues in categorical imperatives. For him, it is inconsistent to assert that, all things considered, an agent ought to do *x*, and yet doing *x* is morally wrong. Kant might be correct. However, it is possible to see how someone might believe that there is no logical inconsistency in this connection. These observations are important in light of Marx's attack of morality in general, or at least some conceptions of it. The general point is that a radical critic need not have an ethical theory to meet this second condition.

The third condition is the least obvious of all these conditions. Why must a radical critic have an alternative in view to criticize successfully the existing order? Throughout history, successful (as well as unsuccessful) revolutionaries have usually had only the haziest idea, if any at all, about the institutions that ought to replace the ones they are intent on tearing down.

The rationale for this condition can be appreciated by pursuing a useful parallel in epistemological skepticism. Among the ancients, the two most prominent forms of skepticism were the Pyrrhonian and the Academic. At least since Hume, we have been led to believe that the Pyrrhonians advocated the suspension of all beliefs about the world; an apocryphal story to the effect that Pyrrho, lost in skeptical doubt, had to be pulled from the path of a chariot by a student has become part of philosophy's folklore. In fact, however, Pyrrho only maintained that we really do not *know* anything; his advice was not that we abandon all belief, except insofar as belief is infixed with certainty. Instead, Pyrrho advised that, once we see that genuine knowledge is impossible, we can achieve a kind of detachment from the world (*ataraxia*) and let our lives be guided by appearances only. His doxastic advice was not to withhold belief but rather to hold all beliefs lightly. The Academics, on the other hand, while denying that certainty, and hence true knowledge, was possible nonetheless maintained that some beliefs are more reasonable to hold than others. Different beliefs have different measures of epistemic warrant attached to them, and as Hume later put it, "A wise man proportions his belief to the evidence." The upshot is that both the Academics and the Pyrrhonians had "doxastic advice" to offer on the basis of their skeptical arguments, though in the final analysis, the Pyrrhonians were more tolerant of our inclinations than were the Academics.

Historically, radical social critics have not been shy about giving advice about how the future ought to be subsequent to the demise of the existing order. In particular, it has usually not been hard to find some suggestions from them about how basic social institutions should be reshaped. However, that does not prove that a radical critique *requires* a sketch of an alternative. Indeed, there are two kinds of objections that might be raised against condition (iii). First, it might be objected that a radical critic need provide no sketch of alternative social institutions. The second objection questions the elaborate requirements (iii) spells out. Let us consider these objections together.

It is hard even to conceive of a radical social critic who offers no advice at all beyond how to get rid of the old order. However, it seems at least possible that advice might not include proposals for new institutions. Getting rid of the old order for some individuals might mean just pulling out. There is a long tradition, in both the East and the West, of withdrawal from the world in the face of human and natural evil. This withdrawal may be solitary or in artificially small groups (e.g., monasteries). It is perhaps significant that these "rejectionists" are nearly always religiously motivated. Furthermore, they usually locate the problem in human nature or at least the human condition. The latter explains defects in social institutions. By contrast, other critics, notably Marx, locate the social evils in the institutions, and deformities in human nature are attributed to the latter. It is doubtful, however, that these rejectionists ought to be called 'radical social critics' instead of 'misanthropes' or perhaps 'whiners'. (Whiners are people who merely complain about undesirable yet ineradicable features of the human condition, such as having to mow the lawn.)

The idea of a radical critique, as it has been articulated above, is really quite recent in historical terms. It is predicated on a scientific understanding (or a pretension to scientific understanding) of the social world. Its foundations are to be located in the first constructions of the "human sciences" in the Enlightenment. It is hard to say who first conceived of applying the methods of the new science to the study of politics and society, though Hobbes and Hume come readily to mind. In any case, by the nineteenth century, the prediction and control that the scientific method promised had clearly captured the imagination of most social thinkers, including, notably, Marx.

Whatever its historical origins, it is clear that a successful radical critique requires a sketch of alternative institutional arrangements in light of the position subscribed to by one of the radical's opponents: the conservative moderate social critic. It is open to the moderate critic to hold that the social ills identified by his radical counterpart are, in one way or another, part of the human condition (or perhaps post-feudal society). At most, they can be ameliorated, but their elimination is a purely utopian ideal that cannot be realized or at least cannot be realized in an industrial society. In addition, conservatives are inclined to argue that serious and systematic attempts to wipe out these social evils are likely to make matters worse. None of this may be true, but the radical critic has to show this—and the only way to do this is to address the Alternative Institutions requirement.

Furthermore, radical social criticism is intended to have action-guiding significance on a society-wide scale. Whether the radical critic favors quick revolutionary destruction of the existing order or the gradual metamorphosis of the offending institutions, rationality requires that he have some idea of where he is going. Given that radical criticism is directed at the basic social institutions of the society, this guiding vision has to be articulated at the level of social institutions. In addition, if social change unleashes dystopian forces, not only will he have failed to achieve his purpose, the results will provide some evidence for the conservative view that significant social change is nearly always a change for the worse.

. These considerations also support the detailed requirements spelled out in (iii) (a) through (c). That the alternative institutions must at least not face the same problems that face existing institutions is obvious. Regarding (b) and (c), if the radical critic has no idea of how alternative institutions might function or if he has no good reason to believe that they can persist as stable social forms, then, for all he knows, conservatives might be right in their pessimistic assessment of the prospects for social change that is both fundamental and beneficial. What is not yet clear is exactly what (iii) (b) and (c) require. Part of the difficulty is that it is unclear how detailed and concrete the description of the alternative institutions must be and how elaborate the function and stability explanations must be. These problems are perhaps best approached at a lower level of abstraction, that is, in a discussion of a particular radical critique. My aim here has only been to argue that this condition, as specified by (a) through (c), must somehow be addressed.

Condition (iv) requires that a radical critic provide a plausible story describing or explaining how society can or will be fundamentally altered. All social systems that endure have mechanisms that tend to preserve their basic institutions. It seems at least possible that these mechanisms are powerful enough to prevent radical social change indefinitely far into the future. A radical critique presupposes that this is not the case. Looked at from another perspective, if the destruction of the existing order or the inauguration of the new society presupposes processes that are unlikely to occur, given existing and foreseeable conditions, the radical vision of what society could and should be like can be justly labeled "utopian"; it has lost its significance for radical social change, and the radical critique must be judged a failure.

On the other hand, if the radical critic can successfully argue that the existing order contains the seeds of its own destruction and that the conditions for the germination of those seeds are, or are about to be, realized, then he has satisfied this requirement. Another way to satisfy this requirement is to show that exogenous shocks (e.g., war) have so weakened key institutions that their destruction is eminently feasible. As was the case with condition (iii), there is a problem in specifying the appropriate level of abstraction at which this tale must be told; surely a detailed blueprint for revolutionary action and inauguration of the new order is not required, but the radical critic must have some idea of how the transition can take place.

It is worth noting that conditions (iii) and (iv) are logically independent. A radical critique may contain a specification of a set of workable alternative institutions, but if the existing order will persist indefinitely, the "problem of the transition" has no solution. Similarly, the radical critic may be able to sketch a plausible story about the destruction of the existing order (perhaps a story in which he is the main character!), but he may have no definite idea about what institutions ought to replace existing ones. In either case, his radical critique will be unsuccessful. Of course, revolutionary *action*, on his part or on the part of others, may be quite successful in destroying the existing order, but, the unity of theory and practice notwithstanding, a radical critique, as it is understood here, is essentially a cognitive enterprise; it consists of a set of propositions, not a series of actions. For example, Marx's radical critique of capitalist society might be successful, even if his own direct revolutionary activity was not.

Conditions (i) through (iv) are necessary conditions for, or presuppositions of, a successful radical critique of a society. What is not yet clear is whether Marx can be said to offer a radical critique of capitalist society as I have specified it. To get this project off the ground, I shall begin by considering a number of systematic challenges to the possibility of a radical Marxian critique of capitalist society. The detailed argument in favor of construing Marx as offering a radical critique of capitalist society is constituted by Part I of this book where I reconstruct that critique according to the guidelines outlined above.

## Three Challenges to the Possibility of a Radical Marxian Critique of Capitalist Society

It might be objected that the conception of radical criticism outlined above is inconsistent with central elements of Marx's thought. Consequently, it is fundamentally mistaken to construe Marx as offering a radical critique of capitalist society. This concern is not an idle one. If Marx did have a radical critique of capitalist society, there is certainly no one place in his writings where it can be found. Though, as I shall argue shortly, it is an overstatement to say that Marx's normative concerns were minimal, it is probably correct to say that those concerns receive less emphasis in his writings than one might expect. So, this objection is worthy of serious consideration. Moreover, it is at least possible that the systematic challenges to construing Marx as offering a radical critique can, in a more generalized form, be represented as challenges to the notion of a radical critique itself, at least as it has been specified above. This would suggest that this notion is of less general philosophical interest than I have supposed.

As it pertains to Marx, there are, as far as I am able to determine, three grounds on which this objection might be made: First, it might be argued that central elements of Marx's theory of history, his Historical Materialism, are inconsistent with the essential action-guiding significance of a radical critique. In particular, according to Marx, radical social change is the outcome of large-scale historical forces and not of intellectual critiques of the existing order. Second, it might be objected that Marx subscribed to a form of ethical relativism that makes negative ethical judgments about capitalist society problematic. For instance, Allen Wood has argued that Marx believed that capitalism was basically a just system, since justice is defined in terms of what is appropriate for the dominant relations of production in a given society. Finally, it might argued on a variety of grounds that the demands of the Alternative Institutions requirement are completely inappropriate for Marx. The paucity of material on post-capitalist society in Marx's writings gives this objection some initial plausibility. In what follows, I shall investigate all of these charges in detail. Following this will be a sketch of the plan for the rest of the book.

### Historical Materialism

Central to the notion of a radical critique is that it has action-guiding significance for social change. According to this objection, for Marx, radical social change is in the works, so to speak, and its explanation will not be attributable to *ideas*. This is the error of the Young Hegelians that he criticizes so sharply in *The German Ideology* (*GI*, MECW, vol. 5, pp. 27–32). Consequently, if a radical critique necessarily has action-guiding significance, and if ideas cannot determine radical social change, Marx's Historical Materialism is inconsistent with a radical critique of capitalism as specified above.

In the following passage from *The German Ideology*, Marx identifies the central failing of the approach of the Young Hegelians:

> Since, according to their fantasy, the relations of men, all their doings, their fetters and their limitations are products of their consciousness, the Young Hegelians logically put to men the moral postulate of exchanging their present consciousness for human, critical or egoistic consciousness [a reference to Feuerbach, Bauer, and Stirner, respectively], and thus of removing their limitations. [*GI*, MECW, vol 5, p. 30]

A change in consciousness, then, is what the Young Hegelians are aiming at; that is sufficient for social change, or, in post-Hegel Germany, perhaps that *is* social change! By contrast, for Historical Materialism, revolutionary social change is a change in the real relations between human beings, especially the relations of production. Furthermore, it is the result of large-scale social forces set in motion by man's real interactions with nature and other men in the course of production. In the case of communism, the latter is not an ideal to be realized but a natural and organic outgrowth of the existing order, which will be fashioned by real people interacting with each other and with their natural environment. Consequently, it is a mistake to think that what Marx is doing is offering a radical critique with action-guiding significance for social change.

There are two difficulties with this objection. First, it overstates the relative unimportance of ideas for Historical Materialism. Second, it overstates the action-guiding significance of a radical critique. Let us consider each of these points in turn. For Marx's Historical Materialism to preclude a causal role for some of the ideas (or propositions) that constitute a radical critique, it is necessary to interpret Marx as an epiphenomenalist of a certain sort. In the philosophy of mind, epiphenomenalism is the view that ideas are the effects of brain processes but are not the cause of either brain processes or even other ideas. They are essentially by-products of something more fundamental, namely, brain processes.

Now what would "social epiphenomenalism" amount to? It would have to be the view that ideas about society that most people would think have action-guiding significance for social change really don't. In short, ideas cannot be causal factors in producing social change, or at least radical social change. If that is Marx's view, then it is fundamentally misconceived to construe Marx as offering a radical critique of capitalist society. The question now reduces to whether or not this is a part of, or is entailed by, Marx's materialist theory of history.

The problem with this objection is that none of the central tenets of Historical Materialism preclude a place for normative beliefs, broadly conceived, in the sequence of events that brings about radical social change. That a society's base determines its superstructure only implies that the *dominant* ideas are the ideas of the ruling class. It further implies that the ideas of the rising class will only become dominant at or about the time that the old order is on the way out. Both of these claims are consistent with the proposition that those ideas are or can be causal factors in producing the new order.

It is clear that Historical Materialism is inconsistent with the view that the propagation of ideas can by itself initiate and sustain social change.[3] According to Marx, revolutionary social change requires a "material base," that is, a contradiction between the forces and relations of production. What is beyond the power of radical criticism is the creation of this material base. Only when these objective circumstances are right can successful revolution take place. Time and again Marx castigates others on the Left for failing to understand the social and material preconditions of radical social change—an understanding they could easily acquire by reading the appropriate works of Marx.

Finally, according to Marx, socialist revolution is supposed to spring from the proletariat's recognition (belief) that the existing order is no longer in their own best interests. This is clearly implied in the following description of the role of Communists in *The Communist Manifesto*: "The Communists are distinguished from the other working-class parties by this only: 1. In the national struggles of the proletarians of different countries, they point out and bring to the front the common interests of the entire proletariat, independently of all nationality" (*CM*, MECW, vol. 6, p. 497). Why should the Communists do this if not to affect the workers' normative beliefs (in this case, beliefs about what is in the workers' best interests), thereby making a major contribution to revolutionary social change?

The temptation to conceive of Marx as a "social epiphenomenalist" is perhaps due to his radical reaction to, and departure from, Hegelian orthodoxy, which assigns a far more significant and comprehensive role to ideas than Marx thought was plausible. As the preceding discussion makes clear, this temptation should be resisted. Social causation, like causation in the natural world, is often exceedingly complex. The role of normative ideas can be overstated, as in the case of the Young Hegelians, or understated, as in the case of the "social epiphenomenalist." And, although Marx's general explanation of social change accords a more subordinate or derivative role to ideas than competing theories, he clearly does not believe that they have no causal role to play. In short, he was not a social epiphenomenalist.

The other difficulty with this objection is that it overstates the action-guiding significance of a radical critique. To say that the latter has this significance is not to claim that it can, by itself, bring about radical social change, even given the appropriate material preconditions. For example, Marx clearly believed that class consciousness among the proletariat is a necessary condition for radical social change. If that is true, the appropriate material preconditions, together with a radical critique would not by themselves be sufficient for radical social change; other conditions are also necessary. But this does not rule out a causal role for a radical critique. But what causal role could a radical critique play?

At most, some elements of a radical critique could provide practical advice about the wisdom, or lack thereof, of specific strategies for revolutionary social change (the Transition requirement), as well as advice about the design of social institutions (the Alternative Institutions requirement). More specifically, a radical critique may exhibit its action-guiding significance through what Lenin called 'agitation' and 'propaganda'. In Leninist theory, agitation is the transmission of simple ideas to the many (e.g., "All power to the Workers' and Soldiers' Soviets"), whereas propaganda is the transmission of complex ideas to the few, most notably, the intelligentsia. At the outer limits, propaganda would include the theoretical writings of leading socialist theoreticians. That part of the radical critique that deals with the transition would be most influential in shaping revolutionary strategy, but after the destruction of the existing order, that part that specifies alternative institutions will be put to the practical test.[4]

Returning to Marx, his *Critique of the Gotha Program* contains the only sustained discussion of post-capitalist society to be found in his writings. As I shall show in Chapters 6 and 7, it is possible to extract from this discussion a characterization of alternative institutions for post-capitalist society. Although Marx's account of post-capitalist society is cast in descriptive rather than prescriptive terms, one of the main purposes of this attack on the Lassallean program was to put the United Workers' Party of Germany on the right track. As Marx says in the covering letter to Bracke that accompanies this document:

> Every step of real movement is more important than a dozen programmes. If, therefore, it was not possible . . . to go *beyond* the Eisenach programme, one should simply have concluded an agreement for action against the common enemy. But by drawing up a programme of principles . . . one sets up before the whole world landmarks by which it measures the level of the Party movement.[5]

This passage is revealing in a number of respects. Although the first sentence downplays the revolutionary significance of a statement of principles that describes the intermediate and long-term goals of the movement, the last sentence makes clear that Marx believed such a statement to be of action-guiding significance.

Lurking in the shadows of the general objection under consideration are worries about the compatibility of a radical critique and some form of determinism presupposed by Historical Materialism. This parallels, at the level of theory, the view that Marx saw himself as a scientist whose sole purpose was to explain and describe class societies in general and capitalism in particular. Recent work on Marx by philosophers has pretty conclusively demonstrated the inadequacy of this view of Marx and indeed of Marx's conception of himself.[6] Additionally, on a more substantive level, the main claims of Historical Materialism can be interpreted in such a way that they are compatible with the action-guiding demands of a radical critique and thus do not presuppose a rigid determinism that renders such a critique problematic.

The purported laws of historical development Marx discovered can be conceived of as large-scale tendencies pulling (or pushing, depending on one's perspective) societies in a certain direction.[7] Though these tendencies are strong, the laws contain implicit *ceteris paribus* clauses to the effect that nothing else interferes. These tendency laws may also presuppose rationality on part of those who bring about radical social change. Such laws allow for the possibility that unpredictable (that is, unpredictable by any theory of society) events such as cataclysmic climatic changes or exogenously caused wars could thwart or hold up the working out of these basic tendencies. If this way of conceiving of the laws of Historical Materialism is right, the rigid determinism that might call into question the possibility of a radical critique is not a presupposition of Marx's theory of history. Indeed, the latter may require, in addition to rationality, that some people (e.g., Communists) in a position to influence decisively the course of historical development have an adequate understanding of these basic tendencies and want to see the latter work themselves out.[8] As Marx and Engels say in *The Communist Manifesto*:

> The Communists, therefore, are on the one hand, practically, the most advanced and resolute section of the working-class parties . . . which pushes forward all others; on the other hand, theoretically, they have . . . the advantage of clearly understanding the line of march, the conditions, and the ultimate general results of the proletarian movement. [*CM*, MECW, vol. 6, p. 497]

To sum up, Historical Materialism can readily be construed in such a way that it does not presuppose a form of determinism that makes a radical critique of the existing order problematical. It also seems clear that the substance of Historical Materialism is not inconsistent with a radical critique.

### Relativism

It is clear that Marx's sociology of morals assigns morality an essentially derivative status. Conceptions of justice, right, and morality in general have their origins in the basic relations of production that characterize a society, and such conceptions serve the interests of the ruling class. Although this account of the origins and function of morality does not appear to entail ethical relativism, it does seem to be congenial to some versions of the latter.[9] Indeed, Allen Wood has argued that Marx was a relativist about justice.[10] Wood claims that, contrary to semi-informed opinion, Marx did not believe that capitalism is a unjust system. According to Wood, (at least distributive) justice is defined in terms of whatever accords with or corresponds to the existing mode of production. Thus, the capitalist does no injustice to the worker by buying his labor power and using it to create surplus value, which he then appropriates. On the other hand, if the capitalist cheats the consumer or pays the worker less than the value of the latter's labor power, he is acting unjustly. Since this does not happen for the most part, capitalists do not in general act unjustly.

Wood's article set off a firestorm of debate about the correct interpretation of Marx on this issue.[11] However, even if Marx was a relativist about justice, it does not follow that he was a relativist about morality generally. Nonetheless, it appears that there is some reason to be concerned about the compatibility of a more general ethical relativism for Marx and a radical critique of capitalist society. The reason for this is that ethical relativism is, at bottom, a profoundly conservative moral doctrine. According to ethical relativism, right and wrong are specified in terms of the principles operative in (or at least subscribed to by) the existing society. An action is morally right if and only if it is prescribed or permitted by the dominant moral code of the society in question. If Marx's sociology of morals is right, currently this is bourgeois morality. If Marx is committed to a thoroughgoing ethical relativism, then, so the objection runs, it does not seem plausible to construe him as offering a radical critique (as it has been specified above) of capitalist society.

There are a number of ways to address this objection. One way would be to construe relativism as specifying right and wrong not in terms of the moral code of the dominant group in a society. A relativist could hold that right and wrong are defined in terms of the principles subscribed to by *any* group in a society. Alternatively, societies can be individuated in such a way that many societies coexist in a given geographical area. For example, capitalism could be said to encompass two distinct societies—bourgeois society and proletarian society. If Marx were a relativist, he could be conceived of as offering a radical critique from the perspective of proletarian morality, or possibly communist morality. This may not be the best way to construe Marx's radical critique of capitalism in the final analysis, but it does provide one way to render the latter consistent with ethical relativism.

Perhaps a less problematic approach would be to call attention to the fact that a radical critique may be normative without being ethical. (See the above discussion of condition (ii), the Normative Theory requirement.) In other words, there may be good reasons for advocating radical social change (e.g., the class interests of the proletariat), but those reasons may not be moral. Perhaps morality does not apply or perhaps it does not issue in categorical imperatives at the level of large-scale social change. Prima facie, anyway, that is a defensible position, and absent any direct textual evidence to the contrary, it is position Marx could have adopted. I have been unable to find any textual evidence to suggest that Marx rejected this position. Thus even if Marx were an ethical relativist, it would not render problematic the idea of a radical Marxian critique of capitalist society.

Besides, it is fairly obvious that Marx did think that there were good reasons, at least from the proletariat's perspective, for condemning the existing order and for advocating radical social change. As will become apparent in Part I of this book, these reasons consist, at the very least, in the fact that certain negative values can be avoided by abolishing capitalism and instituting socialism. Since this "good reasons" approach is all that is needed to get a radical Marxian critique of capitalism off the ground, I shall not attempt to resolve the dispute between those commentators who affirm that Marx was an ethical relativist and those who deny it.

### Doubts About the Alternative Institutions Requirement

The one presupposition of a radical critique that seems most foreign to Marx's thought is the Alternative Institutions requirement. The latter appears to require a rather elaborate specification of alternative institutions that the critic believes should constitute the new order. It has been widely observed that Marx had very little to say about post-capitalist society. Furthermore, he often inveighs against other radicals for offering detailed blueprints for such a society. The objection is that the demand for a specification of the major institutions of communist society is profoundly inappropriate for Marx, reflecting Idealism or worse. Shlomo Avineri sums up what is probably a widely shared view when he says,

> Since the future is not as yet an existing reality, any discussion of it reverts to philosophical idealism in discussing objects which exist only in the consciousness of the thinking subject. . . . for [Marx] communist society will be determined by the specific conditions under which it is established, and these conditions cannot be predicted in advance.[12]

The first sentence is badly overstated, since it would make all human labor an exercise in Idealism. Marx points out that one of the features that distinguishes human labor from animal activity is that a human being can raise up in imagination the product of his labor before he gives it material expression.

> A spider conducts operations that resemble those of a weaver, and a bee puts to shame many an architect in the construction of her cells. But what distinguishes the worst architect from the best of bees is this, that the architect raises his structure in imagination before he erects it in reality. At the end of every labour-process, we get a result that already existed in the imagination of the labourer at its commencement. [*Capital* I, p. 174][13]

The passage from Avineri conflicts with Marx's view that the scientific knowledge of society which he and others (mostly he and Engels) have discovered makes possible for the first time humanity's self-conscious construction of social institutions.[14] This will perhaps be the highest form of dealienating labor ever undertaken. Finally, we *do* know the conditions under which communism will be established—the revolutionary overthrow of capitalism and the rule of the bourgeoisie. The real question is not 'Can we know anything about post-capitalist society?', but rather, 'What can be known and what cannot?'. It is here that the issue is joined with the Alternative Institutions requirement of a radical critique.

The fact is that Marx's theory of history can be used to predict quite a bit about post-capitalist society. As a general point, however one interprets the theory of history, it is abundantly clear that it is possible to know a number of important things about any society if its relations of production are known. As I shall argue in Chapter 6, the relations of production that will characterize post-capitalist society can be specified as follows: (i) The workers control the means of production; (ii) Labor power is not a commodity (i.e., it is not bought and sold on a labor market) and is controlled by the workers. These two conditions allow us to infer at least two things about the economic system of post-capitalist society: Nonworkers do not control the means of production and there is no market for labor power. Given some plausible auxiliary assumptions, this allows us to make further inferences about the economic system of the society. (More on this in Chapter 6.)

Secondly, Marx believed that the distribution of the social product is determined by, and thus can be inferred from, the relations of production. This is why, in the *Critique of the Gotha Program*, he criticizes the Lassalleans for wanting to change the distribution of the social product while ignoring the relations of production (*CGP*, p. 18). On the basis of this fundamental tenet of Historical Materialism and some auxiliary assumptions, Marx infers that the two stages of post-capitalist society will be governed by two different distributive principles: In the first, or lower, stage it will be 'To each according to his labor contribution', whereas in the higher stage, it will be 'To each according to his needs' (*CGP*, pp. 14, 16).

Thirdly, Marx's theory of the state holds that the state is a reification of class conflict; it loses its *raison d'être* with the disappearance of classes. This leads Marx to predict a period of the dictatorship of the proletariat in the immediate aftermath of socialist revolution and the subsequent withering away of the state. On the basis of his theory of history, then, Marx makes some predictions about both the economic and the political institutions of post-capitalist society.

Finally, Marx would maintain that the systematic ills that afflict capitalism will be for the most part abolished in post-capitalist society. So it should be possible to say something about what life would be like without systematic exploitation and alienation and how that might be explained by the economic and political system of post-capitalist society.

All of the above implies that the general contours of the institutions of post-capitalist society can be predicted on the basis of Marx's theory of history; it does not imply that everything can be known about these institutions. Just as feudal and capitalist relations of production can be realized in very different institutional embodiments, different socialist or communist systems will have their own idiosyncratic signatures. The construction of the social institutions of the new order will require creativity as well as ingenuity on the part of the workers; predicting in detail the outcome of such a creative process is surely foolhardy, so there is an important sense in which we cannot know "what life will be like" in post-capitalist society.

Moreover, many of the particular values by which people live their lives either cannot be known in advance or cannot be predicted on the basis of Historical Materialism. With regard to the latter, those values shaped by tradition and national character come readily to mind. For example, the English will likely continue to exhibit an inordinate interest in their household pets, while the French and Italians (but not the Germans) will probably continue to uphold their great culinary traditions. Nonetheless, at a certain level of abstraction, there will be some common values, namely, those necessary to sustain the basic institutions of the new society.

Given that predictions are possible on the basis of Marx's theory of history, what is the explanation for his scant attention to the future? After all, his remarks on post-capitalist society are remarkably desultory. I think there is a relatively straightforward and simple explanation for Marx's relative inattention to the nature of post-capitalist society. This explanation is tactical or pragmatic; as the above quotation from the letter to Bracke makes clear, Marx thought it unwise to spell out too explicitly the medium and long range goals of the workers' movement. There are two obvious reasons for this: First, it simply makes good political sense to try to seek widespread agreement on as little as possible. Relatedly, it is much easier to achieve agreement on what people are against as opposed to what they are for. Surely, Marx was astute enough as a politician to recognize these obvious facts.

Finally, his Historical Materialism gives him good reason to be politically apprehensive about making elaborate predictions about the nature of post-capitalist society. Central to Marx's views about the development of the working class movement is the belief that the proletariat would grow in political maturity. The climax of this process is their recognition that it is possible for them to take their future in own their hands and shape, by a democratic process, new social institutions. The last thing they need is some smart-aleck intellectual telling them how they are going to do this. Not only is this likely to be, and to be perceived as, anti-democratic, it may be counterproductive as well.[15]

These observations might appear to be inconsistent with the implicit action-guiding significance of a radical critique, but they are not. Primarily two aspects of a radical critique have action-guiding significance: what addresses the Transition requirement and what addresses the Alternative Institutions requirement. It is not unreasonable to suppose that they will play their action-guiding roles at different times. Since the demise of capitalism was the first order of business, and on Marx's view, just around the corner, he did not hesitate to offer specific advice on tactics to advance the transition to socialism. This advice was based on "a correct analysis of the situation," as later Marxists like to put it. This sort of analysis depends on an accurate theoretical understanding of the dynamics of class societies in general and capitalist society in particular, and this understanding is provided by the theory of history. This in part explains why some later Marxists have fought so bitterly about what the correct analysis of a given situation is (e.g., in the colossal struggle between Lenin and the rest of the Bolsheviks on the eve of the November Revolution and in Trotsky's struggles against most everyone from the Twenties on). What they are really fighting about is, in Lenin's phrase, "what is to be done."

Although Marx did not offer much in the way of detailed advice about the institutions of post-capitalist society, that is consistent with such advice being implicit in the theory of history, at least when coupled with some plausible assumptions about the circumstances the proletariat will face. Had a successful large-scale socialist revolution actually occurred in his lifetime, we can be reasonably confident that Marx would not have been content to let the chips fall where they may when the workers got around to constructing new institutions.

Presumably, this advice would be informed by his materialist theory of history. For example, the latter would predict that certain modes of distribution of the social product cannot stably coexist with socialist relations of production, and as I shall later argue, that a true market system will produce significant alienation. It may be that in general the advice about alternative institutions that is implicit in the theory of history is largely negative. However, that should not obscure the fact that the complex theoretical understanding of the interrelations among and within social institutions painstakingly achieved by Marx and Engels can and will have action-guiding significance in the construction of the new order.

There is one final challenge to the Alternative Institutions requirement that warrants some discussion. At the end of his chapter on fetishism in *Karl Marx's Theory of History*, G. A. Cohen briefly discusses an obscure suggestion that Marx envisioned communism as an alternative to society and not a form of society or social organization. This is a very difficult idea to get a grip on, but because it represents the most radical kind of challenge to the very possibility of a radical Marxian critique of capitalism, it is worth investigating in some detail.

Let us begin with the famous passage from *The German Ideology* on communism and the division of labor that led Cohen to make his suggestion:[16]

> For as soon as the division of labor comes into being, each man has a particular exclusive sphere of activity which is forced upon him and from which he cannot escape. He is a hunter, a fisherman, a shepherd, or a critical critic, and must remain so if he does not want to lose his means of livelihood; whereas in communist society, where nobody has one exclusive sphere of activity but each can become accomplished in any branch he wishes, society regulates the general production and thus makes it possible for me to do one thing today and another tomorrow, to hunt in the morning, fish in the (afternoon, rear cattle in the evening, criticise after dinner, just as I have a mind, without ever becoming hunter, fisherman, shepherd or critic. [*GI*, MECW, vol. 5, p. 47]

For future reference let us call this the Famous Passage.

This passage implies that the division of labor will be abolished with the advent of communism. Explaining exactly what that means turns out to be very difficult. However, for present purposes, all that is necessary is to provide an accurate interpretation of some of the *consequences* of abolishing the division of labor. As will become evident, accomplishing this is itself a formidable task.

Let us begin with Cohen's interpretation, which is the most radical reading possible. He says,

> This man is not even successively a hunter, fisherman, and critic, though he does hunt, fish, and criticize. For he is in none of these activities entering a position in a structure of social roles, in such a way that he could identify himself, if only for the time being, as a hunter, etc. . . . The abolition of roles is an exacting prescription, but Marx imposed it on future society. . . . He wanted individuals to face one another and themselves 'as such', without the mediation of institutions.[17]

Communism, then, requires the abolition of all social roles and with that, the abolition of all social institutions. No wonder Marx had so little to say about the institutions of communism: There aren't any!

My argument against Cohen's interpretation of this passage proceeds in two stages. First, I'll investigate just what Cohen is getting Marx into; I shall argue that the view he attributes to Marx is intrinsically highly implausible. Then I'll suggest a more modest reading of what Marx might have intended in the Famous Passage. The Principle of Charity in interpretation will thereby rule against Cohen.

The social institutions that constitute a genuine society are themselves constituted by a set of social roles. Not just any set of social roles constitutes an institution; there has to be some kind of internal coherence to the set. But, an institution, or social subsystem as it might be called, is constituted by a set of social roles. For example, capitalist relations of production, which constitute the economic system of capitalist society, relate occupiers of social roles, namely, capitalists and proletarians. To say that someone occupies a social role implies that he can reasonably be expected to engage in certain patterns of purposive behavior in certain circumstances. For example, a capitalist *qua* capitalist can be counted on to try to maximize profits when making offers to buy and sell commodities, and a proletarian *qua* proletarian will sell his labor power for the highest price he can get in bargaining with an employer.

What exactly is the relation between the institution and the relevant behaviors? The latter serve some institutional purpose or function. Teachers communicate certain kinds of knowledge and skills, trade union leaders represent the interests of the membership, and so forth. There may be no pattern(s) of behavior common and peculiar to all teachers or all trade union leaders except insofar as these patterns are functionally defined.[18] Let us say that these behavior patterns realize, or tend to realize, the social function of the institution.

Of course, those who occupy a social role can fail to fulfill the function that the occupation of that role is supposed to realize. This fact points to another implication of what it means to occupy an institutional social role: It entails that the expectation of certain patterns of behavior in certain circumstances is legitimate in the normative sense. That is, a failure of these expectations warrants criticism or requires an excuse. If a doctor fails to be thorough in his examination of a patient, an expectation has been disappointed that is both epistemically and normatively legitimate.

Not every pattern of purposive human behavior is manifestation of a social role. When I play golf by myself, I occasionally kick the ball out of the rough, and I sometimes take a shot over again. Sometimes people who observe me snicker, but they do not object or demand an excuse. Under these circumstances, although I play golf, there is a sense (perhaps more than one!) in which I am not a golfer. On the other hand, if by an act of God, I got on the professional circuit, people would legitimately, both epistemically and normatively, expect me not to do these things.

To sum up, for a person to occupy a social role is for that person to be disposed to engage in a some pattern or patterns of behavior that tend to realize some social goal or function and to generate legitimate expectations among others that these patterns of behavior will be exhibited under certain circumstances. Or, to put the matter more formally,

$X$ occupies social role $S$ if and only if

> i. There is some social function or purpose $f$ such that some patterns of behavior $B_1 \ldots B_n$ tend to realize $f$.

AND        ii. $X$ is disposed to engage in some of these patterns.

AND        iii. Others legitimately (in both senses) expect $X$ to exhibit some of the $B$s in the appropriate circumstances.

Now let us consider what the abolition of social roles amounts to. To say a role has been abolished implies that no one any longer occupies it. This must mean one of two things: (1) The purpose or function the role served has been rendered otiose (e.g., presumably the role of trade union leader will be abolished in this sense when the workers come to control the means of production); or, (2) The purpose remains but it is served without a role-occupier. For example, a technological advance may result in the abolition of a role in the workplace.

We are now in a position to understand the radical thesis that Cohen attributes to Marx, namely, that communism abolishes all social roles and, by implication, all social institutions. It is reasonable to suppose that many of the institutional roles people occupy under capitalism will be abolished, either because technological developments engendered by the development of the forces of production allow the relevant social function to be served by a machine, or because there will be no social function to be served. Regarding the latter, any communist has his rogues' gallery of social roles under capitalism that he believes would be abolished under communism. However, some functions or purposes now served by role-occupiers will continue to need to be served under communism. Since these purposes would persist under communism, Marx would be committed to the view that they would be served without role-occupiers. This in turn entails that one of the following is true: (1) All such functions can be served by nonhumans; or, (2) Some of those functions will be served by humans but not *qua* role-occupiers, since no human occupies any social role. Given the above definition of what it means to occupy a social role, the latter in turn entails that one of the following is true: (a) Human behavior will not exhibit patterns of behavior and thus will generate no epistemically legitimate expectations—in short, human behavior will be unpredictable; or, (b) Some human behavior will exhibit patterns, but it will generate no normatively legitimate expectations.

If this is what abolishing all social roles amounts to, it is wholly unreasonable to believe that all social roles, and by implication all social institutions, could be abolished under communism. Clearly, not all functions served by role-occupiers under capitalism will either disappear or be served by nonhumans under communism. (The role of parent comes readily to mind.) This entails that some of these functions will be served by humans but not *qua* role-occupiers. This in turn entails that (a) or (b) above is true. But neither of these is the least bit plausible. In general, the fact that people engage in coherent and predictable patterns of behavior which generate normatively legitimate expectations and which serve a variety of social functions makes social life both possible and intelligible. It is not even clear that an association of individuals who occupy no social roles is consistently thinkable.[19]

Cohen's interpretation saddles Marx with a view that has implications no responsible social thinker would accept. This constitutes a strong prima facie case against interpreting Marx in this manner. This case can be further strengthened by providing a more modest reading of the passage in dispute.

As the first sentence of the Famous Passage makes clear, Marx's intention in this passage is to contrast some features of life that are consequences of the division of labor with what life would be like under communism without the division of labor. The key to understanding this passage is this first sentence: "For as soon as the division of labour comes into being, each man has a particular exclusive sphere of activity, which is forced upon him and from which he cannot escape." If the remainder of the passage is intended to contrast this aspect of life in societies with the division of labor (notably capitalism) with life under communism, then Marx is saying two things about communist society: (i) People will not have a particular exclusive sphere of activity (i.e., a social role in the workplace) forced on them; (ii) They will not be forced to remain in any such sphere. It is consistent with this that they occupy social roles as defined above; it is just that they will not be forced to occupy one role. On this reading, the latter part of the Famous Passage predicts two things: that people will occupy many more (productive) roles than under, for example, capitalism and that their choices about these roles will be free. This is a much more modest reading of what Marx is saying, though it remains a profoundly radical thought. This reading of the Famous Passage does not commit Marx to the very implausible consequences that follow from Cohen's interpretation.

Cohen might respond to this interpretive argument by pointing out that it does not fully explain the last clause of the Famous Passage where Marx denies that communist man "will become hunter, fisherman, shepherd or critic." Indeed, in the following passage, which Cohen quotes in his discussion, Marx seems much less ambiguous in predicting the abolition of social roles:

> with a communist organisation of society, there disappears the subordination of the artist to local and national narrowness, which arises entirely from division of labor, and also the subordination of the artist to some definite art, thanks to which he is exclusively a painter, sculptor, etc., the very name of his activity adequately expressing the narrowness of his professional development and his dependence on the division of labor. In communist society there are no painters but at most people who engage in painting among other activities. [*GI*, MECW, vol. 5, p. 394]

Cohen gives the following reading of the last sentence of this passage:

> We deny that the last sentence says: 'In a communist society there are no full-time painters but at most part-time painters.' People do paint, but the status of painter is not assumed even from time to time.[20]

Summing up his interpretation, he says,

institutions represent 'fixation of social activity, consolidations of what we ourselves produce into an objective power above us'. It is no great exaggeration to say that Marx's freely associated individuals [i.e. communism] constitute an alternative to, not a form of, society.[21]

If social roles constitute institutions and if Marx is saying that all social roles will be abolished, as the immediately preceding quotation seems to indicate, then Cohen is right to maintain that Marx envisioned the disappearance of all social institutions under communism. What can be said in response?

Two things. First, Marx is engaged in a bit of semantic overkill. Second, the passage in question supports equally well the interpretation that Marx envisions a radical restructuring of social roles, but not their abolition. Let us begin with the first point.

As Marx (claims to) understand the German equivalent of the term 'painter', it connotes a narrow professional development and dependence on the division of labor. Surely this is an exercise in creative semantics. Interpreted literally, this means one cannot be a painter without having a narrow professional development. To put it another way, a broadly developed painter is a contradiction in terms. This is clearly overstated as a claim about the *meaning* of the *term*,[22] a judgment that is supported by the foregoing analysis of what it means to occupy a role.

On the other hand, it is arguably not an *empirical* overstatement about being a painter in bourgeois society (and now we come to the second point). Being a painter in bourgeois society does, as a matter of empirical fact, presuppose the division of labor. Good painters in our society nearly always "go professional" and do it full-time for a living. The passage, then, can be plausibly interpreted to be making a claim about what it "means," in a non-semantical sense, to be a painter in bourgeois society.

On this reading, to say that in communist society there are no painters is a hyperbolic and misleading way of saying that role-occupiers will not be "subordinated to local and national narrowness" and not have the narrowness of training characteristic of painters in bourgeois society. But it does not imply that there will be no one occupying the role of painter in communist society, even part-time. Some individuals will continue to exhibit the appropriate patterns of behavior and to generate legitimate expectations among others about their behavior. And this behavior will continue to serve a social function, for example, giving artistic expression to the culture of which the artist is a part.

Possibly, the nature of the social function that artistic endeavor serves will be transformed in the new society, as will the kinds of behavior artists exhibit. (Compare the Soviet conception of the role of the artist with that of bourgeois society.) In this sense, the *role* of a painter may be transformed, which is what Marx might have had in mind in saying there will be no painters in communist society. But, strictly speaking, there will still be painters under communism: There will still be legitimate expectations about certain patterns of behavior and painting will still serve some useful social purpose. The social necessity for role-occupiers in the arts generally is far from obvious, to say the least. But once we move from the fringes of social organization, where painters camp out, to aspects of life that require more highly structured social roles, the need for role-occupiers becomes palpable. Consider, for example, *dentists*. Though I suppose I could get used to a part-time dentist, I would be apprehensive, to say the least, if the person holding the drill wanted to face me "as such," without the mediation of social roles and institutions.

To summarize, though the texts are not unambiguous, there is a more modest reading of the relevant passages that does not commit Marx to a frankly outlandish position: Communism is not a form of society but an alternative to it. The Principle of Charity requires that the latter be rejected and the former accepted.

Besides, perhaps too much is being made of a couple of passages. The most important argument against the view that communism involves the abolition of all social roles is more indirect and more comprehensive. If a plausible reconstruction of Marx's account of the institutions of post-capitalist society can be given, then Cohen's suggestion must be rejected. That burden will be fully discharged in Chapters 6 and 7 of this book. An examination of this suggestion will have proved worth the effort, however, since it provides a first look at what is involved in specifying a social institution. This turns out to be of first importance when we turn our attention to post-capitalist society in Chapters 6 and 7.

A complete reconstruction and critical evaluation of Marx's radical critique of capitalist society would obviously be a monumental undertaking. My interest in this book is in the first and third requirements. The core of radical criticism consists of its diagnosis of systemic social ills and its vision of the new society—the good society—which is to take the place of the existing order. Part I of this book reconstructs these two elements of Marx's radical critique of capitalist society. Chapters 2, 3, 4, and 5 address the Critical Explanations requirement and Chapters 6 and 7 address the Alternative Institutions requirement.

Chapter 2 is about alienation; it reconstructs Marx's critical explanations of the various manifestations of alienation in capitalist society in such a way that these manifestations are shown to be rooted in capitalism's basic institutions, notably the economic system. Chapters 3, 4, and 5 are about exploitation. Chapter 3 shows that there are fatal difficulties with Marx's own argument for the charge of systematic exploitation against capitalism. Chapters 4 and 5 constitute a systematic statement and critical evaluation of other ways of arguing for this contentious charge that have appeared in the secondary literature over the past two decades. Those who have made these arguments sometimes represent them as what Marx really had in mind, but this interpretive question is only of secondary interest. What is most important is to give Marx the best case possible, whether or not that case best captures his intentions.

Chapters 6 and 7 sketch Marx's vision of post-capitalist society with an eye toward the demands of the Alternative Institutions requirement. Though

Marx had some very definite ideas about the main institutions of post-capitalist society, (namely, the economic system and the political system), his vision of that society will be shown to be significantly incomplete in the following three respects:

1. Marx has no account of the constitutive social roles of the state or how state power will be wielded in the first phase of post-capitalist society. By contrast, he does have such an account for the economic system, as I shall show in Chapter 6.
2. His argument for the claim that the state will "wither away" in the second phase of post-capitalist society is incomplete at best.
3. Perhaps most importantly, there is not much in the way of argumentation for the implicit claim that post-capitalist society will not suffer the social ills (exploitation and alienation) characteristic of capitalist society. Marx seems to have assumed that the destruction of capitalist institutions would result in the elimination of these problems, but such an assumption is unwarranted. After all, it might turn out that the new institutions will also produce these social evils. He makes almost no effort to explain why this will not happen.

That Marx's vision of post-capitalist society is significantly incomplete will come as no surprise to those familiar with the texts. Indeed, this is a commonly voiced complaint. Nonetheless, it is of some interest to identify the exact nature and extent of this incompleteness. A much more serious objection would be that Marx's vision of post-capitalist society cannot be realized. That is, the Alternative Institutions requirement cannot be—and not merely has not been—satisfied. This implies that Marx's radical critique of capitalist society is not just incomplete; rather, it is fundamentally and fatally flawed.

There are two strategies a critic might use to establish this. One would be to argue that Marx's vision of post-capitalist society is unrealizable because it is internally inconsistent. However, the very incompleteness of Marx's vision alluded to above would appear to insulate him from a charge of inconsistency. Of course, the critic could add elements to Marx's account (in the guise of reconstruction) to create such a problem, but that would obviously be an unfair tactic.

A second strategy would be to argue that Marx's account of post-capitalist society, such as it is, presupposes the disappearance of permanent features of the human condition, for example, changes in an unchangeable human nature or impossible increases in the level of material wealth. One apparent problem with this approach is that claims such as these would seem to be difficult to sustain on a sufficiently long historical view. For example, how can one know what human nature will be like or what technology will bring in, say, two hundred or five hundred years?

Despite the apparent difficulties with each of these strategies, Part II of this book uses both of them to argue that Marx's vision of post-capitalist society cannot in fact be realized. That vision has two phases or stages, and these will be tackled independently. Chapter 9 employs the first strategy to argue that the first, or lower, phase of post-capitalist society cannot be realized. Chapter 8 employs the second strategy to argue that the second, or higher, phase cannot be realized. The opening section of Chapter 8 examines the two general strategies in more detail and indicates how the pitfalls alluded to above can be avoided. The ultimate conclusion of this book is that Marx's radical critique of capitalist society must be judged a failure because the Alternative Institutions requirement cannot be satisfied.

Chapter 10 considers market socialism as an alternative to Marx's vision of post-capitalist society; I shall show that the former represents a repudiation of Marx's radical critique of capitalist society and that, not coincidentally, it faces the same systematic social problems as capitalism, though the question of whether it is to be preferred to capitalism will not be settled. The second section of this last chapter consists of some general observations about social criticism in light of the project of radical critique, and it reexamines my claims about the intellectual burdens that any social critic must shoulder.

# I

## RECONSTRUCTION

# 2

# Alienation

The guidelines implicit in the discussion of the Critical Explanations requirement and the Alternative Institutions requirement for a successful radical critique provide the framework for the reconstruction of Marx's radical critique of capitalism contained in the next six chapters. The purpose of the next four chapters is to reconstruct Marx's systematic explanations of the ills or defects of capitalist society in such a way as to address the Critical Explanations requirement.

Before passing on to a discussion of alienation, let us consider in a bit more detail the demands imposed by this requirement. According to the latter, the sources of the defects of capitalist society have to be located in its basic structure or its fundamental processes. Otherwise, it is open to Marx's critics to argue that radical social change is not needed to correct the defects in question. Forestalling this challenge is crucial to the radical critic's program. However, talk about basic structures and fundamental processes seems rather obscure and metaphorical. What does it really amount to? Implicit in this talk is a certain conception of explanation in the social sciences that needs to be articulated and at least provisionally defended.

What a theorist takes to be a basic structure or a fundamental process is relative to certain theoretical commitments. For example, in Marx's theory of history, the economic system is, at one level, basic or fundamental because a wide range of social phenomena are explained by appeal to it. For present purposes, it would be circular to establish the economic system's fundamental status by showing that Marx's critical explanations are traceable to the economic system. Fortunately, his theory of history is intended to explain a much broader range of phenomena than the defects of capitalist society, so there is independent justification for regarding the economic system as fundamental. For Marx, then, a sufficient condition for satisfying the Critical Explanations requirement is to explain the defects of capitalist society by appeal to the economic system.

What, then, is the economic system? This analytical question is more difficult than it might appear to be. A complete answer will emerge in Chapters 6 and 7, but a first approximation might go something like the following: An economic system is simply that social institution by which humans regulate their interactions with themselves and nature for the purpose of production. For Marx, different economic systems have different essential features. That is how types of economic systems are individuated. These essential features are also expressive of certain theoretical commitments. In other words, what counts as an essential feature of an economic system is determined by whatever explains the range of phenomena the theory is supposed to explain. Although it may be tempting to think of essential features in terms of defining (i.e., necessary and sufficient) conditions, this is probably not required, and it is certainly foreign to the spirit of Marx's thought.

In light of the above, it would be sufficient to meet the Critical Explanations requirement if it can be shown that the defects of capitalist society can be explained by appeal to essential features of its economic system. Before passing to a consideration of what Marx thought those features are, a word needs to be said about the kind of explanations I shall employ. It is a hotly debated question in the philosophy of science whether or not all explanations are causal. Certainly some of them are, so to avoid this controversy, I shall, where possible, offer only causal explanations in my reconstruction of Marx.

In addition, I shall assume that social structures or aspects (features) thereof can be said to cause certain phenomena (kinds of events) or states of affairs. This is a common enough assumption in the sciences generally and the social sciences in particular. For example, in economics the fact that firms are assumed to be profit maximizers is used to explain, among other things, product innovation and differentiation. More generally, economists (including Marx) explain a wide variety of phenomena by appeal to structural features of the economic system. It may be that, in the final analysis, only events can cause other events, in which case I assume that talk about structures can be replaced by talk about events. Shamelessly, I won't argue for this assumption; I adopt it because my main concern is to get Marx's story out, and often these structural explanations seem to capture best Marx's intentions. My only concession to current disputes about explanation in the social sciences is that I eschew functional explanations of the defects of capitalist society, except insofar as they can be construed as causal. To sum up, Marx can be thought to satisfy the Critical Explanations requirement if a causal story about the defects of capitalist society can be told that appeals to essential features of the economic system.

What, then, according to Marx, are the essential features of the capitalist economic system? In *Capital* Marx gives two alternative characterizations of the essence of the capitalist economic system: One is structural in that it calls attention to social relations among individuals vis-à-vis the process of production; the other is modal in that it focuses on the purpose for which production takes place. Let us consider the structural characterization first.

In volume II of *Capital* Marx says,

> Whatever the social form of production, labourers and means of production always remain factors of it. . . . For production to go on at all they must unite. The specific manner in which this union is accomplished distinguishes the different economic epochs of the structure of society from one another. In the [capitalist] case the separation of the free worker from his means of production is the starting point given. [*Capital* II, pp. 34–35]

Earlier, in volume I he says,

> The capitalist epoch is therefore characterized by this, that labour power takes in the eyes of the labourer himself the form of a commodity which is his property. [*Capital* I, p. 167n1]

These two passages encapsulate, in a fairly terse manner, the structural characterization of capitalism: The free worker sells his labor power as a commodity. As Marx elsewhere notes (*Capital* I, pp. 165, 166), the worker is free in two senses: (1) As the owner of his labor power, he is neither enslaved nor enserfed; (2) On the other hand, he is also "free from," that is, bereft of, control or ownership of the means of production.

This structural characterization of capitalism can be summarized in the following two conditions: (i) The workers do not control the means of production; and (ii) Labor power is a commodity owned by the workers.

The other characterization of capitalism Marx offers is specified in terms of the purpose for which production takes place. In this connection, Marx contrasts production for exchange with production for use. In a system of production for use, production and consumption are not mediated by buying and selling in the marketplace. For example, in a primitive community, production is directly determined by the needs of the community as expressed by custom, tradition, and the demands of the ruling class for particular use-values. On the other hand, production for exchange, or commodity production, is undertaken with the market in view. What makes something a commodity for Marx is that it is produced for exchange and not for direct use (*Capital* I, p. 48). In a system of commodity production, most goods and services are commodities. This distinction provides the backdrop for Marx's other characterization of capitalism:

> Capitalist production is distinguished from the outset by two characteristic features.
> *First*, it produces its products as commodities . . .
> . . . The *second* distinctive feature of the capitalist mode of production is the production of surplus value as the direct aim and determining motive of production. [*Capital* III, pp. 879, 880]

Surplus value under capitalism takes the form of the exchange value that remains after the factors of production, including labor power, have been paid for. What the second condition says, then, is that the drive for surplus value, in the form of profit, determines production under capitalism. These two conditions are logically distinct; production could be for exchange without being for maximizing surplus value and vice versa. Something like the former seems to have characterized the later stages of the long decline of feudalism; markets for goods and services had become more extensive before capitalism came on the scene. Nevertheless, the aim of production was not to maximize surplus value; rather, it was to acquire particular use-values through exchange.

On the other hand, it is at least logically possible for a system to be guided by the desire to maximize surplus value, even though commodity production is not widespread. A slave society without markets could, in principle, exhibit this feature. However, Marx believed that, as a matter of empirical fact, this would not happen. Until surplus value took the form of exchange value, its form was that of use-values. The amount of surplus it was rational to want was, in consequence, limited by what the oppressor could consume. Once surplus value takes the form of exchange value (which can serve as *capital*), this brake on the accumulation of surplus is removed (*Capital* I, p. 226).

To summarize, let us restate more clearly the two essential features of capitalism given in the above quotation as follows:

i'. (Most) goods and services are produced as commodities; that is, for exchange in the market.

ii'. The motive for production by those who control the means of production is profit; that is, production is for profit, not for use.

Given that Marx has two alternative characterizations of capitalism, what can be said about the relation between them? G. A. Cohen has produced a pair of interesting Marxian arguments to show that, as a matter of empirical fact, any social system that satisfies the first characterization will satisfy the second and vice versa.[1] What Cohen has not addressed is the question of why Marx would offer two different characterizations.

Some insight into this question can be gained by considering Marx's theoretical goals or purposes. Marx's theoretical goals were manifold: He wanted to explain the genesis of capitalist society (i.e., how and why capitalist society arose from feudalism) and how and why capitalism will give way to socialism and communism. The achievement of these explanatory goals would serve to flesh out important details of the larger project of his Historical Materialism. Let us call these 'genetic explanations'. In addition to explanations of this sort, Marx also wanted to explain how capitalism functions as a going concern, that is, how it reproduces itself. Let us call these 'reproduction explanations'. Providing explanations of this sort is one of the main tasks of *Capital*. Finally, he wanted to give critical explanations in the sense required by the Critical Explanations requirement. The point of these explanations is to show that many of the undesirable features of nineteenth century industrial society were deeply rooted in the basic economic system.

Given this diversity of theoretical purposes, it is not surprising that Marx appeals to (elements of) different conceptions of what capitalism *is* in his various explanations. It would make for a nice architectonic if each theoretical purpose could be mapped onto one characterization of capitalism; Marx doesn't do it, but as far as I can tell, this creates no problems.

The interest of the current project is in the critical explanations. Successful critical explanations that appeal to elements of either of the characterizations of capitalism given above (i.e., [i], [ii], [i'], or [ii']) will satisfy the demand of the Critical Explanations requirement for *radical* diagnosis of society's illnesses. Such explanations go to the roots of things—in this case the fundamental features or processes of capitalist society.

I shall proceed on the methodological assumption that all of the defects Marx found in capitalist society can be brought under headings of alienation or exploitation. In the last section of Chapter 5, I shall explain why Marx's radical critique of capitalism is not, or at least need not be, based on a charge of injustice. In the first section of Chapter 8, I shall indicate why the fact that capitalism is responsible for significant poverty does not appear in this reconstruction. Beyond these two points, I have no a priori argument for this methodological organizing assumption, though I believe the discussion of the remainder of Part I of this book vindicates it. Without further ado, let us turn to critical explanations of the various forms or manifestations of alienation in capitalist society.

The topic of alienation brings together a number of important themes in Marx's radical critique of capitalism. My purpose in this discussion is not to give a complete account of everything that Marx had to say on this topic. Quite a few works in the recent literature contain elaborate general discussions of Marx's account of alienation.[2] My concern is to identify specific defects of capitalist society that fall under this heading and to see if their causes can be traced to the fundamental features or processes of capitalism that have been identified above. In one respect, the accounts I will develop will be relatively atheoretical; I shall identify defects, ills, or harms of capitalist society without making a systematic effort to explain exactly why they are bad things. A full account of the Normative Theory requirement mentioned in Chapter 1 would address exactly this issue. Limitations of space and my critical interests in Marx's radical critique preclude a full treatment of this topic.

The term 'is alienated from' is a relational term: Persons are one of the relata, but the other term varies with the context. Marx speaks of workers (or in some cases people in general) under capitalism as being alienated from: (1) the product of their labor, (2) their laboring activity, (3) their species being, and (4) their fellow human beings (*EPM*, MECW, vol. 3, p. 277). In what follows I shall consider each of these kinds of alienation in turn. In each case, three questions will guide the inquiry:

a. In what does the relevant form of alienation consist? Put another way, what exactly does it mean to say, for example, that under capitalism, man is alienated from his species being?

b. What exactly is the harm or defect that Marx is calling attention to? As noted above, the answer to this question will be left at a relatively atheoretical or intuitive level.

c. What is Marx's explanation for the relevant form of alienation and can the latter be traced to the essential features of capitalist economic system identified above?

## Alienation from the Products of Labor

Labor just is the purposive investment of human skill and energy in the creation of a product. By its very nature, it involves the externalization and realization of human capacities in things that have use value. This fact about labor cannot constitute grounds for criticism. However, under capitalism, something more is involved:

> The *alienation* of the worker in his product means not only that his labour becomes an object, an *external* existence, but that it exists *outside him*, independently, as something alien to him, and that it becomes a power on its own confronting him. It means that the life which he has conferred on the object confronts him as something hostile and alien. [*EPM*, MECW, vol. 3, p. 272]

What does this mean? One obvious sense in which the product is alien to the worker under capitalism is that he does not own it. By putting his labor into it, he has created it, but it belongs to another—the capitalist; moreover, it passes to the control of yet another when it is sold on the market. However, some aliens are friendly; why is this one not? In the *1844 Manuscripts*, the answer is not very clear. It is primarily in the later works, especially *Capital*, that a more comprehensible and detailed account emerges. This account depends crucially on the fact that, under capitalism, the worker's product is a commodity and, relatedly, that capitalism is a system of commodity production.

Marx begins volume I of *Capital* with an elaborate discussion of the nature of commodities and commodity production. Recall the contrast between commodity production, or production for exchange, and production for use: To describe an economic system as a system of production for use is to say that most products are produced with their ultimate use in view. For example, the family unit under serfdom produced most of their food, clothing, shelter, and so forth, for their own use (*Capital* I, p. 82). Even means of production were produced with their particular uses in view. By contrast, in a system of commodity production, or production for exchange, articles are not produced with their ultimate use in view. Producers, as well as those who control the means of production, do not know and do not care about the use to which their products will be put. All that matters is exchange value. The market rules in that articles do not get produced unless price cover costs. The goal of production under capitalism is not the creation of use-values; instead, those who control the means of production seek exchange value. And, although

Marx recognizes that production for exchange has occurred sporadically throughout history (with the possible exception of primitive communist societies), he points out that only under capitalism is production for exchange widespread and pervasive.

Marx believed that this fact is of fundamental significance for understanding the capitalist mode of production. He says, "the mode of production in which the product takes the form of a commodity, or is produced directly for exchange, is the most general and most embryonic form of bourgeois production" (*Capital* I, p. 86). This passage occurs in the famous section on the fetishism of commodities, and it is here that we find the explanation of how the alienation of the producer from the product creates an alien, hostile force.

The key to fetishism is the notion of mystification. In the most famous passage on the fetishism of commodities, Marx says,

> A commodity is therefore a mysterious thing, simply because in it the social character of men's labor appears to them as an objective character stamped on the product of that labor; because the relation of the producers to the sum total of their own labor is presented to them as a social relation, existing not between themselves but between the products of their labor . . . the existence of the things *qua* commodities, and the value-relation between the products of labour which stamps them as commodities, have absolutely no connexion with their physical properties and with the material relations arising therefrom. There it is a definite social relation between men that assumes, in their eyes, the fantastic form of a relation between things. [*Capital* I, p. 77]

The upshot of this passage is that relations among producers masquerade as relations among products of labor. When objects are produced as commodities (i.e., for exchange), they seem to have value in and of themselves. The exchange value of commodities appears to be an autonomous phenomenon, but it is in fact a reified expression of relations among producers. This social character of individual labor is not evident to the producers when articles are produced, not directly for use, but for exchange in the market. The social character of their labor is only revealed when commodities "meet" (i.e., exchange) in the market. The market serves as an automatic and impersonal device for the coordination and allocation of social labor, which entails that the large-scale social organization of production is not subjected to the conscious control of the producers, or anyone else for that matter.[3] The fact that the reality behind the fetishistic relations among products does not dissipate even after it has been exposed explains why commodity fetishism has the character of a mirage in that it persists even after its explanation is known.[4] It is also part of the explanation for why a correct understanding of the capitalist mode of production is not a sufficient condition for revolutionary social change. The foregoing explains the most important sense in which the worker's product is an alien thing.

Many commentators on Marx stop here in their exposition of commodity fetishism and do not press on to ask the obvious question: 'What is so bad about commodity fetishism?'. It is true, if Marx is right, that things are not as they appear in the social world. But physics teaches that this is true of the natural world. For example, a table appears solid, but contemporary physics tells us that it is mostly empty space. It does not seem that the mere fact that appearance and reality diverge constitutes grounds for criticism.

Because Marx's concerns in *Capital* are not primarily normative, he does not address this question directly. It must be inferred from what he says. Let us pursue Marx's discussion of the fetishism of commodities in *Capital* I a bit further. Immediately following his account of the fetishism of commodities under capitalism, he discusses three ways in which production can be carried on without fetishism. One purpose of this is to make clear by way of contrast what fetishism is; another purpose is to indicate that commodity production is not the only form of production. He canvasses the solitary production of Crusoe (one of Marx's few forays into Crusoe economics), production for the lord and for the family unit under feudalism, and finally communist production, though the latter is not named as such (*Capital* I, pp. 81–83). What all these modes of production have in common is that what regulates and determines production is completely apparent, or to use Marx's term, *transparent*, to the producer.

Crusoe's needs, and the quantity of labor required to satisfy them, directly determine his production; the wants and needs of the feudal lord determine the goods and services in kind that the serf must deliver; the needs of the family directly determine production in the peasant household; and, finally, in communist society, production would be determined by a consciously formulated plan to meet the needs of the community of producers. All societies other than capitalism, then, have a kind of transparency in their mode of production in two respects: (1) Needs directly determine what gets produced; and (2) The true social relations among individuals in production and distribution are immediately evident to all.

This permits a kind of social self-understanding that is denied to everyone—capitalist and worker alike—under capitalism. Under capitalism, the domination of commodities seems to both workers and capitalists to be an ineluctable feature of the universe and not a feature of a particular, historically transitory mode of production. One reason why Marx would regard this lack of social self-understanding as a bad thing is that it helps to prevent the destabilization of capitalist relations of production by preventing the workers from conceiving of different ways in which production may be organized.[5] The fact that workers' products, as commodities, reflect in a mystified way social relations among the producers makes those products truly alien beings in the sense that the social relations they represent, and at the same time mask, are not understood.

Where is the hostility, though? For this it is necessary to turn to the forces unleashed when commodity production takes place on a widespread basis. As capitalist relations of production came to predominate throughout the world, market forces came to have an enormous impact on people's lives. As

Marx says in *The German Ideology*, "thus, for instance, if in England a machine is invented which deprives countless workers of bread in India and China, and overturns the whole form of existence of these empires, this invention becomes a world historical fact" (*GI*, MECW, vol. 5, p. 51).

It was no part of the intention of the worker who invented this machine to disrupt the system and threaten the livelihood and even the lives of Indians and Chinese. He may not even find out that this has happened. That it does happen, however, is an inevitable consequence of a worldwide system of commodity production.

Commodities are loose cannons on the deck of society. It is superficial to blame the problems of the Indians and Chinese on the greed of individual capitalists; harms of this sort are endemic to a system that sets in motion large-scale social forces beyond anyone's conscious control, and such forces continue to manifest themselves today. At the time of this writing, American steel and textile production is shrinking dramatically as cheaper, imported products flood the market. Steel and textile workers who have been working at the same factory for decades are being thrown out of work and onto the street, and there is an important sense in which no one is to blame. It is the capitalist system, a system of commodity production, that produces these tragic unintended consequences.

Marx expresses this point vividly in *The German Ideology*: " . . . trade . . . rules the whole world through the relation of supply and demand—a relation which, as an English economist says, hovers over the earth like the fate of the ancients and with invisible hand allots fortune and misfortune to men, sets up empires and wrecks empires, causes nations to rise and to disappear . . ." (*GI*, MECW, vol. 5, p. 48).

Adam Smith thought of the Invisible Hand as a largely beneficial device for coordinating production on a large scale. For Marx, by contrast, it was a hand not connected to any brain, a social force that tosses lives about, helping some (usually capitalists) and harming others (other capitalists and most proletarians). It is in this sense that the worker's product confronts him as a hostile force. Market forces operate through the circulation of commodities (and money).[6] The alien nature of these forces consists in the fact that commodity production links producers together in ways they can neither understand nor control.

A final way that the workers' products manifest themselves as hostile forces consists in the inherently imperialistic nature of the capitalist mode of production. Given that the purpose of production under capitalism is to maximize surplus value, capitalists have an incentive to expand their markets and with it, the capitalist mode of production, to ever more distant places. As Marx and Engels say in a particularly colorful passage in *The Communist Manifesto*, "the cheap prices of its [the bourgeoisie's] commodities are the heavy artillery with which it batters down all Chinese walls, with which it forces the barbarians' intensely obstinate hatred of foreigners to capitulate. It compels all nations on pain of extinction, to adopt the bourgeois mode of production" (*CM*, MECW, vol. 6, p. 488). The capitalist mode of production, then, possesses a mechanism that requires its inmates to recruit others to their ranks by producing inexpensive commodities that serve both to undermine non-capitalist relations of production and to promote capitalist relations of production throughout the world.

To sum up, there are three objectionable features of capitalist society that are the result of the worker's alienation from his product:

1. The worker's product, *qua* commodity, is a mystified manifestation of social relations among producers. Its fetishistic quality prevents the worker from understanding the true nature of the social system of which he is a part and prevents him from conceiving of different ways production might be organized. In other words, not only must the worker participate in recreating the system that oppresses him, he will also be unaware that this is what he is doing.
2. The worker's product, *qua* commodity, is the lifeblood of a system that unpredictably wreaks havoc with people's lives by way of the operation of market forces. That is, large-scale social forces not subject to anyone's conscious control shape the fate of capitalists and workers alike, often to their detriment.
3. The worker's product, *qua* commodity, is the decisive weapon in the imperialistic war that the capitalist mode of production wages on other modes of production. The worker, then, creates the very weapons that are used to force others to share his fate.

An examination of Marx's explanations of these ills shows that they all satisfy the demands of the Critical Explanations requirement, as it has been articulated in the introduction to this chapter. That is, all his explanations appeal to fundamental features of the capitalist economic system, notably, the fact that capitalism is a system of commodity production, or production for exchange; this figures prominently in all three explanations. In both (1) and (2), it is obvious that it is the commodity nature of production under capitalism that explains both mystification and the capricious and unfortunate consequences of the operation of market forces. Finally, the explanation of the imperialistic nature of the capitalist mode of production appeals to the commodity nature of production, together with the fact that the goal of production is to maximize surplus value.

In addition, all of these critical explanations have or can be given microfoundations in that they can be articulated at the level of individual actions or at least kinds of actions. All of the explanations considered above have that character in that they are addressed to facts about the lives of individuals. The provision of microfoundations does not entail a strict reducibility of the social to the nonsocial; all it rules out are unanalyzed claims about the interactions of macrophenomena, such as what one finds in much of contemporary macroeconomics or in some functional explanations in Marxist theory of history. No such explanations have been employed above.

## Alienation from the Activity of Laboring[7]

Alienation manifests itself not only in the worker's relation to the product but also in relation to the activity of labor itself. In his discussion of this form of alienation, Marx comes closest to the contemporary notion of alienation as involving hateful work. In a dense passage in the *1844 Manuscripts* he begins an extensive list of undesirable features of alienated labor under capitalism as follows:

> What, then, constitutes the alienation of labour?
>
> First, the fact that labour is external to the worker, i.e. it does not belong to his intrinsic nature; that in his work, therefore, he does not affirm himself but denies himself, does not feel content but unhappy, does not develop freely his physical and mental energy but mortifies his body and ruins his mind.... It is therefore not the satisfaction of a need; it is merely a *means* to satisfy needs external to it. Its alien character emerges clearly in the fact that as soon as no physical or other compulsion exists, labour is shunned like the plague.... Lastly, the external character of labour for the worker appears in the fact that it is not his own, but someone else's, that it does not belong to him, that in it he belongs, not to himself but to another.... it is the loss of his self. [*EPM*, MECW, vol. 3, p. 274]

It is clear that a normative or valuational conception of human labor pervades this entire passage: There is some way labor is supposed to be, and it is systematically denied or perverted under capitalism. That is part of what it means to say that labor is alien under capitalism. Marx's claim that labor is not the satisfaction of a need is extremely important in this connection. It suggests that there is an original need to labor that is being systematically frustrated under capitalism. The explanation for this need will have to await the discussion below about species being. It is sufficient to point out here that one of the defects of capitalism is its systematic frustration of the need to engage in, how shall we say, truly human labor. Finally, as the last sentence of this quotation indicates, its alien character is also manifested in the fact that this labor belongs to someone other than the worker.

Let us try to lay out more carefully the specific harms or defects of alienated labor under capitalism and their respective explanations. According to the above passage, the frustration of the need to engage in truly human labor that the worker suffers manifests itself in at least three ways: It harms the body, it stunts the mind, and it becomes a mere means to needs external to it. Let us consider the first two together. That work under capitalism harms the body and stunts the mind are consequences of the intense and highly fragmented character of the production process in the workplace. The physical dangers workers face and the sheer physical exhaustion work engenders were well-established facts of capitalist society by Marx's time. The same is true of the highly fragmented nature of work, which makes the worker little more than an appendage of the machine. In *Capital* I especially, Marx documents all of this in excruciating detail.

His account of these evils of nineteenth century factory work is embedded in a larger theoretical examination of how the capitalist system operates. The key to understanding these features of labor under capitalism is to be found in the "boundless thirst for surplus value." At the beginning of Section 2 of Chapter X of *Capital* I, Marx contrasts capitalism with pre-capitalist economic formations on just this point:

> It is, however, clear that in any given economic formation of society, where not the exchange-value but the use-value of the product predominates, sur- plus labour will be limited by a given set of wants which may be greater or less, and that here no boundless thirst for surplus-labour arises from the nature of production itself. . . . But as soon as people, whose production still moves within the lower forms of slave-labour, corvee labour, &c are drawn into the whirlpool of an international market dominated by the capitalist mode of production, . . . the civilised horrors of over-work are grafted on the barbaric horrors of slavery, serfdom, &c. It was no longer a question of obtaining from him a certain quantity of useful products. It was now a ques- tion of production of surplus-labour itself. [*Capital* I, pp. 226–27]

Similar considerations explain the highly fragmented character of work under capitalism, which stunts the mind. The productive efficiency of orga- nizing work in this dehumanizing way was widely remarked on by earlier political economists. In his sarcastic discussion of Proudhon in *The Poverty of Philosophy*, Marx cites Adam Ferguson as the first to note this (*PP*, MECW, vol. 6, p. 181). This theme has been taken up and elaborated in great detail in the writings of twentieth century critics of capitalism, especially in the Sixties. Marx's contribution to this discussion is that he saw that this technical fact about production has far-reaching implications only under a particular mode of production—widespread commodity production driven by boundless thirst for surplus value—in a word, capitalism. In a system where the sole aim of production is to maximize surplus value, this technical fact determines how work is organized.

Another feature of labor under capitalism that makes it inhuman or a perversion of truly human labor is that it "is merely a means to satisfy needs external to it." Any kind of labor has an "internal" goal, that is, its particular purpose. The goal of shearing sheep is the removal of wool; the goal of raising crops is the production of food. However, such goals may or may not actually guide production. If some purpose outside of the process of production guides it, then labor is a mere means to other (i.e., external) needs. This recalls the familiar contrast between production for use and production for exchange.

In pre-capitalist systems of production for use, both the laborer and the oppressor have as their goals the production of use-values. Particular labor is not separated from its intrinsic purpose, the creation of particular use- values. And, the producer does not conceive of it as so separated. Of course, production under, say, feudalism is not the realization of a harmony of in- terests between serf and lord; the serf, after all, is being exploited. But production is guided by the need for use-values, and labor is a means to satisfy the need for those use-values.

By contrast, under capitalism, a regime of commodity production, labor has for the worker an overriding purpose or goal that is external to the labor process: wages, which he exchanges for things that satisfy other needs. The "external" character of labor is revealed in the fact that this money is something he would be perfectly content to get in the absence of working. Additionally, the capitalist's goal in giving (or, more accurately, selling) the worker the opportunity to labor is the realization of exchange value. For the worker, work is a mere means to what he really cares about, viz., wages, and for the capitalist, the ultimate purpose of the entire process of production (which is constituted by the labor of others) is the realization of exchange value as profit. For both, then, the purpose of labor is exchange value, something external to the labor process.

It may be that the need for truly human labor is nothing more than the need to fashion use-values to meet other needs. The fact that under capitalism, it is logically possible that the individual worker's goal could be realized without laboring implies that this need for truly human labor is not being met.

The above account goes no further than explaining what Marx might mean when he says that labor under capitalism is merely a means to satisfying needs external to it, but this "metaphysics of labor" does not specify exact harms or evils suffered by the workers, or society at large, as a result of the perverse nature of labor under capitalism. What is the harm and why? The most obvious suggestion is that "perverted labor" is a bad thing because it prevents the satisfaction of the need for truly human labor, and satisfaction of the latter is something intrinsically good. This makes sense out of the complaint that alienated labor is "unnatural," where the latter term has negative value connotations.[8] For now, let us assume that something like this is what Marx had in mind.

A more complete account of the harms of alienated labor emerges if one looks to what explains the purely instrumental character of "perverted labor" under capitalism. Marx's explanation for this is that the capacity to labor, labor power, is a commodity under capitalism. As the introduction to this chapter indicates, one of the distinguishing features of capitalism is that labor power takes the form of a commodity, which entails that it is bought and sold in the marketplace.

There are quite definite harms that are traceable to the commodification of labor power. To see what they are, recall that to say that something is a commodity is to say that the purpose of its production is sale in the market place. Since, by definition, a commodity is also a use-value, it follows that it is actually sold in the marketplace. If labor power is a commodity, it follows that the purpose of its production is sale in the marketplace; it also follows that it is actually sold. Let us consider what each of these implications amounts to.

How is labor power produced? No mystery here. The worker eats, rests, reproduces his kind, and so forth. Now for Marx to claim that labor power is a commodity is, according to his own definition of a commodity, to imply that the purpose of all of these activities is to recreate labor power for sale to the capitalist. Surely this is an overstatement. Workers have to eat, sleep, and so forth, in order to sell their labor power, but that is not the "goal" of these activities, in the sense that that is the end in view, to use Dewey's phrase. Contrast this with the production of rolled steel. Here the sale of the product can be accurately described as the goal of production, since those who control production have that as their end in view. On the other hand, eating, sleeping, procreating, and so forth, are engaged in for their own sakes, or because there is an original need for food, rest, sex, and so forth, or both. It is just false to say that the purpose of these activities is the sale of labor power.

Despite this overstatement on Marx's part, there is an element of truth in what he says that goes beyond the fact that there are biological prerequisites for labor. In the transition from feudalism to capitalism, the working day and the working year became much longer. There were fewer holidays and the rhythms of industrial life did not permit the erratic work schedules characteristic of the feudal manor. Perhaps more importantly, as capitalism geared up, the drive to maximize surplus value mandated a progressively longer and more intense workday. This soon reached the point where inhuman labor (in Marx's normative sense) occupied most of the worker's life. As a result, the rest of a worker's life had become much more subordinated to the job. For this reason, the sale of labor power becomes *a* purpose, if not *the* purpose of the production of labor power. In this sense, the worker comes to conceive of himself as just another commodity. Certainly the capitalist treats him that way. This is imperfectly captured[9] in a passage quoted earlier when Marx says that "labour power takes *in the eyes of the labourer himself* the form of a commodity" (*Capital* I, p. 167n1, emphasis added). Because labor power was not sold on a widespread basis prior to the predominance of capitalist relations of production, this attitude could not be prevalent in pre-capitalist modes of production.

A clear illustration of this general point can be found in contemporary higher education. Young people these days seek out education largely to improve their position in the "job market." They seek out education for its own sake about to the extent that nineteenth century proletarians ate their food for the variety of tastes and textures it had. To be sure, even in the darkest days of the Industrial Revolution, some people (usually members of the ruling class) ate for the joy of eating; and today some undergraduates (also usually members of the ruling class) seek education for its own sake. But clearly, this is the exception and not the rule, at least for the working class.

More generally, the commodification of labor power causes previously autonomous spheres of life to be conceived of by the workers themselves more as instrumentalities in the production of labor power and less as spheres of life that have intrinsic interest. Additionally, the increasing subordination of these spheres of life to the production of labor power is not merely a matter of changed perception—the realities change as well. As food becomes mere fuel, American fast food chains have begun to batter down the French walls of great cuisine. As education becomes job training, students take business ethics instead of studying Marx.

Let us turn to the second implication of the commodification of labor power: the fact that it is bought and sold. What the worker receives in exchange for his labor power is wages. The wage constitutes the purpose of labor for the worker under capitalism. As indicated earlier, this is a kind of perversion of the natural purpose of labor, which is to satisfy the need to labor. How does the fact that labor power is sold for wages harm the worker or society at large?

One obvious harm that results from this treatment of labor power is that it contributes to the widespread attitude in bourgeois society that a person's value *qua* person, is determined by his income. It is, after all, difficult to separate in imagination a person from his labor power. The worth of a person is thought to be determined by the value of that person's labor power. (Just ask any medical doctor in the United States.)

Secondly, when a person's labor power can be bought and sold, that person is not for a time a self-determining being. This is the significance of Marx's claim that the worker's labor is not his. During the workday, the capitalist tells the worker what to do and in some cases imposes restrictions on him off the job as well. This is one respect in which labor under capitalism is dehumanizing.

There is one final respect in which the worker's labor is alien and harmful. When the worker's product is sold in the market as a commodity, particular, concrete embodied labor takes the form of abstract labor—exchange value. The fact that labor takes this alien form of exchange value is responsible for two other harms that workers suffer under capitalism. To see what they are, it is necessary to take a brief detour through the Labor Theory of Value.

On one interpretation, the Labor Theory of Value asserts that the (exchange) value of an object is identical to the quantity of labor in it.[10] I say, 'is identical to' in part because Marx sometimes speaks of the value of a product as "embodied," "crystallized," or even "congealed" labor.[11] According to Marx, what the capitalist buys from the worker is not his labor, but his capacity to labor, that is, his labor power. Given that the value of something is the quantity of labor required to produce it, the value of labor power is the quantity of labor necessary to produce a given quantity of that capacity. Furthermore, given that things that exchange have equal value, this quantity of labor is exactly what the worker gets from the capitalist in the form of the subsistence wage. However, when the labor power the capitalist has bought is discharged in the process of production, it is creative, in that more value is produced than the value of the labor power. Roughly, it is this surplus that the capitalist rakes off a profit. So far, this is just standard Marxian economics. What does this have to do with alienation?

Its significance for alienation is this: The existence of profit is a necessary condition for the reproduction of the capitalist system. According to the Labor Theory of Value, profit, as surplus value, represents embodied labor. This labor forges anew each day the chains that bind the proletariat by reproducing capitalist relations of production.[12] This particular evil of capitalism is logically parasitic on other defects of capitalism, since if the capitalist system was actually in the best interests of the workers, this would hardly be an objection to it. But, given that there are other defects, this is an additional one. Prison systems, for example, do not visit on their inmates the indignity of requiring them to build their prison walls ever higher each day.

Finally, the fact that the worker sells the capitalist (not his labor but) his labor power, masks the fact that for part of the day the worker creates value for which he is not compensated, that is, for part of the day he works for free. This is the fetishism of the commodity of labor power. By contrast, in pre-capitalist systems of production for use, the expropriation of surplus value was transparent to the exploited and the exploiters alike. Thus, not only does the worker's labor serve in the daily re-creation of capitalist relations of production, but the worker is also completely unaware of this secret of surplus value.

What I have attempted to do in this section is to identify ills or defects of capitalist society that can be attributed to alienated labor and to explain their causes. A summary of this section, which lists these problems and which more explicitly addresses the Critical Explanations requirement would perhaps be useful at this juncture.

According to Marx, alienated labor consists in, or is responsible for, the following objectionable features of capitalist society:

1. Labor is physically harmful and mindless for many proletarians.
2. Labor is a mere means to needs external to it and is not an expression of the original need for truly human labor.
3. The worker conceives of himself and/or his labor power as a mere commodity.
4. Previously autonomous spheres of life become distorted by being harnessed to the production of labor power.
5. People come to think that the value of a person is determined by the value of his labor power.
6. Alienated labor under capitalism involves the domination of the worker by those who control the means of production, which under capitalism is the capitalist.
7. Alienated labor, in the form of surplus value, is required to sustain or recreate capitalist relations of production.
8. The mystification induced by the wage labor contract prevents the worker from recognizing number 7.

If the explanations of these facts are to satisfy the Critical Explanations requirement, they must appeal to some of the essential features of capitalism that constitute the two alternative characterizations of the latter identified in the introduction to this section, namely, the following:

i. The workers do not control the means of production.
ii. Labor power is a commodity owned by the workers.
i'. (Most) goods and services are produced as commodities, that is, for exchange in the market.

ii'. The motive for production by those who control the means of production is profit. That is, production is for profit, not for use.

The explanations for all of numbers 1 through 8 satisfy the Critical Explanations requirement, and all have or can be given microfoundations.

Regarding number 1, the technical fact that mindless and exhausting work is the most productively efficient is of significance under capitalism only because the goal of production is to maximize profits. Note that greed among all or most capitalists is not necessary to explain this aspect of labor under capitalism. Market forces, as manifested through competitive pressures, insure the survival of only those firms that produce most efficiently. Furthermore, competition is a necessary feature of any system of commodity production.[13] The explanation for number 1, then, appeals to two essential features of capitalism: (i') and (ii').

Numbers 2 through 5 all appeal to conditions (ii) and (i'), the fact that labor power is a commodity and the fact that commodity production is widespread. The essential perversity of labor under capitalism, reflected in the fact that it is done for the sake of some external need, is directly traceable to the commodification of labor power, which in turn presupposes the widespread production of commodities.[14] The same two conditions appear in the explanation of number 5: Only in a world where both the products of human labor and the capacity for labor have a price, which represents exchange value, is the value of a person conflated with the exchange value of his or her labor power. Finally, numbers 3 and 4, of course, directly appeal to (ii). All of these explanations have beliefs, attitudes, or situations of individuals as their explananda.

Number 6 is explained by (i) and (ii), given the additional premise that a person cannot be separated from his labor power: When the capitalist buys labor power, he rents its owner.

The explanation of number 7 appeals to (ii) and (i'). Expended labor has to take the form of exchange value, and so production must be for exchange, in order that capitalist relations of production can be reproduced. The worker's labor, in the form of surplus exchange value, is necessary for the reproduction of capitalist relations of production. In this way, the worker (helps to) re-creates the system that oppresses him.

Finally, the explanation of number 8 presupposes the labor/labor power distinction and the fact that what the worker is selling is a commodity—his labor power—the fetishistic nature of which obscures from him a correct appreciation that it is his unpaid labor that sustains capitalist relations of production.

## Alienation from Species Being

The term 'species being' (*Gattungswesen*), a remarkably awkward expression, is borrowed from Feuerbach, who used it to apply to man in general or mankind as a whole. Marx uses it infrequently and then primarily in his early writings. In explicating Marx's claim that, under capitalism man is alienated from his species being, let us begin with his account in the *1844 Manuscripts* of what it means to say that man is a species being. In the most perspicuous passage explicating this notion, Marx says,

> Man is a species-being, not only because in practice and in theory he adopts the species (his own as well as those of other things) as his object, but—and this is only another way of expressing it—also because he treats himself as an actual living species; because he treats himself as a *universal* and therefore free being. [*EPM*, MECW, vol. 3, p. 275]

The other passages are much worse. As a true representative of the species *Philosophicus germanicus* Marx explains the obscure by the more obscure. Let us see what sense can be made of this.

We start with the ordinary meaning of the term 'species'. In explaining the meaning of a species term (e.g., 'human', 'dog', 'dolphin'), taxonomists give us a list of characteristics common and peculiar to all those individuals to which the term applies. Perhaps, then, what lies behind this talk of species being is a set of characteristics common and peculiar to all humans, that is, a conception of human nature. It used to be widely believed that Marx denied that there was such a thing as human nature. And in fact he does criticize other thinkers' (e.g., Feuerbach's and Bentham's) conception of human nature.[15] But this does not imply that he believed there is no such thing as a human nature—it could just be that others have gotten it wrong.

Indeed, it would be pretty implausible to deny that there is such a thing as human nature.[16] 'Human' is, after all, a natural kind term. And, in the *1844 Manuscripts* Marx goes on at length about the differences between humans and other animals. Given that, there are really only two issues worth debating: One is whether or not he has picked out the right set of characteristics; the other is the use to which the concept of human nature is put. Our interest in this chapter lies with the second of these; the first would be taken up in a discussion of the Normative Theory requirement.

In what follows I shall lay out Marx's conception of human nature. This will make it possible to explicate what Marx means when he says that man is a species being. This in turn will permit an explanation of what it means to say that man is alienated from his species being and why that might be a bad thing. Finally, we shall investigate Marx's explanation(s) for this alienation.

Marx's conception of human nature centers on man's interaction with the physical world in the process of production. He argues that this interaction is fundamentally different from the manner in which all other animals interact with the world, and perhaps more importantly, that many features of the human condition can be explained on this basis.[17] Although all animals interact with nature to sustain themselves, man's interaction is universal in that all of nature is potentially of use to him in meeting his needs.

The universality of man appears in practice precisely in the universality which makes all nature his *inorganic* body—both inasmuch as nature is (1) his direct means of life, and (2) the material, the object, and the instrument of his life activity. . . . For labour, *life activity, productive life* itself, appears to man in the first place merely as a means of satisfying a need—the need to maintain physical existence. Yet the productive life is the life of the species. . . . The whole character of a species—its species character—is contained in the character of its life activity. [*EPM*, MECW, vol. 3, pp. 275–76]

The last sentence of this quotation strongly suggests that the nature of a species is its life activity. Since, for man, labor is life's activity, this implies that the capacity to labor is at the core of Marx's conception of human nature. ("At the core of" is a somewhat fuzzy expression, but it will do for now.) Human labor is distinguished from the "needs-meeting" activity of animals in two respects: (1) It is universal (i.e., it has the potential to take any element or aspect of nature as its object), and (2) It is purposive. These both entail that labor is free. This needs some explanation.

In a passage quoted in Chapter 1 (*Capital* I, p. 174), Marx says that what distinguishes the worst architect from the best bee is that the former raises up in imagination an idea of the product of his labor before it is actually produced. This is what makes labor purposive.[18] Given that all of nature is potentially an object of labor, it follows that it is within man's power to choose what purposes to (try to) realize. It is in this sense that labor is free. Put another way, (positive) freedom is control, and man's ability to control his environment requires that he be able to conjure up "ends in view" or purposes that he then attempts to realize. It also follows from (1) and (2) that it is possible for labor not to be directly determined by immediate physical needs and to be thereby an expression of the need to labor discussed in the last section. Otherwise, either not all of nature would be potentially an object of labor or man's needs-meeting activities would be directly determined by his needs, as in the case of animals.

Human nature, then, consists in the capacity to engage in a indefinitely large number of needs-meeting activities through (purposive) labor. The first clause of the first quotation in this section suggests, though it does not actually state, that *only* man is a species being, though obviously he is not the only being who is a member of a species (and thus has a nature). He is, however, the only being who recognizes his nature. When Marx says that "man . . . adopts the species as his object" and "treats himself as the actual living species," he means that man conceives of himself as a being who labors.

This interpretation gathers further support from Marx's discussion (shortly after the passage quoted immediately above) of labor as conscious life activity: "Conscious life activity distinguishes man immediately from animal life activity. It is just because of this that he is a species being, i.e. that his own life is an object for him. Only because of this is his activity free activity" (*EPM*, MECW, vol. 3, p. 276). To say that his own life is an object for him must mean that his life activity (i.e., his capacity for free, universal labor) is an object for him. To say that it is an *object for* him entails that he is aware of this fact about himself. That is, he conceives of himself as a being who has this capacity. To sum up, human nature consists in the capacity to engage in free, universal production. To say that man is a species being is to say that he (alone) is aware of his nature. At least, that is the best sense I can make out of it.

Now, what would it mean to say that man is alienated from his species being? In one sense,[19] 'alienation' implies loss. To say that *y* has been alienated from *x* implies that *y* has been lost to *x*. It can also mean that the alienated thing is not really lost but that it is no longer understood by the person from whom it is alienated. For example, I have become alienated from some of my undergraduate papers by losing them; there are others that I still have, but I can't believe they are mine.

This gives us a handle on interpreting Marx's claim that, under capitalism, man—or at least the proletarian—is alienated from his species being. It could mean either of two things: It might mean that he has lost his human nature, in the sense that he no longer has the capacity for free, universal production. That is, it may be that the proletarians have become so *degraded* that they have lost the capacity for free, universal production, in the same way that an accident victim who has been blinded has lost the capacity for sight. Given Marx's conception of human nature, this entails that these proletarians are not . . . *really human*. On the other hand, the claim that proletarians have been alienated from their species being might mean that, although they still have the relevant capacity, they does not recognize this capacity as theirs. The second alternative appears to be more modest: The proletarians really do have the capacity to engage in universal and free production, but the capitalist system masks this fact from them.

These interpretations can be rendered consistent and indeed complementary by noticing an ambiguity in the term 'the proletarians'. The latter can be understood either distributively or collectively. If we understand it distributively, the first interpretation is plausible; if we understand it collectively, the second interpretation is plausible. That is, it may be that individual proletarians have lost the capacity for free, universal production as a result of living and working in a capitalist system (though it is of course possible that, as individuals, they can regain it). On the other hand, collectively, they may have this capacity, though this is effectively masked from them. Let us pursue this second point first.

There is no question that capitalism has universalized production to an historically unprecedented extent. The range of both producer and consumer goods fabricated under the capitalist mode of production is truly astonishing. On a collective level, Man has produced according to his nature. On the other hand, individuals have not. This immense productivity has been possible up until now only because of the division of labor both within the firm and across society. Division of labor within the firm has, of course, been characterized by the extreme fragmentation of task (one-sidedness) and the riveting of the worker to a particular task (unfreedom). As was explained in the second section of this chapter, the division of labor across society, which is a consequence of commodity production, integrates production in a thoroughly mystified way.

Let us suppose that individual proletarians have lost the capacity for truly human labor. As noted above, on Marx's conception of human nature, that means they are not truly human.[20] Some of the strangeness of this claim dissipates if we understand Marx's conception of human nature to be partly normative. Roughly, his conception of human nature in part describes how humans ought to be and not merely how they in fact are. Just as in the last section, truly human labor was distinguished from the "perverse" labor characteristic of capitalist society, here we can distinguish truly human beings (of which capitalist society has few) from proletarians. If this judgment sounds excessively harsh, it is worth noting that it is ultimately a judgment about capitalist society—or at least that is what a radical critic would say.

The harms or defects of capitalist society resulting from the workers' alienation from their species being and the respective explanations are in part implicit in some of the above discussion, but a more explicit and detailed account of both would be useful. Let us begin with the "degrading" of the proletarians that constitutes one manifestation of alienation from species being. This consists in the dehumanizing labor proletarians must perform under capitalism. Some of the harms this inflicts were discussed in the last section. That man is a species being whose nature consists in the capacity for free, universal productive labor greatly magnifies the tragedy this involves.

This point can best be appreciated by way of contrast with a more bourgeois conception of human nature, say, Bentham's. For Bentham, what man seeks is pleasure and the absence of pain (both broadly construed to include pleasures and pains of the mind). A bourgeois apologist for capitalism might say, "Look, it's true that labor under capitalism is not much fun; most people really don't like their jobs. But, there are compensations. The standard of living is rising, even for some proletarians. That it takes a lot of disagreeable work, even inhuman labor, is just part of the price that must be paid." Behind this response is a conception of labor as purely instrumental and, in some sense and in the final analysis, just not that important. Labor has been devalued by conceiving of it as purely instrumental, since the valuable ends it serves can in principle compensate for its dehumanized character.

By contrast, if Marx is right, the dehumanization of labor is a much deeper tragedy. Given his conception of human nature, the dehumanization of labor is the dehumanization of persons. One way of looking at what the alienation from species being amounts to is that it magnifies enormously the importance of the harms and defects discussed in the previous section on the alienation from labor. Absent Marx's conception of human nature, it is open to his critics to admit the points he wants to make about alienated labor but deny their larger significance or argue that the evils of alienated labor can be traded off for other goods. On the other hand, if Marx is right, this strategy looks much less promising: Degraded labor logically implies degraded humans.

To explain this alienation from species being, it is necessary to explain two things: (1) the dehumanization of the worker, and (2) the mystification that prevents the proletariat from recognizing their capacity for free, universal labor. Let us begin with number (1). Given Marx's conception of human nature, to explain the dehumanization of the worker, it is sufficient to explain the dehumanization of labor. The key to this, of course, is the division of labor. Labor under capitalism is divided in two respects: within the firm and across society. As noted in the last section, it is a technical fact about most production that extreme fragmentation of task increases productive efficiency. This fact is significant under capitalism only because capitalism is a system of production for exchange in which the goal of production is the maximization of exchange value. Competition requires all capitalists to adopt production methods that are maximally efficient, which, given the technical fact alluded to above, explains why workers are forced into narrow, repetitive tasks in nearly all sectors of the economy.[21]

Another aspect of dehumanized labor is explained by the societywide division of labor characteristic of capitalism. Suppose that a capitalist firm could efficiently organize production in such a way that workers were not required to do narrow, repetitive tasks. Still, they would be making only one, or at most a few, products. Productive efficiency demands that (groups of) workers be tied to the production of one or a few products. By contrast, in pre-capitalist modes of production, individuals (and groups) produced in a much more universal manner. Feudal serfs, for example, produced their own food, clothing, shelter, and so forth. On the other hand, even the most versatile modern lawyer under capitalism can do at most only a narrow range of things. He may work on a bankruptcy case in the morning, a personal injury case in the afternoon, a tax case in the evening, and a divorce after dinner. But he won't be able to fix a leaky faucet.

The societywide division of labor, which is peculiar to capitalism, is a direct consequence of the fact that capitalism is a system of commodity production where the goal of production is to maximize surplus value. Only under such circumstances are groups of workers forced to specialize in both of the ways just alluded to.

Let us turn now to the mystification of the proletariat's capacity for free, universal labor. Here the explanation is straightforward: The fetishism of commodities prevents them from understanding the ways in which their individual, one-sided labor is socially integrated. It makes it seem that commodity production is the only way individual labor can be integrated (or made universal).[22] For this reason, they fail to recognize their collective capacity for free, universal labor, which is implicit in the level of development of the forces of production they have collectively achieved under capitalism, and which is latent in them as individuals. The fact that capitalism is a system of commodity production (and the goal of production is maximization of exchange value), then, is responsible for this form of the proletariat's alienation from its species being.

The above explanations of the harms or evils of the proletarians' alienation from their species being all appeal to essential elements or processes of capitalism, again especially the fact that capitalism is a system of commodity production. Thus they satisfy the Critical Explanations requirement. Furthermore, they are either cast at the level of, or can easily be extended to the level of, the effects of social phenomena on individuals.

Thusfar the discussion has proceeded on the assumption that it is the proletarian who is alienated from his species being. There are two other aspects of alienation from species being which are much broader, in the sense that they can afflict any member of capitalist society. One concerns religion and the other the role of money in capitalist society. Let us consider the second first.

A section of the third manuscript in the *1844 Manuscripts* is called, "The Power of Money." It contains a scathing critique of the power of money in bourgeois society. Like other forms of alienation, it involves mystification, and what is mystified is man's species being:

> The distorting and confounding of all human and natural qualities . . . the *divine* power of money—lies in its *character* as men's estranged, alienating and self-disposing *species-nature*. Money is the alienated *ability of mankind*.
>
> That which I am unable to do as a *man*, and of which therefore all my individual essential powers are incapable, I am able to do by means of *money*. . . . If I have the *vocation* for study but no money for it, I have *no* vocation for study, that is no *effective*, no *true* vocation. On the other hand, if I have really *no* vocation for study but have the will *and* the money for it, I have an *effective* vocation for it. [*EPM*, MECW, vol. 3, p. 325]

Unlike most of the other passages from the *1844 Manuscripts* quoted in this chapter, this one does not require much comment and discussion. The main point about species being is that the universality of man's activities comes to reside in money. What is bad about all of this is that it distorts individual human beings and our perceptions of them. For example, a rich member of the bourgeoisie appears artistically cultivated because he has bought many works of art, while a poor person who would benefit from art is denied access to it.

It might be thought that this defect is not peculiar to capitalist society; after all, money has been around quite a lot longer than capitalism has. However, what is distinctive about money under capitalism is its ubiquity and its overriding importance in human affairs. Prior to capitalism, there were many things of great value (e.g., noble lineage) that either could not be bought with money or could be bought with money only with great difficulty or under exceptional circumstances. Relatedly, only under capitalism is social power so directly a function of money; under other social forms, power is not so directly a function of money.

The explanation for this state of affairs is no mystery: Capitalism is a system of production for exchange (commodity production). Production for exchange on a widespread basis cannot take place in a barter system; money takes on its important role because of its functional necessity for a system of commodity production.[23]

One final aspect of man's alienation from his species being concerns religion. Marx's (theoretical) remarks on religion are rather sketchy, and his primary interest was to link religious criticism to political criticism. In two of the works where religion is discussed at greatest length—"On the Jewish Question" and the introduction to the *Contribution to the Critique of Hegel's Philosophy of Law*—Marx does not mention the concept of species being. His references to religion in the *1844 Manuscripts* are always glancing. What follows, therefore, is somewhat speculative but, I think, a plausible reconstruction of Marx's thought.

An important part of Feuerbach's (and others') critique of religion was adopted by Marx—the claim that religion represented a form of alienation. More specifically, man projects the perfection of his own qualities onto the supernatural.[24] God, then, is an alienated form of Man. Marx criticized Feuerbach for not seeing that an adequate understanding of this phenomenon is to be found in the ensemble of social relations among men (*Theses of Feuerbach* [IV and VI], MECW, vol. 5, p. 4). Still, part of Marx's critique of religion can be plausibly construed as consisting of the complaint that belief in a Supreme Being is a manifestation of man's alienation from his species being, in the sense that man projects onto God (without recognizing it) his human nature—the capacity for free, universal production—by attributing to God both omniscience and omnipotence.

Religion is no more than an instrumental or logically parasitic evil in that it is a manifestation of more fundamental problems in civil society. Marx expresses this point with great eloquence in the following passage: "Religion is the sigh of the oppressed creature, the heart of a heartless world, just as it is the spirit of spiritless conditions. It is the *opium* of the people" (*Contribution to the Critique of Hegel's Philosophy of Law*, MECW, vol. 3, p. 175).

A critique of religion consists in the exposure of its causes and its function in civil society. It is not easy to tie the explanation of religion to specific essential features of capitalism, as it is done for the other ills of capitalist society discussed in this chapter. However, it is probably fair to say that Marx believed that civil society would have to be abolished or overcome (*aufhebung*) to eliminate the conditions that make religion both possible and necessary. In this sense, his critique of religion is radical.

### Alienation from Others

Marx's account of the alienation of persons from other persons under capitalism has three facets: the alienation of the state from civil society, the estrangement that characterizes the relation between the worker and the capitalist, and the generalized interpersonal alienation that reflects the commercialization of life in capitalist society.

Let us begin with the state and civil society. A complete account of Marx's views on the state and civil society (especially in relation to Hegel) will not be attempted here. The ensuing discussion is narrowly focused on the exact nature of the alienation of the state from civil society, why this is a defect of capitalist society, and what its explanation is.

'Civil society', a term Marx takes over from Hegel, roughly corresponds to what we would call "the private sector" in the capitalist economic system. Marx says that it "embraces the whole commercial and industrial life of a given stage . . . [and] only develops with the bourgeoisie" (*GI*, MECW, vol. 5, p. 89). The use of the word "embraces" makes it unclear whether civil society consists of more than that. For instance, does it include other inter-personal relations among individuals in capitalist society? For present pur-poses, I am not sure that it matters. It is clear, however, that civil society is distinct from the state. And, according to Marx, the state is an alienated form of civil society.

To understand the nature of this form of alienation, it is necessary to make a brief foray into Marx's theory of the state in capitalist society. Ac-tually, "theories" would be a more accurate word, since two such views can be found in his writings.[25] One holds that the state is merely the executive committee of the ruling class whose main purpose is to manage the "common affairs" of that class. (See, e.g., *GI*, MECW, vol. 5, pp. 90, 355; and *CM*, MECW, vol. 6, pp. 486, 505.) Alongside of this theory we find suggestions of a more complex view. On this account, the state, as represented by the bureaucracy, takes on a life of its own as it were, and achieves a certain autonomy from the ruling class. In both of these theories, the concept of alienation plays a key role.

Let us begin with the first theory. The separation of the state from civil society involves the alienation of individuals from themselves and other people:

> Where the political state has attained its true development, man . . . leads a twofold life, a heavenly and an earthly life: life in the *political community*, in which he considers himself a *communal being*, and life in *civil society*, in which he acts as a *private individual*, regards other men as means, degrades himself into a means, and becomes the plaything of alien powers. . . . In the state, . . . [man] is the imaginary member of an illusory sovereignty ["On the Jewish Question," MECW, vol. 3, p. 154]

The main idea is that capitalist society is really a dual world of civil society and the state. The latter is a distorted and mystified expression of the former. This distortion is manifested in two ways: (1) The state is supposed to pursue the common good when in fact it represents the interests of the ruling class; (2) In the democratic state, each man is supposed to be equal and equally sovereign, but in fact, the bourgeoisie rule. Mystification is achieved through the dominant ideology (more on this shortly) and the formal equality of all citizens.

Why exactly does this make capitalist society defective? What exactly is the harm? As usual, Marx is not very explicit, but it is not hard to hazard a guess. Civil society needs the state in order to persist as a stable social form. The fact that this stability cannot be achieved without mystification is an indication that civil society is itself defective. In the above quotation, Marx briefly mentions some of the defects of civil society, in particular that men treat each other and themselves as (mere) means. More generally, since civil society just is the private sector, the defects of civil society include all the defects of the capitalist economic system.

The alienation of man from man also expresses itself politically in the primary function of the bourgeois state—the protection and enforcement of the right to private property.[26] "The right of man to private property is, therefore, the right to enjoy one's property and to dispose of it at one's discretion . . . without regard to other men, independently of society, the right of self-interest. . . . It makes every man see in other men not the *realization* of his own freedom, but the *barrier* to it" ("On the Jewish Question," MECW, vol. 3, p. 163). Private property is the legal expression of man's separation from man. The above quotation suggests that private property is, in one sense, inherently antisocial, in that it protects and reinforces individuals in their egoism. It is the great atomizer of civil society. In addition, the right of private property permits individual capitalists to pursue their interests in ways that harm the proletariat and society at large. The state's role is to enforce and reinforce this right to private property, as, for example, when the police crack down on strikers.

As with other forms of alienation, there is mystification involved in this context as well. This mystification is provided primarily by ideology. The ideology of bourgeois society proclaims the natural (God-given, for John Locke) right of private property. Bourgeois political economists, especially the vulgar variety, put out the story that the system of private property is in the best interests of all. In this way, the state is (falsely) represented as acting for the common good, when in fact it represents the interests of the ruling class.

In sum, the guarantee of the right of private property, which is the primary function of the bourgeois state on the first theory, is a causal factor in the production of two kinds of harms or defects of capitalist society: (1) the harms inflicted by the economic system on individuals and society at large, and (2) the perpetuation of egoism that accompanies any system that recognizes a right of private property in the means of production.

The explanations for these ills of capitalist society are rooted in the essential features of capitalism. The harms inflicted by the economic system are simply whatever harms are traceable to these essential features. The perpetuation of egoism is probably a functional necessity for capitalist relations of production. Some regard functional explanations as prima facie suspect, but the theory of ideology has a role to play in explaining how the functional necessity for egoism contributes to its incidence. More specifically, the ethos of individualism and laissez-faire (which includes respect for private property) that came to prominence with the rise of capitalism is one of the most important causal factors in reinforcing this egoism.[27]

Marx's second theory of the state is more subtle and sophisticated. On this theory, the state is not merely a passive creature of the bourgeoisie. It takes on a life of its own as it were, and develops its own interests, which may not always coincide with those of the ruling class and only accidentally with those of society at large.

The semiautonomous character of the state is represented or made manifest by the rule of the bureaucracy. In the *Eighteenth Brumaire* he says of the French bureaucracy:

> Every *common* interest was straightaway severed from society, counterposed to it as a higher, *general* interest, snatched from the activity of society's members themselves and made an object of government activity, from a bridge, a schoolhouse and the communal property of a village community to the railways, the national wealth and the national university of France.[28]

The evils of bureaucracy are not, of course, confined to the French system.

In commenting on Marx's observations about the bureaucracy, Shlomo Avineri has said: "The sociological significance of Marx's analysis of bureaucracy lies in his insistence that bureaucratic structures do not automatically reflect prevailing social power relations but pervert and disfigure them. Bureaucracy is the image of prevailing social power distorted by its claim to universality."[29] Detailed discussions of the evils of bureaucracy and its causes are hard to find in Marx's writings. The above-quoted passage from the *Eighteenth Brumaire* is more of an aside in an historical narrative that the opening statements of an elaborate theory. However, we do know that the bourgeois state itself is explained, largely if not completely, in terms of its functional necessity for capitalist relations of production. Whether or not this explanation can be extended to the bureaucracy and its evils (and articulated at the microlevel) is unclear. Nevertheless, we can be reasonably confident that Marx believed that the evils of bureaucracy were deeply rooted in such a way that reforming the state would not significantly ameliorate them. Instead, what is required is the abolition of civil society itself. This might explain why he did not devote much attention to the bureaucracy in his theoretical writings; if the bureaucracy is a distorted expression of social power inherent in capitalist relations of production, the latter should be the focus of attention for the radical critic and for revolutionary action.

The second form of alienation mentioned in the beginning of this section is the estrangement of the worker and the capitalist. In the *1844 Manuscripts* Marx presents this form of alienation as a direct consequence of the forms of alienation discussed in the previous three sections of this chapter (*EPM*, MECW, vol. 3, p. 277). The argument here is hard to follow, though two points are salient: First, there is an unremitting conflict of interest between the worker and the capitalist, since the capitalist realizes a profit (surplus labor in the form of surplus exchange value) only to the extent that he is able to exploit the worker. We shall return to this point in the discussion of exploitation in subsequent chapters. Second, by creating surplus value for "the other," the worker's labor, and by implication, the worker, is an indispensable element in the re-creation of capitalist relations of production. Why this is a problem and what its explanation is were discussed in the first section of this chapter on the alienation of the worker from his product.

The final form of alienation between man and man to be canvassed in this section is the generalized interpersonal alienation that reflects the commercialization of life in capitalist society. In a little-discussed commentary on James Mill's *Elements of Political Economy* Marx gives us one of the clearest and most radical statements of his critique of alienation in capitalist society to be found in the entire corpus. In contrast to the *1844 Manuscripts* and other early writings, he is much clearer in his identification of specific ills or defects of capitalist society and their causes. Although this commentary on Mill ranges over nearly all manifestations of alienation, two of them receive more attention here than elsewhere. One concerns the character of interpersonal relations implicit in the credit relation; the other concerns the inherent alienation that relates those who produce for exchange. Let us begin with the former.

Marx begins his discussion of credit by considering the case of a rich man who lends money to a deserving poor man—an exception but a genuine possibility—credit under the most favorable circumstances, according to the bourgeois apologists. The purely instrumental nature of this exchange is revealed in the fact that the poor man's life, talent, and activity serve to guarantee the loan: "One ought to consider how vile it is to *estimate* the value of a man in *money*. . . . Credit is the *economic* judgment on the *morality* of a man."[30]

Other examples of this phenomenon abound. When a small business is sold, it is customary for accountants to assign a monetary value to *good will*, the reputation the firm has built up over the years. The new owners may not be men of good will, but they can buy it from the previous owner; indeed, they must buy it, since no owner, unless he is a fool, would throw it in for free. The inherent "falseness" of this was discussed above in the section on alienation from species being; here the point concerns the perversity of buying and selling human virtues.

Relatedly, the man who cannot get credit is not only poor, he is judged morally unworthy because he lacks (credit) worthy character traits. His inability to get credit is a manifestation of poor character. Of course, it is often hard to tell who is *deserving* and who is not. Thus begins the game of snooping on the part of the potential creditor and dissimulation on the part of the prospective debtor ("Comments on Mill," MECW, vol. 3, p. 216).

Now, borrowing and lending have been around for a long time—indeed longer than money. Why is the credit system a defect of *capitalist* society? Marx points out that only under capitalism is the credit system fully developed;

indeed, it is only under capitalism that it makes sense to speak of a credit *system*. The reason for this is that only under capitalism has money (without which borrowing and lending on a widespread basis would be impossible) assumed its preeminence. This is a direct consequence of the fact that capitalism is a system of production for exchange. When commodities are produced on broad scale, barter is impractical. Presumably, credit transactions have always had (at least embryonically) the vile characteristics Marx identifies, but it is only under capitalism that these transactions are common and pervasive. In consequence, only under capitalism do people widely relate to each other in this manner.

Toward the end of his commentary on Mill, Marx pushes his analysis farther—to production for exchange itself. In a few remarkable pages he comes to identify exchange, and production for exchange, as inherently vicious. To see why this is, imagine two individuals, *A* and *B*, each of whom needs two things, *x* and *y*, in some given quantity to live or to live well. *A* produces *x* and *B* produces *y*. If *A* produces more *x* than he needs for his own consumption and *B* produces more *y* than he needs for his own consumption, that is, if each produces a surplus, an opportunity for exchange exists. Suppose further that each is aware of this beforehand and that some production by both *A* and *B* takes place for the purpose of mutual exchange. Marx says that this surplus production

> does not mean rising above selfish need. On the contrary, it is only an indirect way of satisfying a need.... Production has become a *means for gaining a living*, labour to gain a living....Our production is not man's production for man as a man, i.e., it is not *social* production.... Each of us sees in his product only the objectification of his *own* selfish need, and therefore in the product of the other the objectification of a *different* selfish need, independent of him and alien to him. ["Comments on Mill," MECW, vol. 3. p. 225]

Production for exchange, then, by its very nature, is defective or objectionable in a number of respects: (1) It is inherently selfish; (2) It is a mere means to an end, and not something undertaken for its own sake (i.e., as an expression of the need to labor) or as something undertaken to meet the need of another; (3) It is part of a system of production in which people's selfish needs confront each other in the form of their products.

The alienating character of labor under capitalism and the fetishism of commodities is implicit in all of this. Our main interest in this section, however, is in the character of the social relations among individuals implicit in a system of production for exchange; Marx's view is quite clear: It is one of mutual using or exploitation.

> Far from being the *means* which would give you *power* over my production, [your need, your desire] are instead the *means* for giving me power over you.... The *social* relation in which I stand to you, my labour for your need, is therefore also a mere *semblance*, and our complementing each other is likewise a mere *semblance*, the basis of which is mutual plundering. The intention of *plundering*, of *deception*, is necessarily present in the back-

ground, for since our exchange is a selfish one, on your side as on mine, and since the selfishness of each seeks to get the better of that of the other, we necessarily seek to deceive each other. ["Comments on Mill," MECW, vol. 3, pp. 225–26]

Strong stuff. I take Marx to be arguing for the view that the deep explanation for egoism in capitalist society is to be located in the fact that capitalism is a system of production for exchange. It might be objected that egoism was not unknown in pre-capitalist societies and that it has obvious survival value.

I suspect Marx would respond by pointing out two facts: (1) Although exchange, and production for exchange, (not to mention the use of money) has occurred in all pre-capitalist societies, excepting perhaps primitive communist societies, what distinguishes capitalism is that this is the predominant form of production. And only under capitalism is the *individual* (as opposed to some group to which the individual belongs) the focus of nonaltruistic behavior. The explanandum under discussion is not selfish, egoistic behavior per se, but its ubiquity. (2) The survival value of selfish behavior presupposes a social system that rewards it and punishes alternative behavior patterns. Capitalism is just such a system. Not all actual or possible social systems have this characteristic.

There is an independent argument for Marx's position: In the mutual alienation of owned objects that constitutes exchange, there is always a more favorable alternative for each side which can be achieved only at the other's expense, to wit, the price could be less, and at the limit, zero. Exchange is not entirely a zero-sum game, since both parties can benefit from it, but within the parameters that make exchange possible, one person's loss is another's gain. The inherently conflictual nature of exchange fosters concern for the self because it is the self who gets taken advantage of if the other party is skillful. We each search out the other's "reservation price," as economists call it. This is, in effect, the worst bargain the other person will accept. The social structure of the exchange situation puts people in a position where they feel compelled to act in this way. If this is right, the conflict of interest between the capitalist and the worker is a special case of the conflict of interest inherent in all exchange.

Marx's explanation for the ubiquity of selfish, egoistic behavior in capitalist society accords well with his view that humans transform themselves by means of the way they produce. They find themselves in a set of relations of production, which are not of their own making and yet which profoundly influence their lives. Marx's explanation of the ubiquity of egoism also explains why he derides bourgeois apologists who claim that selfish, egoistic behavior is part of the fabric of the human condition. At most it is part of the fabric of the capitalist mode of production. On the other hand, all this leaves somewhat unclear the nature of truly human production and its interpersonal consequences. Marx briefly discusses the former at the end of his comment on James Mill. We shall return to this in our discussion of the nature of labor and production in post-capitalist society.

**Some Preliminary Conclusions**

In surveying Marx's account of alienation as it has been developed to meet the requirements of a radical critique, there are a number of general points worth remarking on. One is that Marx is able to explain an extraordinary range of objectionable features of life in capitalist society by appeal to a small number of essential features of the capitalist economic system. At the crossroads between science and human values, Marx shows himself to be an exceptionally good scientist. Although all of these essential features get called on to appear in the explanans of the various critical explanations, one or perhaps two do most of the work: The fact that capitalism is a system of commodity production and the fact that labor power is a commodity.

The general theoretical significance of these features of capitalism, especially commodity production, has been largely underappreciated by Marx scholars.[31] Part of the reason for this is that the relation between alienation on the one hand and commodity production and the commodification of labor power on the other in Marx's writings has been partially obscured by the fact that these topics commanded his attention at different times in his career. In the early writings, where alienation is a prominent topic, he had not clearly identified the general theoretical importance of commodity production for understanding capitalism. In the *German Ideology* and the *1844 Manuscripts*, for example, private property and the division of labor appear to be accorded greater significance; both presuppose commodity production, though Marx did not seem to be aware of this at the time, or if he was, he did not make much of it. Commodity production first gets extensive treatment in the *Grundrisse* and *Capital*. By that time, alienation was not as prominent in Marx's critique of capitalist society (exploitation was), and Marx's theoretical normative interests were in abeyance to some extent. However, the earlier account of alienation is never repudiated, and the later writings serve to augment and further articulate that account.

Another common theme in Marx's discussions of alienation is that many of the forms of alienation involve some kind of mystification. Essential features of capitalist society mask what is "really going on." More specifically, they mask defects of this form of social organization, usually (but not always) from the people whose interests are adversely affected by the system. Obviously, this is an instrumental evil, at least for the proletariat, insofar as mystification imparts stability to capitalist relations of production. However, there may be more to it than that. Sometimes Marx seems to think that the sheer "falseness" involved in a manifestation of alienation is a bad thing in and of itself. A good society is not plagued by systematic misconceptions about its nature. Its members understand their human nature, their production, and their fellow humans. This connects up a with a number of views advanced in the recent literature, which posit that central intrinsic values for

Marx logically presuppose self-knowledge and knowledge of social relations among individuals.[32] A necessary condition for a good society is that it is a "transparent society."

The account of alienation developed in this chapter also hints at a form of naturalism in Marx's value theory. In two facets of alienation, the concepts of truly human labor and truly human beings are prominent. Marx's condemnation of capitalism is sometimes best captured by the anachronistic use of psychoanalytic terms such as 'perversion'. The line between facts and values is blurred, but that is probably a strength and not a defect of the normative theory implicit in Marx's critique of capitalism.[33]

An important consequence of this discussion of alienation is that it imposes some very significant burdens of proof on Marx's thought when it comes to the Alternative Institutions requirement. A necessary condition for satisfying the latter is that the alternative institutions must not reproduce the defects of existing institutions. For example, Marx's critique of interpersonal relations under capitalism is directed at the fact that capitalism is a system of production for exchange. Implicit in this critical explanation is the view that any system of production for exchange will "commercialize" interpersonal relations. Assuming that post-capitalist society avoids this problem presupposes that post-capitalist society will not be a system of production for exchange, that is, it will not have a market economy. Indeed, in Chapter 6 I shall argue that Marx's repeated appeals to commodity production in his critical explanations of alienation constitute part of the motivation for Marx's rejection of markets—and his commitment to central planning—in his conception of, or proposals for, post-capitalist society. But this gets ahead of our story.

# 3

# Exploitation:
# Marx and Böhm-Bawerk

One of the most important charges that Marx levels against capitalist society is that the worker is systematically exploited by the capitalist. The 'systematic' part is important; this exploitation is not accidental or incidental. Rather, it is built into the very structure of the capitalist economic system. Or so Marx claims.

Marx's position is controversial. While defenders of capitalism won't deny that exploitation occasionally takes place in capitalist society, they deny the proposition that all the workers are systematically exploited in capitalist society. That is, they deny, and Marx affirms, the idea that capitalism is exploitative by its very nature. The main purpose of this chapter is to evaluate Marx's position on this question.

My strategy will be to begin with a close look at Marx's actual path of argumentation in *Capital* I to the conclusion that the capitalist necessarily and systematically exploits the worker. Part of my intention in tracing out this path is to identify a number of substantive theses in economic theory that Marx believed and to indicate how he thought they hang together to support the charge of exploitation. Some of the arguments for these theses are obscure and not very convincing, but later in this chapter I shall show that he has available other, more plausible arguments.

However, the chain of argumentation to the conclusion that the workers are exploited depends crucially on the labor theory of value, which many contemporary Marxists have rejected. Unfortunately, some of the problems with Marx's original argument have not been correctly understood. In particular, the objections to the labor theory first offered by the nineteenth century Austrian economist Eugen Böhm-Bawerk have been ignored or inadequately appreciated by nearly all contemporary Marxists.

The result has been that attempts to rehabilitate Marx's charge of exploitation end up facing the same problems confronting Marx's original argument. This will be shown in detail in Chapter 4, which consists of a systematic critical evaluation of recent attempts to establish the Marxian charge of exploitation. Chapter 5 consists of a discussion and critical evaluation of the economist John Roemer's reconstruction of Marx's account of exploitation. Since Roemer is not trying to capture Marx's exact intentions, he does not face the problems identified in Chapter 4, though, as I shall show, he does face some distinctive burden of proof difficulties. The larger purpose of Chapters 3, 4, and 5 is to explicate and evaluate all the ways in which Marx or a Marxist might try to sustain a charge of systematic exploitation against capitalism.

## Marx's Argument of *Capital* I

Volume I of *Capital* opens with a lengthy and often obscure discussion of the nature of a commodity. Marx begins by making the standard (for classical economics) distinction between use value and exchange value. In short order, Marx establishes a principle about exchange value, which is of fundamental importance for his economic theory: the Law of Value (hereafter the LV). Simply stated, the LV says that commodities that exchange in the market are equal in value. Marx states the LV and suggests an argument for it in the following passage:

> A given commodity, e.g., a quarter of wheat is exchanged for x blacking, y silk, or z gold, &c. . . . But since x blacking, y silk, or z gold, &c. each represents the exchange-value of one quarter of wheat, [they] must, as exchange-values, be replaceable by each other, or equal to each other. Therefore, first: the valid exchange-values of a given commodity express something equal. [*Capital* I, p. 44]

Strictly speaking, exchange values are ratios. The exchange value of, say, a bushel of wheat can be specified in terms of certain quantities of other commodities that could be acquired for that bushel in the market. What Marx says is that these ratios "express something equal," which I have interpreted to mean that they have equal (quantities of) value. Marx conceives of value as an objective quantity possessed by things with use value (though that quantity takes the "form of value" only in a system of commodity production). The relation between value and exchange value is given by the LV: Commodities that exchange in the market have equal quantities of value. This ignores fluctuations due to the vagaries of supply and demand, but the basic idea is that the values of commodities that exchange in the market are, when equilibrium obtains, equal. This, of course, raises the obvious question, 'What is value?'. Marx begins his attempt to answer this question as follows:

> The two things [i.e., commodities that exchange] must therefore be equal to a third, which in itself is neither the one nor the other. Each of them, so far as it is exchange-value must therefore be reducible to this third. [*Capital* I, p. 45]

Admittedly, what this means is obscure, but let us pursue it a bit further. The task Marx sets for himself is to find out what this "third thing" is. He proceeds by a process of elimination to rule out every possible candidate (or so he thinks) except one: human labor in the abstract. He says,

> there is nothing left but what is common to them all; all are reduced to one and the same sort of labour, human labour in the abstract.
>
> Let us now consider the residue of each of these products; it consists of . . . a mere congelation of homogeneous human labour, of labour-power expended without regard to the mode of its expenditure. All that these things now tell us is . . . that human labor is embodied in them. When looked at as crystals of this social substance common to them all, they are—Values. [*Capital* I, p. 46]

This passage is remarkably unilluminating except for one thing: It suggests very strongly that Marx believed that the value of a commodity just *is*, i.e., is identical to, a certain quantity of labor. He uses forms of three terms that are frequently repeated in his economic writings when he is talking about the value of a commodity: 'embodied labor', 'congealed labor', and 'crystallized labor'. (A fourth is 'ossified labor'.)

A few paragraphs later Marx notes that some people might think that his view implies that unusually inefficient labor would make a commodity more valuable than it would otherwise be. He rejects this by claiming that the value of the commodity is not the embodied labor *tout court*, but the socially necessary embodied labor, which is labor of average skill and intensity. Not all embodied labor, then, need be socially necessary. With this qualification in hand, we get what might be called,

*The Value Identity Thesis*: The value of a commodity is identical to the quantity of socially necessary labor contained in it.

The Value Identity Thesis (hereafter, just the 'Identity Thesis') and the LV directly imply:

*The Labor Time Corollary*: Commodities (or more exactly, bundles of commodities) that exchange in the market in equilibrium contain the same quantity of socially necessary labor.

My attribution of the Identity Thesis to Marx is somewhat controversial, so some interpretive arguments are in order. The following considerations make it plausible to suppose that Marx believed it:

1. The Identity Thesis makes sense out of Marx's ubiquitous talk of value as congealed, crystallized, embodied, or ossified human labor. These vivid metaphorical expressions clearly convey the meaning of the Identity Thesis and are highly misleading if he did not believe it. Another example: At the end of the paragraph from which the preceding quotation comes, he quotes his own *Critique of Political Economy* as follows: " 'As values, all commodities are only masses of congealed labor time' " (*Capital* I, p. 47).

2. The chain of reasoning leading from the LV to the Labor Time Corollary (hereafter the 'LTC') is best explained as leading through the

Identity Thesis. If value just *is* socially necessary labor time, and commodities that exchange have equal value, then it follows that the commodities that exchange have equal quantities of socially necessary labor time in them. The Identity Thesis is an obvious lemma linking the LV and the LTC.

3. It is clear that Marx believed that the term 'value of a commodity' denoted an objective feature of that commodity and not, for example, a subjective appraisal of use-value. If Marx did not subscribe to the Identity Thesis, the nature of value (i.e., what it *is*) remains very mysterious.

Further, less direct considerations in favor of this attribution are found in the following digression.

Before proceeding with Marx's chain of argumentation to the conclusion that the worker is exploited under capitalism, a brief digression on the aims of classical economic theory would be in order. One of the primary tasks of classical economics was to construct a general theory of exchange. Such a theory seeks to answer some fundamental questions about the phenomenon of exchange. One of these is the ontological question, 'What is value?', to which Marx's answer is the Identity Thesis: The value of a commodity is identical to the quantity of socially necessary labor in it. Analogously, thermodynamics asks the question, 'What is heat?', to which the current answer is 'Heat (just) is the motion of molecules'. Note that if this is right, the Identity Thesis is not merely a stipulative definition; indeed, it is not a definition at all. It is a substantive ontological claim about the nature of value. The two terms have the same reference but not the same meaning.

If we were looking for a definition, strictly speaking, of the term 'value' that all classical economists would be likely to accept, it would probably be something like the following functional definition: The value of a commodity is whatever determines it's equilibrium price. Since 'equilibrium price (of a commodity)' means 'exchange ratio with other commodities when supply and demand are in balance', it follows that the value of a commodity, by definition, is whatever determines its exchange ratio with other commodities when supply and demand are in balance. The Identity Thesis tells us that this "whatever" is the amount of socially necessary labor.

There is a useful analogy in recent developments in the philosophy of mind. Some Materialists believe that mental events are simply nothing but brain processes. The former are identical with and reducible to the latter. We may think that there are two kinds of things, for example, pains and the firings of C-fibers in the brain, but the Reductive Materialist maintains that there is in fact only one kind of thing—brain processes. Another kind of materialist, the Eliminative Materialist, maintains that there really are no mental events; there are only brain processes. What we *call*, for example, a thought is in fact a brain process, but strictly speaking, there are no thoughts, just as there are no cases of Zeus throwing thunderbolts, though there is lightning.

There is a striking parallel in the history of economic thought. If my reading of Marx is right, Marx had a reductive value theory. Value just *is* embodied socially necessary labor. Subjective value theory (not to be confused with marginal productivity theory), which postdates Marx, answers the fundamental ontological question in an eliminativist spirit: In one sense, there is no such thing as exchange value; there are only subjective estimates of use value. Everything true and theoretically significant that an economist wants to say about exchange ratios can be said by talking only about subjective estimates of use value—if this theory is correct. *Both* theories provide an answer to the fundamental ontological question.

This brings us to a second fundamental question classical economics sought to answer, to wit, 'Why does a particular quantity of one commodity exchange for a (usually different) particular quantity of another commodity?'. In other words, what explains why commodities have the exchange ratios that they do? This is the central question of what has come to be called price theory. Marx's answer to this question is a consequence of the LV and the Identity Thesis, namely, the LTC: Commodities that exchange have equal quantities of socially necessary labor. This is intended to explain why exchange ratios are what they are. To sum up, in the tradition of classical economics, Marx sought a general explanation of the phenomenon of exchange in both its qualitative and quantitative dimensions. That is, he sought answers to two general questions: (i) 'What is value?', and (ii) 'What explains particular exchange ratios?'. The Identity Thesis answers (i), and together with the LV, implies an answer to (ii).

Where is the famous Labor Theory of Value? In the paragraph immediately following the qualification about socially necessary labor noted above, Marx gives the first full statement of what has come to be called the Labor Theory of Value (hereafter, just the LTV): "We see then that that which determines the magnitude of the value of any article is the amount of labor socially necessary, or the labor time socially necessary, for its production" (*Capital* I, p. 47).

The Labor Theory of Value is usually stated as something like the following: The value a commodity is determined by the quantity of socially necessary labor time required for its production. A problem with this formulation is evident in light of the above definition of the term, 'the value of a commodity'. If that definition is right, then the LTV means the same thing as 'Whatever determines the (equilibrium) exchange ratios of commodities is determined by the quantity of socially necessary labor time required for production'. Understood in this way, the LTV is misleading, obscure, and ambiguous. It is misleading because it suggests that value and embodied socially necessary labor time are distinct things, and the latter determines the former. The two are not distinct if my attribution of the Identity Thesis to Marx is correct. It is obscure because if value is something other than embodied socially necessary labor, but is determined by it, the following questions come to mind for which there are no easy answers in Marx: (i) What, then, is value (if not embodied labor)? (ii) *Why* is value determined by socially necessary labor time? Note that if the Identity Thesis is right, (i) does not come up and (ii) has a ready answer, to wit, 'Because that is what value *is*'. Finally, this formulation of the LTV is ambiguous because of the highly ambiguous term 'determine'. Does it mean 'causally determine', 'logically determine', or, more vaguely, that there is some determinate relation between value and embodied labor? The big insider trading scandal in commentary on Marx's economics is the trading on the ambiguity of this little term.

For these reasons, I shall avoid the standard formulation of the LTV. But what, then, is the LTV? At the risk of being offensive and seeming idiosyncratic, I shall claim that the Labor Theory of Value just is the conjunction of (i) the Identity Thesis; and (ii) the Labor Time Corollary.

Here's why: As I have argued above, Marx believed the Identity Thesis, and he clearly believed the LTC. So, the only question is whether or not these two should be identified with the LTV. The fact that, on this reading, the LTV consists of two statements and not one is an argument in favor of this identification and not against it. How many real theories contain one statement? More importantly, these two statements do all the work one would want a theory of value to do. Together they solve (if Marx is right), the quantitative and qualitative problems of value as conceived of by classical economics. The Identity Thesis answers the ontological question and the LTC explains equilibrium exchange ratios. What more—and indeed what less— can be asked of a theory of value? In the final analysis, the best argument for this identification is that it allows a clear representation of Marx's thought. That is the burden I shall try to discharge in the remainder of this chapter.

Let us return to our story leading to the conclusion that the proletariat are exploited. In Chapter VI of *Capital* I Marx recognizes that the conjunction of the LTV and the LV create a potentially serious problem in his explanation of the workings of the capitalist mode of production. That problem is the Kantian question, 'How are profits possible?'. The reasoning is straightforward: Profits represent surplus value, which according to the Identity Thesis is embodied labor. Insofar as he does not labor, the capitalist is effectively a middleman who, according to the LV, buys commodities (the inputs) at their value and sells commodities (outputs) at their value. According to the LTC, equal quantities of labor are being swapped. But, since the capitalist *qua* capitalist[1] does not labor and laboring is the source of value, it would seem that profits are impossible. Furthermore, since things that exchange have equal value (the LV), the capitalist cannot be systematically cheating by buying inputs below their value and/or selling outputs above their value.

It is important to note that this problem does not arise in Marx's system from the LV alone. The Identity Thesis is also required. If the value of a commodity were not a quantity of socially necessary labor, then the surplus value that profit represents would not be a quantity of socially necessary labor. If the Identity Thesis were false, the source of profits could be elsewhere than in laboring, and it is at least possible that the capitalist is responsible for the existence of those profits. This possibility evaporates when the Identity

Thesis is added. It follows from the latter that labor and labor alone is the source of profits; so, the capitalist is buying and selling embodied labor; he is making profits, which represent command over embodied labor, but there is no systematic cheating going on. Thus the question, 'How are profits possible?'

Marx regards his solution to this general problem as one of his most important contributions to political economy.[2] He argues that there is one commodity, which when purchased at its value, can be employed to create new value. This commodity is of course labor power, the capacity to labor. The value of labor power is determined in the same way as the value of any other commodity: It is the quantity of socially necessary labor required to produce it, which is represented by the subsistence wage. The worker is not being cheated, at least in any straightforward sense, since he is being paid the full value of what he is selling—his labor power.

Labor power, when it is actualized in the course of production, is creative in that it produces more value than what is needed to produce it. That means that the value produced by the discharge of the labor power is greater than the value of the labor power, which is, of necessity (by the LV) purchased at its value. So, a gap opens up between the value of the labor power and the embodied labor (value) in the finished product. And, herein lies the profit that the capitalist, as owner and marketer of the finished product, is able to rake off. The value received by the worker in wages is less than the value he actually puts into the finished product.

Suppose, for example, that it takes the worker six hours to produce goods, which when sold, realize enough value to sustain himself and his family and to offset the cost of the means of production used up in the creation of the product. Does the capitalist tell him to take the rest of the day off and go fishing? By no means. The worker must work, say, twelve hours, for the capitalist. This extra six hours is used to create surplus value, which the capitalist receives as profit. Effectively, for part of the day, the worker works for free, though this fact is mystified by the wage contract that specifies a certain wage in return for a certain number of hours worked. If one assumes that what the capitalist buys is the worker's labor, instead of his labor *power*, there is no way to account for the existence of profit, since (by the LV) the worker would get the full value of his labor. The central proposition in Marx's Theory of Surplus Value (hereafter, the TSV) is that the profit that accrues to the capitalist is the difference between the value the worker creates in the course of producing the product over a given period and the value of the worker's labor power, which the latter receives in the form of the subsistence wage.

This analysis constitutes the basis for Marx's claim that the worker is systematically exploited by the capitalist. However, exploitation consists in more than the mere fact that the capitalist appropriates surplus value that the laborer creates. Central to Marx's conception of exploitation is the fact that the worker does not control the means of production and thus, in one sense, is forced to deal with the capitalist. Now there has been a great deal of controversy in the secondary literature in recent years about what this forcing amounts to and, more generally, about exactly what conditions are necessary and sufficient for exploitation in Marx's sense. We shall turn to some of these disputes in due course, but for now I want to keep them in the background and proceed without a rigorous definition of exploitation. Intuitively, the idea is reasonably clear: The worker is exploited because he is forced to create surplus value for the capitalist.

This account of exploitation, and the larger economic analysis in which it is embedded, have a number of things going for it. The Identity Thesis implies that what the capitalist gets when he realizes profits belongs, in a primitive nonlegalistic sense, to the worker. If the appropriation of surplus value is not theft, it's surely like theft in that it consists in the forced and uncompensated[3] appropriation of something intimately connected with the being of another. This takes on further significance in light of Marx's view that the capacity to labor is central to what makes us human. Secondly, it locates the explanation for exploitation in the structural characterization of capitalist relations of production: Nonworkers and only nonworkers control the means of production and the workers (have to) sell their labor power as a commodity. This implies that as long as capitalist relations of production prevail, there will be exploitation.

Unfortunately, the economic analysis that supports this claim that the worker is systematically exploited by the capitalist has a number of serious difficulties. To see what they are, it might be helpful to restate the main theoretical claims and to summarize the arguments that connect them. There are three independent theoretical claims Marx advances; the LTC is a direct consequence of the LV and the Identity Thesis, and the LTV is just the conjunction of the LV and the LTC. The three independent claims are the following:

1. *The Law of Value* (LV): Commodities that exchange in the market have equal value.
2. *The Identity Thesis*: The value of a commodity is identical to the quantity of socially necessary labor required to produce it.
3. *The Theory of Surplus Value* (TSV): The profit that accrues to the capitalist is the difference between the value of the labor power he employs and the value embodied in the product he sells.

At this juncture, the argument for the LV has not been discussed. We return to it below. The argument for the Identity Thesis (and the LTC) can be represented as follows:

1. Commodities that exchange in the market have equal value (the LV).
2. Value is a magnitude.
3. The only magnitude that commodities that exchange in the market have in common is a quantity of (socially necessary) labor time. (See *Capital* I, pp. 45–46.)
∴4. The value of a commodity is identical to the quantity of (socially necessary) labor time required to produce it (the Identity Thesis).
∴5. Commodities that exchange in the market have equal quantities of

(socially necessary) labor. (This is the LTC, from [1] and [4].)

The argument for the TSV is perhaps best construed as an inference to the best explanation.[4] Stated informally, it goes something like this: The Identity Thesis implies that the surplus value realized by the capitalist as profit is embodied labor. Given that the capitalist buys inputs and sells outputs at their values (the LV), the best explanation for the existence of profit is that the capitalist buys the worker's labor power, and the value of that labor power is less than the value created by the discharge of labor power in the process of production. It is worth noting that Marx's argument for the TSV presupposes both the LV and the Identity Thesis.


**Böhm-Bawerk's Objections**

A critic has at least six potential lines of attack on Marx's chain of reasoning that leads to the TSV, which is the penultimate step in the argument for the charge of exploitation. He can argue that one or more of Marx's arguments for the LV, the Identity Thesis, or the TSV is defective. By contrast, the argument from the Identity Thesis and the LV to the LTC is safe. More powerful objections would take the form of showing that one or more of the three central theses is false. By the late nineteenth century, Marx's critics pressed a number of these lines of attack. Marx's most systematic critic at this time was the Austrian economist, Eugen Böhm-Bawerk. Böhm-Bawerk argues that these three Marxian theses are inconsistent with some obvious facts of life under capitalism.[5] His critique of Marx has sunk into undeserved obscurity in recent years. This is unfortunate if only because Böhm-Bawerk's discussion is a model of what good criticism should look like. He carefully restates Marx's argument, identifies what he takes to be its defect, considers potential counterarguments, and traces the problems to fundamental features of Marx's vision of economic life. The tone is respectful and without acrimony throughout.

An assessment of Böhm-Bawerk's objection is obviously important for evaluating Marx's charge of exploitation against capitalism. In the remainder of this section I explain this objection in more detail. The next section considers how Marx might have responded, and the final section of this chapter vindicates Böhm-Bawerk. This final section also lays some important groundwork for Chapter 4.

According to Böhm-Bawerk, the primary problem in Marx's system is that it contains a massive inconsistency, which is brought on by a conflict between the chain of reasoning outlined above and an obvious and elemental fact of life under capitalism. To see what this is, it is necessary to follow out some of the details of the workings of the capitalist system according to Marx's theoretical system or economic vision. For analytical purposes, Marx divides the capital owned by an individual capitalist into two categories: constant capital and variable capital. The former represents investments in raw material, equipment, buildings, and so forth. Although these things have value (after all, they have been produced by labor) that is transmitted to the finished product as they are used up, they are "sterile" in that they create no new value. The variable capital is the capital used to purchase labor power; by that fact, it is more fecund. According to the Identity Thesis, profit just is embodied (socially necessary) labor. It is an implication of the TSV that this profit comes from the variable capital, which is invested in labor power.

It is at this point that a major difficulty appears. It seems to follow from the LV, the Identity Thesis, and the TSV that the rate of return on investment (rate of profit) of a firm or an industry is proportional to the variable capital, that is, the amount of capital invested in labor power. For technical reasons, different industries use a different mix of capital goods and labor. For example, companies that operate oil refineries have most of their capital tied up in physical plant and machinery. By contrast, a large truck farming operation may use most of its capital as working capital to pay wages to migrant laborers. Marx refers to this mix of constant and variable capital as the 'organic composition' of the total capital. Since only labor power can produce (new) value, the more labor-intensive an industry is (i.e., the greater the proportion of variable capital to total capital), the more surplus value it should produce (assuming, as Marx does, that the rate at which the workers are exploited remains constant[6]). But, this is contrary to the obvious facts. Capitals of the same amount tend to yield the same return on investment (rate of profit) regardless of their organic composition.

This is where matters stood at the end of volume I of *Capital*. Marx himself was aware of this problem (see *Capital* I, p. 290) and promised a solution to it in volume III. Twenty-seven years elapsed from the appearance of volume I to the appearance of volume III. Marx had died, and the job of seeing through the publication of volumes II and III fell to Engels. In the interim, Marx's followers tried over and over to solve the "problem of the average rate of profit." It even happened that a prize essay competition was instituted on the question.[7] None of the proposals was satisfactory, however, and the appearance of volume III, wherein the solution was supposed to be found, was eagerly awaited.

The appearance of volume III was something of a disappointment to the underlaborers in Marx's garden. The explanation of the tendency for the rate of profit to equalize was straightforward and unobjectionable enough: Capitalists shift resources from branches of production where profit rates are low to those where profit rates are high; the resultant competition brings about a tendency toward equality in the rate of return on investment. In volume III of *Capital* Marx clearly recognizes that all of these facts taken together require that in actuality commodities that exchange in the market do not have equal values, and, by implication, do not contain equal quantities of socially necessary labor. Rather, some bundles of commodities tend to be sold *above* their values and others tend to be sold *below* their values. The LV, and by implication the LTC, are "modified" to account for this; the (equilibrium) *price* at which goods exchange is what Marx calls the price of production, which is defined as the cost of the item (reckoned in labor units) plus the profit (at the average rate). The "deviations" of price from value cancel each other out in the aggregate so that the total profit represents total surplus value and the average price of all commodities represents average value. In industries where the proportion of variable capital is lower than the average (i.e., where there is a higher proportional investment in "non-creative" machinery, etc.), commodities sell at a price over their value, and conversely, for industries with a higher investment in labor, commodities sell at a price under their value. Notice that Marx never thinks to "adjust" the Identity Thesis. Value still just is embodied labor. Hanging onto the latter requires Marx to do exactly what he did, namely, modify the LV and the LTC. What justifies or warrants this decision is not yet evident.

Nonetheless, although things no longer exchange according to their (labor) values, Marx claims that there is a complex and determinate relation between prices and values: The values of commodities "stand behind" and indirectly determine equilibrium prices. Or, as Marx puts it, values are "transformed" into prices (of production). Marx's own solution to the problem of the relation between price and value (the so-called Transformation Problem) is acknowledged to be defective. But, other solutions to this problem have been proffered by Marx's followers in this century and the last.

Böhm-Bawerk's central objection is that Marx has contradicted himself. Put quite simply, in volume I of *Capital*, he asserts that under capitalism, commodities that exchange have equal value (ignoring minor fluctuations in supply and demand); in volume III he explicitly denies this. He admits, in effect, that the LV, the Identity Thesis, and the TSV cannot be rendered consistent with the fact that industries with different organic compositions of capital tend to return the same rate of profit. No wonder the above-mentioned prize essays failed—these writers were trying to do the impossible!

By itself, pointing out this inconsistency is no great feat, though Böhm-Bawerk does it in an especially clear and careful manner. What distinguishes his account is that he traces out the consequences of this inconsistency throughout the chain of Marx's reasoning. Let us begin with the LV, which says that commodities that exchange in the market have equal value. What is problematic about giving up the LV and admitting that commodities that exchange do not have equal value? The problem for Marx in admitting that, strictly speaking, the LV is false is that he has given up a key premise in his argument for the Identity Thesis. The supposition that commodities that exchange in the market have equal value is what set Marx off to discover what magnitude they have in common, which he concludes is labor. If he admits that these bundles of commodities do *not* have equal value, then he has lost a crucial premise in his argument for the Identity Thesis. As a result, (as far as *Capital* I is concerned), the Identity Thesis is unsubstantiated.

Moreover (and the importance of this is hard to overstate), his explanation of profit is predicated on the assumption that capitalists buy and sell commodities at their values. The crucial distinction between labor and labor power and the entire complicated exposition of how labor power gives rise to surplus value, which is raked off by the capitalist as profit—this "great secret of political economy," as Engels calls it—is motivated by the supposition that capitalists, like everyone else, exchange things at their values. *The question, 'How are profits possible?', which motivates all this, has a bite only if there is reason to believe that commodities that exchange have equal value.* But now we are told that commodities that exchange in the market in the capitalist mode of production do not, and indeed never did, have equal values. The problem is that there is now no reason to believe the TSV, which identifies the source of profits in labor power. Without the TSV, Marx has no grounds for maintaining that the workers are systematically exploited by the capitalists.

In at least one respect, the above representation of Marx's argument is misleading. It is doubtful that Marx came to believe in the Identity Thesis and the LTC on the basis of an argument from the LV. Versions of the Identity Thesis were commonplace in classical political economy from Adam Smith onward. It was part of the intellectual furniture of the universe for the classical tradition. Nonetheless, when we look for Marx's actual justification for the Identity Thesis and the LTC, we find that it does proceed from the LV as indicated above (cf. *Capital* I, pp. 44–47).

So, one way of looking at the problem raised by the "unresolved contradiction," to use Böhm-Bawerk's phrase, is that it renders unsound the argument for the Identity Thesis. This does not prevent Marx from continuing to assume it, but the fact remains that he has no apparent justification for this assumption.[8] However, without the Identity Thesis, Marx has no justification for the TSV. The claim that the capitalist's profit is embodied labor created by the discharge of labor power is simply unwarranted. The upshot of all of this is that Marx's claim that the workers are systematically exploited by the capitalists is unsubstantiated.

### Can Marx Avoid the "Unresolved Contradiction"?

There is something very peculiar about the unresolved contradiction. Marx seems to have made a colossal and obvious error. When a thinker of the first rank makes a mistake like this, it is incumbent on an interpreter to look for explanations. Perhaps he wrote the offending passages absentmindedly or after an all-night chat with Bakunin about some of the finer points of their disagreements. However, a more satisfying explanation would be that Böhm-Bawerk has misconstrued Marx in some way. This gains some plausibility from the fact that Marx himself seems to have recognized the problem in volume I of *Capital* (*Capital* I, p. 290). On the other hand, the idea that Böhm-Bawerk has misconstrued Marx looks implausible if one disinterestedly compares what Marx actually says with Böhm-Bawerk's representation and criticisms of his argument. It certainly seems as if Böhm-Bawerk has got Marx's number. Resolving this puzzle will take us pretty deep into some methodological questions in the philosophy of economics, but, as I hope the subsequent discussion will show, it will cover its costs.

Let us begin with Marx's approach to this problem in volume III. In Chapter X of volume III he just drops the LV, as if it were a simplifying assumption of the sort commonly employed by economists. Böhm-Bawerk interprets Marx to be offering a substantive universal law that is supposed to explain exchange ("Any pair of bundles of commodities that exchange [in equilibrium] have the same value") and ultimately ground the charge of exploitation, whereas Marx seems to have regarded the LV as a convenient simplifying assumption that can be dropped at the more complex level of analysis that volume III is intended to provide. Perhaps the root of the problem is that Marx's purpose in introducing the LV and the Identity Thesis (and the LTC) was not to justify the charge of exploitation; arguably, his main purpose was much more strictly scientific—he wanted to explain how the capitalist economic system functions as a going concern, that is, how goods and services get produced and capitalist relations of production get reproduced.

To this end, Marx employs what subsequent economists have called reproduction models. A reproduction model is an abstract representation of an economic system. Only certain features of the real world are represented in the model, and the model often makes some extreme simplifying assumptions (e.g., economists since Ricardo have often employed models in which only one good—corn—is produced). Explanation of certain features of real-world economic systems is supposed to be achieved by this relation of representation. Let us call the model, together with the claim that it accurately represents certain key aspects of an actually existing economic system, a *model-theoretic explanation.*[9]

In volume III of *Capital* Marx makes clear that the LV was intended to apply only to a situation (i.e., a reproduction model) in which the producers own their means of production and produce for the market, which of course is not the capitalist mode of production. He says,

> The *punctum saliens* will best be brought out if we approach the matter as follows: Suppose the labourers themselves are in possession of their respective means of production and exchange their commodities with one another.
> ... The exchange of commodities at their values, or approximately at their values, thus requires a much lower stage than their exchange at their prices of production, which requires a definite level of capitalist development. [*Capital* III, pp. 175, 177]

Let us call a system of the sort Marx describes a regime of simple commodity production. There has been some dispute about whether or not Marx thought that simple commodity production, as a full-blown economic system, ever existed in reality.[10] However, it is not necessary for this to be true for the model to have explanatory power. After all, classical mechanics supposes a perfect vacuum; even if none exists, classical mechanics still provides adequate explanations.

For the purposes at hand, it is not necessary to pursue the details of Marx's model-theoretic explanation of the workings of the capitalist mode of production. Indeed, that is the primary topic of volumes II and III of *Capital*. The important thing to notice is that under this conception of what Marx is up to, it is possible to understand why he posits the LV at one stage in his discussion (in his general treatment of commodity production in volume I of *Capital*) and then simply drops it when he constructs a more complicated model (by volume III). In this sense, Böhm-Bawerk's charge that there is an unresolved contradiction in the Marxian system seems misleading at best. Marx, however, is not blameless in this affair. In the opening chapter of volume I of *Capital*, where the LV is introduced, it turns out that he is not talking about the capitalist mode of production; instead, he is talking about one feature of the capitalist economic system *considered in abstraction from capitalist relations of production*. For all of us who have struggled through this chapter, it is likely that this comes as something of a surprise. Nonetheless, this has to be what is going on if Böhm-Bawerk's charge is to be avoided.

A similar move is available with respect to both parts of the LTV (the Identity Thesis and the LTC). One purpose of a model-theoretic explanation of an economic system is to represent the system as a circular flow of social wealth.[11] For mathematical techniques to be applicable, it is necessary to choose a unit of value or numeraire. Since labor is a common factor in all commodities,[12] labor time is the obvious candidate for the unit of value or aggregator of social wealth. Thus the assumption is that equal quantities of labor are exchanged and that value is embodied labor.

Since the statements in a model-theoretic explanations are, strictly speaking, about the model, they are not straightforward empirical claims about actually existing economic systems. How, then, are they to be criticized? One way to do this is to show that statements about the model are inconsistent. A model can hardly represent reality if it is internally inconsistent. Another way to criticize these statements, or the model of which they are a part, is to show that an important range of phenomena that the model is supposed to represent or explain cannot be represented or explained by that model. A third form of criticism is to show that certain statements of the model are unnecessary, in the sense that an alternative model can be constructed that does not make the relevant assumptions and yet represents the same features of the system as the other model. Considerations of simplicity argue against the model with the unnecessary statements in it.

In recent years, economists, including those sympathetic to Marx, have pressed the latter two kinds of criticisms against Marx's models. Two of these will be briefly mentioned here. The first is the problem of heterogeneous labor. It is an obvious fact that labor is not qualitatively homogeneous. Some labor is highly skilled, and it earns a higher wage as a result. If labor is to serve as an aggregator for a reproduction model, skilled labor must, in some sense, be reducible to unskilled labor. Marx is aware of this requirement and claims that the higher wage imputed to skilled labor can be explained in terms of the labor expended in acquiring the skill and the labor of the teachers. One problem with this is that it leaves out of account natural talents and abilities. I could spend twice as much time with Louis Armstrong's teacher, but my wages as a trumpet player would still be at least a little less than Armstrong's.[13] The other move against the LTV is to show that it is unnecessary to explain prices and profits. This is one of the conclusions of Piero Sraffa's *The Production of Commodities by Commodities*.[14]

Let us suppose that these criticisms are well-taken. On the other hand, in light of the above methodological discussion, it seems reasonable to believe that Böhm-Bawerk, in claiming that Marx has an unresolved contradiction in his system, has not appreciated Marx's method. For Marx, the LV (as well as the LTC) is treated as an assumption that holds only under a regime of simple commodity production. When the model is complicated and the LV and the LTC are relaxed, there remains a determinate relation between value (and, by implication, embodied labor) and equilibrium price, though it is more complicated than originally assumed. Böhm-Bawerk's claim that there is an unresolved contradiction in Marx's system is based on a misunderstanding of what Marx was up to. At any rate, I suspect that this would be the judgment of many contemporary economists.

### Vindicating Böhm-Bawerk

Unfortunately, the above appreciation of Böhm-Bawerk is incomplete and seriously misleading. I want to suggest that the latter's critique of Marx is far more profound than the above account would indicate. In this section I shall argue that there is in fact an unresolved contradiction in Marx's system and that this contradiction can be used to show that the Identity Thesis is false. Since the latter is a crucial premise in Marx's argument for the charge that the capitalist systematically exploits the worker, that argument must be rejected as unsound. Finally, as I shall show in the next chapter, insights provided by Böhm-Bawerk's critique of Marx can be used to undermine an entire class of reconstructions of Marx's argument for the charge of exploitation.

Let us begin with some concessions. It is reasonably clear that in *Capital* Marx does construct a model of the capitalist mode of production. Given that he intended to explain thereby certain features of the latter, it is safe to say that he was offering a model-theoretic explanation. As such, he can be conceived of as making simplifying assumptions that can be evaluated along the lines suggested above.

On the other hand, it seems that key statements in Marx's model, notably the LV and both parts of the LTV, are intended to be true of the real world (and not just the model). The main evidence for this is that Marx offers independent arguments for both the LV and the Identity Thesis in *Capital* I (which together entail the LTC). Furthermore, Marx uses the LV and the Identity Thesis to argue for the TSV, which says that the surplus value that the capitalist realizes as profit is explainable by the creative properties of labor power. The TSV in turn is used to ground the charge of exploitation.

All this strongly suggests that Marx believed that these simplifying "assumptions" (the LV and both parts of the LTV) really are true.

If these claims were merely convenient model-theoretic assumptions of the sort made by contemporary economists, surely Marx would not base one of his main normative judgments against capitalism on them. If, for example, the Identity Thesis represents a pragmatic decision to treat embodied labor as a value numeraire so as to make mathematical techniques applicable, the nature of the argument for the charge of exploitation against capitalism is completely obscure.

Furthermore, just as in the writings of contemporary economists, we would expect to find no explicit argument at all for these assumptions. Or, if Marx was being self-conscious about his methodology, he might have said at some point that these assumptions are justified by the explanatory adequacy of the various models as they emerge in *Capital*. But this is not what we find in *Capital* I where they are introduced. Instead, early in *Capital* I, Marx offers an independent argument for the LV—an argument for its *truth*—not its convenience; in addition, he uses the LV to argue for the Identity Thesis. The LV and both parts of the LTV (the Identity Thesis and the LTC) are used in his model-theoretic explanations, but that is not all that they are used for; in short, they are not mere assumptions.

If this is right, it suggests an important methodological difference between Marx and much of contemporary economics that warrants some discussion. Two features of contemporary economics that non-economists find most puzzling are the severe abstraction and the unrealistic assumptions that characterize the models. For example, neoclassical economics regularly assumes perfect information among all participants in the market.[15] Furthermore, there is almost never any argument given for why (or, more accurately, how) we should accept these assumptions. Sometimes it seems that economists are like Lutherans: Just as the latter have their doctrine of Justification by Faith, the former seem to have a doctrine of Justification by Assumption. Perhaps the implicit argument is something like, "Give me these assumptions and I will construct a reproduction model." These assumptions are to be understood as, at best, rough approximations of the way things are, and the models they describe are intended to be rough representations of economic reality.[16]

This method may not, in and of itself, be a bad way to proceed. As noted above, the physical sciences have had much success using unrealistic simplifying assumptions (e.g., the assumption of a perfect vacuum in classical mechanics). By contrast, it seems that the key claims in Marx's models were intended to be descriptive of reality at some level and not merely rough approximations thereto. For example, Marx maintains that the LV does hold for certain segments of both ancient and modern economies.[17]

However, one problem with this account of Marx's models is that it does not explain why he seems to have regarded some of the statements of the model as simplifying assumptions that can be discarded at higher or more complex levels of analysis. If they are simplifying assumptions, aren't they intended as mere approximations to the truth? Or, to put the point more paradoxically, this account does not explain the sense in which these statements are intended to be true of the real world.

This problem can be solved by interpreting these statements as describing *tendencies*. To say that there is a tendency for something to happen (be the case) does not entail that it will happen (be the case). Statements describing tendencies can be understood to contain an implicit *ceteris paribus* clause. For example, Marx's LV really says, "*All else equal*, commodities that exchange in equilibrium have equal value." The LTC says, "*All else equal*, commodities that exchange in equilibrium contain equal quantities of socially necessary labor."[18]

Though it makes perfect sense (whether or not it is true) to say, "All else equal, commodities that exchange in the market have equal value" (and *mutatis mutandis* for the LTC), it does not make any sense to say this of statements that are intended as approximations. For example, it makes no clear sense to say, "All else equal, all market participants have perfect information." Tendency laws with *ceteris paribus* clauses are not approximations of reality; the tendency is either "there," or it is not.[19] Literally, tendency laws say what would happen if all else were equal; to put it another way, they can be said to describe what *is* happening behind the "noise" of disturbing influences.

This account of Marx's model-theoretic explanations has the following not inconsiderable virtues for interpreting Marx:

a. It explains why Marx felt compelled to argue for the LV and, on that basis, for the Identity Thesis and the LTC in *Capital* I. (By contrast, contemporary economists often do not argue for their assumptions.)
b. It explains why Marx could drop the LV, and by implication the LTC, (in their unqualified form) in *Capital* III. At that stage of his analysis, all else was not equal.
c. It explains why he maintains that modified versions of the LV and the LTC continue to hold in full-blown capitalism. (The working out of these tendencies are modified by circumstances inconsistent with all else being equal.) The claim that a modified LV and LTC hold under capitalism looks like a bald assumption, but if they are universal laws with an implicit *ceteris paribus* clause, these modified laws are something he is entitled to, assuming that his argument for these laws at the beginning of *Capital* I succeeds.

These concessions to Marx seem Neville Chamberlain-like in their scope and extent, as far as Böhm-Bawerk is concerned. Though Böhm-Bawerk was Austrian, he might feel like a Czech. If we grant all this, haven't we resolved the unresolved the contradiction? The answer is no and here's why: Any time someone finds himself committed to a set of apparently inconsistent propositions, he has the option of giving up or reinterpreting one of the statements. This is just what Marx does with the LV and the LTC. He claims, in effect, that these laws, without their *ceteris paribus* clauses, hold only when considered in abstraction from capitalist relations of production and that under capitalism only some qualified version of the laws hold. But this sort of move is satisfactory only if Marx can successfully argue that these laws do hold when all else is equal. In short, genuinely resolving a contradiction, as opposed to simply restoring consistency, requires more than giving up or reinterpreting one of the statements in the inconsistent set. And the full scope of Böhm-Bawerk's critique of Marx can only be appreciated by examining his objections to Marx's attempts to justify these laws considered in abstraction from "disturbing influences."

Let us begin with the LV. Marx's only explicit argument for the LV is scattered through the first three sections of Chapter I of *Capital* I.[20] We can do no better than Böhm-Bawerk's clear and concise summary of it:

> Marx had found in old Aristotle the idea that "exchange cannot exist without equality, and equality cannot exist without commensurability" [*Capital* I, p. 65]. Starting with this idea he expands it. He conceives the exchange of two commodities under the form of an equation, and from this infers that "a common factor of the same amount" must exist in the things exchanged and thereby equated [*Capital* I, pp. 44–45], and then proceeds to search for this common factor to which the two equated things must as exchange values be "reducible" [*Capital* I, pp. 45–46].[21]

As Böhm-Bawerk points out,[22] conceiving of exchange as an *equation* is by no means forced upon us by the nature of the phenomenon, as Marx seems to believe. There are at least two other ways of interpreting or describing the phenomenon: (1) Exchange could be viewed as the swapping of *equivalents* in the juridical sense that both parties agree to the equity of the transaction without there being any identity involved; (2) Subjective value theory, of which Böhm-Bawerk was an early proponent, understands the exchange phenomenon as involving as essential *difference* in the personal valuations, or estimates of use value, made by the parties to the exchange. In other words, when *A* and *B* exchange eight bushels of wheat for one hundredweight of iron, the possessor of the wheat values the iron he does not have more than the wheat he does have, and the possessor of the iron values the wheat he does not have more than he values the iron he does have. Whether or not this is right, it does show that Marx's conception of exchange as an equation of value is *not* self-evident and forced on us by the phenomenon itself.

Is there any other way to argue for the LV? Let us turn to Marx's resolution of the problem posed by different organic compositions of capitals (the "contradiction") in Chapter XI of volume III. Five paragraphs into that chapter, Marx poses the problem:

> The really difficult question is this: how is this equalisation of profits into a general rate of profit brought about, since it is obviously a result rather than a point of departure? [*Capital* III, p. 174]

To which he gives as direct an answer as one could ask for:

The whole difficulty arises from the fact that commodities are not exchanged simply as *commodities*, but as *products of capitals*, which claim participation in the total amount of surplus value proportional to their magnitude. [*Capital* III, p. 175]

To describe something as the product of capital is to presuppose capitalist relations of production. The latter constitute a "disturbing influence" vis-à-vis the LV. This is confirmed by what comes after the above-quoted passage, which is Marx's description of the workings of a regime of simple commodity production. His account of the simple commodity production model ends with the following:

The exchange of commodities at their values . . . requires a much lower stage than their exchange at their prices of production, which requires a definite level of capitalist development. [*Capital* III, p. 177]

These two passages confirm the suspicion that the *ceteris paribus* clause in the LV is supposed to screen out capitalist relations of production.[23] The crucial question now is, 'What reason is there to believe that the LV holds (would hold) without the *ceteris paribus* clause (or, what comes to the same thing, with that clause doing no work) in a regime of simple commodity production?'.

In Chapter XI of *Capital* III Marx offers no explicit argument for this. He simply asserts that it does hold. This seems like justification by assumption, but other elements of Marx's thought suggest some reasons for this claim. One reason for maintaining that exchange involves a swapping of things with equal value (as well as equal quantities of socially necessary labor) in a regime of simple commodity production can be found in a recurring theme of *Capital*: Economic relations among commodities are phenomenal manifestations of social relations among individuals. Recall that in a regime of simple commodity production all individuals own their own means of production and produce for the market. By implication, there is no wage labor. In consequence, no one is realizing value at another's expense, so to speak. Reproduction under these circumstances requires us to suppose that things with equal value are being exchanged (the LV)—at least given an objective conception of value.

The LTC can be given a similar justification.[24] By hypothesis, there is a stable division of labor in a regime of simple commodity production, which is manifested in stable exchange ratios of various commodities. That eight bushels of wheat exchange for one hundredweight of iron is a manifestation of how labor must be apportioned among various branches of production to meet the needs of the community. The fact that labor is divided the way it is calls for explanation. Positing the LTC means that equal amounts of (socially necessary) labor are continually being exchanged. This explains the economic cohesion of a society of independent yet interdependent producers. If unequal amounts of labor were being exchanged, the division of labor would be unstable. In a recent discussion of simple commodity production, Hindess, Hirst, et al. say,

> In exchanging the products of their labor as equivalents in labor time, in-
> dependent producers blindly and unknowingly are supposed to reproduce
> the proportions of social labor objectively necessary for a certain composition
> and scale of production. Marx . . . explain[s] the necessity of equivalent ex-
> change in terms of its reproducing the division of labor. Equivalent exchange
> is conceived as a particular form of a universal 'natural law'. To breach the
> law of value is to problematise the division of society's labor and in turn the
> existence of society itself.[25]

What we have, then, is a functional explanation for the division of labor
in a regime of simple commodity production. The use of the terms 'blindly
and unknowingly' make this explanation seem much more mysterious than it
really is. It is easy enough to describe a feedback mechanism, based on some
modest assumptions about the disutility of labor and the rationality of economic actors, to explain why exchange ratios, conceived of as values or as
embodied labor, stabilize where they do. That is, no one would regularly
exchange the more valuable for the less valuable or a commodity with more
embodied labor for one with less embodied labor. Thus the argument for the
LV and the LTC.

Marx, then, is being construed as offering a model-theoretic (functional)
explanation of the hypothetical stability of a regime of simple commodity
production by appeal to the LV and the LTC. The latter two, without their
qualifying clauses, are asserted to be literally true of such a regime. Capitalism
can then be conceived of as a kind of overlay on such a system—in which all
else is not equal. Here we have a totally new argument for both the LV and
the LTC, which is derived from methodological considerations about how
model theoretic explanations do their explaining.

However, Böhm-Bawerk has a decisive objection against the LV and the
LTC, conceived of as holding in an unqualified form only under a regime of
simple commodity production. To see what it is, let us consider in more detail
Marx's particular model of simple commodity production (*Capital* III, pp.
175–177). He asks us to imagine a two-person economy where *A* produces
one kind of commodity, call it *I*, and *B* produces another kind of commodity,
call it *II*. Their respective products require different quantities and qualities
of means of production. Following Marx, let us assume that both produce
their own means of production and that *A*'s production involves more means
of production, relative to *B*'s production; or, to put it in capitalist terminology,
*A*'s production is relatively capital-intensive and *B*'s production is relatively
labor-intensive. Suppose further that *A* and *B* labor the same number of hours
in a day at the same level of intensity. How does value flow through the
system? For ease of exposition, we use terminology appropriate to capitalism.

The value of variable capital used up is the same in both cases for a given
period of time—the value quantity necessary to replenish the workers' labor
power for, say, a day. The quantity of surplus labor (quantity of profit) is
also the same—the difference between the total value added by the day's
labor and the value of the variable capital used up. A day's income, then, is
equal to the wage plus the "profit." Or, as Marx puts it,

To put it the capitalist way, both of them receive the same wages plus the same profit, or the same value, expressed, say, by the product of a ten-hour working-day. [*Capital* III, p. 176]

But, although the quantity of profit (surplus value) produced in a day is the same, the respective *rates* of profit (surplus value divided by the value of the total capital employed) would be different, and this is because the production of *I* involves more constant capital. (Recall that the production of *I* is capital-intensive.) Marx readily admits that the rate of profit would be different, but he maintains that this would be a matter of indifference to *A* and *B*, since the greater value of a unit of *I* compensates for the greater value outlay for constant capital.

If labourer [*A*] has greater expenses, they are made good by a greater portion of the value of his commodity, which replaces this "constant" part, and he therefore has to convert a larger portion of the total value of his product into the material elements of this constant part, while labourer [*B*], though receiving less for this, has so much less to reconvert. In these circumstances, a difference in the rates of profit would be immaterial.... [*Capital* III, p. 177]

It is a consequence of Marx's account that differences in the organic composition of "capital" in this model of simple commodity production have no significance for exchange ratios; all that counts is total embodied labor. Indeed, he has to hold this, if he is to maintain that capitalist relations of production are the "disturbing influence" that makes all else *not* equal in the capitalist version of commodity production.

But it is this very implication that Böhm-Bawerk maintains is false. That is, in a regime of simple commodity production, different organic compositions do have a systematic effect on exchange ratios. For vividness, Böhm-Bawerk asks us to imagine that it takes five years for *A* to produce the means of production used in the production of commodity *I* and an additional year to turn out a quantity of commodity *I*. By contrast, *B* takes the production of *II* through all of its stages in a month. Böhm-Bawerk says,

Now Marx's hypothesis assumes that the prices of the commodities I and II are determined exactly in proportion to the amounts of labor expended in their production [the LTC], so that the product of six years' work in commodity I only brings as much as the total product of six years' work in commodity II. And further, it follows from this that the laborer in commodity I should be satisfied to receive for every year's work, with an average of three years' *delay* of payment, the *same* return that the laborer in commodity II receives *without any delay*; that therefore delay in the receipt of payment is a circumstance which has no part to play in the Marxian hypothesis, and more especially has no influence on ... the crowding or understocking of the trade in the different branches of production, having regard to the longer or shorter periods of waiting to which they are subjected.[26]

This is plainly implausible. In short, no such system would reproduce itself. The point is that while Marx does take into account the fact that the production of $I$ is capital-intensive and thus has a higher price than it would have if it were less capital-intensive, he believes that this higher price is wholly accounted for by the greater quantity of embodied labor "transferred" to the final product (*Capital* III, pp. 176–177), as the Identity Thesis requires.[27] Böhm-Bawerk contends that this is false. For reproduction to take place (in other words, even in equilibrium), unequal quantities of embodied labor must be exchanged in a regime of simple commodity production.

But if unequal quantities of embodied labor are exchanged and yet equal *values* are exchanged, it follows that value is not identical to socially necessary labor time (i.e., the Identity Thesis is false). On the other hand, if the Identity Thesis is true, then the LV must be false. Indeed, Marx is facing the same problem here that he faced in the original contradiction, which is why Böhm-Bawerk says there is an *unresolved* contradiction in the Marxian system. The foregoing shows that he is right about this.

Is there any way to resolve the contradiction on Marx's behalf? There is, but the cost is too high. If the Identity Thesis is to be saved—and it must be for Marx's argument for the claim that the capitalist exploits the worker— Marx would have to admit that the LV, without the *ceteris paribus* clause, does not hold in a regime of simple commodity production; the *ceteris paribus* clause must "screen out" not only capitalist relations of production, but also different organic compositions involved in the production of different commodities. However, to make this work, Marx would have to argue that the LV, without the *ceteris paribus* clause, does hold under some circumstances. But what might those circumstances be?

Let us consider a yet simpler model, which can be called "really simple commodity production." The most famous instance of this model is Adam Smith's deer-beaver model. In Smith's own words,

> In that early and rude state of society which precedes both the accumulation of stock and the appropriation of land, the proportion between the quantities of labour necessary for acquiring different objects seems to be the only circumstance which can afford any rule for exchanging them for one another. If among a nation of hunters, for example, it usually costs twice the labour to kill a beaver which it does to kill a deer, one beaver should naturally exchange for or be worth two deer. It is natural that what is usually the produce of two days or two hours' labour, should be worth double of what is usually the produce of one day's or one hour's labour.[28]

Notice that Smith assumes the absence of any capital goods (no "accumulation of stock"). The hunting is done bare-handed. As this quotation indicates, a stable exchange ratio would emerge. Equal values—and equal quantities of socially necessary labor—are being exchanged. That is, the LV and the LTC, in their unqualified forms, hold. The (stability of the) exchange ratio is thereby explained, but this is a model without capital goods. Actually, it is not necessary to suppose the complete absence of capital goods; all that is required is equal organic compositions. For example, suppose it takes as long to construct a beaver trap as it does to fashion a bow.

What the contrast between Smith's model and Marx's model shows is that the organic composition of capital systematically affects (equilibrium) exchange ratios. This forces one of two options regarding the LTC, depending on how much work the *ceteris paribus* clause is supposed to do. If that clause does not screen out different organic compositions of capital, the LTC is just false. Even when all else is equal (i.e., even when capitalist relations of production are abstracted), equilibrium exchange ratios are not determined by the quantity of embodied socially necessary labor time. The second option saves the LTC by claiming that the *ceteris paribus* clause screens out differences in organic compositions of capital, as well as, for example, capitalist relations of production.

It might seem that the logic of the situation does not dictate either option, as long as consistency is maintained. But that is not exactly the case. Recall that one of the main tasks of classical economics is to answer the question, 'What explains equilibrium exchange ratios?'. *If we are looking for a (complete) explanation of equilibrium exchange ratios, the LTC is just false.* At most, embodied labor is just one factor in the determination of those ratios; relative organic compositions of capitals is another—and there may be others.

This does raise the question of why it was necessary to pursue Marx back to Smith's forest to catch him. Why won't the original contradiction do the job? After all, Marx admits that in capitalist equilibrium, the LTC does not hold. Equal quantities of socially necessary labor are not being exchanged. The reason is that it is plausible, though just barely, to maintain that capitalist relations of production distort an underlying tendency toward equal labor exchange. Marx has been interpreted as saying, in effect, "We have to look at 'pure' commodity production, the embryonic form (he uses this very term) of the capitalist mode of production, to see the LTC at work." If the LTC held without the *ceteris paribus* clause under such circumstances, then it might be plausible to claim that the LTC holds, with that clause, under capitalism.

These results about the LTC have parallel implications for the Identity Thesis. Recall what the Identity Thesis says:

> *The Identity Thesis*: The value of a commodity just is the quantity of socially necessary labor contained in it.

But the term 'the value of a commodity' means 'what determines equilibrium exchange ratios of commodities'. So, the Identity Thesis means the same thing as,

> ($IT_1$) What determines equilibrium exchange ratios of commodities is the quantity of socially necessary labor in them.

This reveals an ambiguity in the Identity Thesis because of the ambiguity of that infamous term, 'determines'. How might it be disambiguated? The most natural way to read it makes it simply another way of asserting the LTC, 'Commodities that exchange in the market in equilibrium have equal quantities of socially necessary labor in them'. On this reading, the LTC and the

Identity Thesis are one and the same! (Proponents of the customary under-standing of the LTV can take heart.) But, since the LTC is false, it follows that the Identity Thesis is false.

Let us consider two other readings of (IT):

(IT$_2$) One factor that systematically affects exchange ratios is the quantity of socially necessary labor contained in commodities.

(IT$_3$) There is some determinate relation (which perhaps can be mathe-matically modeled) between exchange ratios and the quantity of socially necessary labor time.

The first cannot be accepted, at least on Marx's behalf, for two reasons: First, although it is a statement of identity (the 'is' is the 'is' of identity and not the 'is' of predication), it is not an Identity *Thesis*. That is, it does not permit an answer to the ontological question, 'What is value?'. Secondly, it leaves the door open for other factors as determinants of equilibrium exchange ratios, factors that would deprive labor of the special status it needs for the TSV and the charge of exploitation. The second also cannot be accepted. After all, it is not even a statement of identity. For that reason, it too does not permit an answer to the ontological question, 'What is value?'. Nor can it serve to ground the charge of exploitation in the way Marx intended.

The importance of this ontological question is hard to overstate; an answer to that question is required by a (general) *theory* of value, which is, after all, what Marx is offering. To maintain that a theory of value does not require an answer to this question would be akin to claiming that a general theory of thermodynamics does not require an answer to the question, 'What is heat?'. Without answers to these ontological questions, one literally does not know what one is talking about.

I conclude that the most plausible reading of the Identity Thesis has it meaning the same thing as the LTC, which, suitably reworded, is a statement of identity. Given that the LTC is false, so is the Identity Thesis. Therefore, the LTV is false.

Our main interest in this chapter has been in the chain of reasoning Marx employs to support the charge that the capitalist systematically exploits the worker. A crucial link in that chain is the Identity Thesis, which says that the value of a commodity is identical to the quantity of socially necessary labor contained in it. If this thesis were true, it would mean that what the capitalist acquires when he realizes a profit belongs, in a primitive and nonlegalistic sense, to the worker. Given the coercive or quasi-coercive relation between the capitalist and the worker, this goes a long way toward proving that there is a kind of systematic theft going on in the capitalist-worker exchange. It also implies that there is an irreconcilable conflict of interest between the two great classes of capitalist society. Both of these implications can serve as key elements of a radical critique of capitalist society.

Despite the importance of the Identity Thesis, Marx's argument for it in volume I of *Capital* is remarkably weak. A somewhat stronger indirect argument for it can be reconstructed out of parts of volume III of *Capital*, but this argument ultimately fails as well. As yet, we have no good reason to believe that the workers are systematically exploited by the capitalists.

### Appendix: G. A. Cohen's Reading of the LTV

My reading of the LTV is not uncontroversial and indeed not consistent with everything Marx said about it. G. A. Cohen provides an alternative interpretation of Marx's understanding of what the LTV asserts, together with supporting quotations. He says,

> Suppose a commodity has a certain value at a time t. Then that value, says the labor theory, is determined by the amount of socially necessary labor time required to produce a commodity of that kind. Let us now ask: required to produce it *when*? The answer is: at t, the time when it has the value to be explained. The amount of time required to produce it in the past, and, *a fortiori*, the amount of time actually spent producing it are magnitudes strictly irrelevant to its value.[29]

Cohen's reading about what the LTV asserts raises some delicate questions about the difference between interpretation and reconstruction. It may be that his way of construing what the LTV says is substantively best in that it avoids certain difficulties facing another construal, but there are reasons for denying that this is what Marx had in mind. This is true despite the fact that Marx occasionally formulates the LTV in a way favorable to Cohen's reading.

On my interpretation of Marx, the LTV is equivalent to the Identity Thesis ("The value of a commodity just is the quantity of socially necessary labor contained in it").[30] The best evidence for this reading of the LTV is Marx's repeated use of metaphorical talk of value as embodied, congealed, or ossified labor. This metaphorical talk is pervasive in Marx's economic writings and cannot be explained on Cohen's reading of the LTV. For example, at the end of the paragraph where the LTV is first introduced in *Capital* I, Marx quotes his own *Critique of Political Economy* as follows: " 'As values, all commodities are only masses of congealed labor time' " (*Capital* I, p. 47).

Cohen argues against my reading of the LTV on the grounds that it cannot account for exchange ratios under two kinds of circumstances: (i) when abnormally efficient or inefficient labor is used, or (ii) when technology changes in such a way that the standard amount of labor required to make the product changes.[31] However, neither kind of case creates a problem for the embodied labor reading of the LTV. As a theory of exchange ratios, the LTV is intended to be a complete explanation for exchange ratios only under conditions of equilibrium in a regime of simple commodity production. It says that, assuming equilibrium and assuming a regime of simple commodity production, commodities that exchange contain equal quantities of socially necessary labor. If complications are introduced, for example, capitalist relations of production or changes in the amount of socially necessary labor required to produce an item, then embodied labor is not the only factor that systematically affects exchange ratios; in other words, embodied labor completely determines exchange ratios only *ceteris paribus*—and *ceteris* isn't *paribus* if (i) or (ii) holds.

Perhaps the most important reason against Cohen's interpretation and in favor of the embodied labor reading is that only the latter makes sense out *Marx's* argument for the charge of exploitation against the capitalist. If Marx believed, as Cohen interprets him, that embodied labor is irrelevant to the value of a commodity, then it is not obvious how *he* (Marx) reasoned to the conclusion that the capitalist's profit is the result of something akin to theft. Moreover, if the value of a commodity is the labor that would be required to reproduce the object, it is harder to understand Marx's indignation about the capitalist's profit. It is an odd kind of booty that the capitalist is hauling away when he gets part of the value of the product.

Indeed, Cohen maintains (correctly, as far as I can tell) that his interpretation of the LTV cannot in fact support the charge of exploitation; on his view, what is needed to sustain this charge is the much weaker claim that the worker and only the worker creates what has value, namely, the product. The argument erected on this premise, what Cohen calls the 'Plain Argument', will be discussed in considerable detail in the next chapter. But, it is worth noting that Cohen makes no effort to show that *Marx* offered the Plain Argument in *Capital* I or anywhere else, nor does he offer a representation of Marx's actual path of argumentation to the conclusion that the capitalist systematically exploits the worker.

The view I defend, as Cohen himself emphasizes, is the popular view of what Marx had in mind, and the criticisms I raise in this chapter against my interpretation of how Marx understood the LTV can be reformulated to handle Cohen's variant reading. Unfortunately, proving that would require a restatement of most of the arguments of this chapter. Since Cohen himself raises decisive substantive objections against this variant reading, no purpose would be served by reformulating my substantive arguments. The purpose of this brief appendix is to call attention to this alternative understanding of the LTV and suggest some reasons why it is probably not Marx's considered view.

# 4

# Parasite Exploitation

## Different Conceptions of Exploitation

The LTV is no longer accepted by most of those who write on Marx. The reasons for this are various; some recognize that it cannot account for the value imputable to unimproved natural resources—land, in the economists' broad sense of the term—or the value returns to natural talents and abilities. Some economists maintain that a theory of distribution (a theory of who gets what in an economic system) can be constructed without any theory of value at all. Whatever the reasons, there is cause to mourn the passing of the LTV from the perspective of exploitation theory. If value just is embodied labor, then the capitalist *qua* capitalist is getting embodied labor created by someone else. Furthermore, even if the capitalist labored in the past to create capital, his return on his investment over and above replacement costs must still be attributable to laborers other than himself. This means that the profit that accrues to the capitalist results from something akin to theft; in taking his "cut," the capitalist is taking something that belongs, in a clear and unambiguous (yet pre- or nonlegal) sense, to the laborer. The LTV and the TSV together provide a persuasive and analytically elegant basis for the charge of systematic exploitation against capitalism. The associated critique of capitalist society is obviously radical, since the explanation for exploitation is to be found in the structure of capitalist relations of production.

However, in light of the problems with Marx's economics detailed in the preceding chapter, a new argument is needed to defend Marx's conclusion that the workers are systematically exploited under capitalism. That conclusion is, of course, central to Marx's radical critique of capitalist society, so it is important to see if it can be sustained in some other way. In the recent secondary literature on Marx, a number of new arguments have been advanced in support of this conclusion. In this chapter and the next I shall state and evaluate these arguments. This task is considerably complicated by the fact that different conceptions, or definitions, of exploitation are involved. Some way of classifying these conceptions is needed to organize the discussion. John Roemer provides a useful point of departure in his identification of two conceptions of exploitation, what he calls 'property relations exploitation' and 'surplus value exploitation'.[1] Let me begin with a brief discussion of these; then I shall propose a more inclusive classification scheme than Roemer supplies.

Both kinds of exploitation identified by Roemer take their inspiration from what might be called 'surplus reproduction models' of the capitalist economic system. The basic idea of a reproduction model is that an economic system can be conceived of as a system that functions to sustain its members by producing goods and services adequate to maintain a given standard of living. Surplus value comes from the production of new wealth over and above what is needed to sustain the producers in the society at that standard of living.

This raises some obvious and immediate questions for positive economics, namely, how is the surplus distributed and to whom? Marxists like this way of viewing economic systems, since it implies that what the latter do is to create a pot of extra wealth for contending classes to fight over; and, capitalist relations of production being what they are, this is a fight the capitalists will largely win and the workers will largely lose. Exploitation is to be found somewhere in the conditions of this fight or its outcome. This way of conceiving of economic systems also accords with Marx's view that economic systems (e.g., slavery, feudalism, capitalism) can be differentiated by their respective modes of surplus extraction (*Capital* I, p. 209).

A necessary condition common to virtually all forms of exploitation is that the exploited are forced in some way or other in their dealings with the exploiters. There has been some recent controversy about whether or not this condition is met in the case of the capitalists and the proletariat.[2] Since my concern is with different conceptions of exploitation, I shall largely bypass this debate over whether or not this common necessary condition is met.

The main difference between surplus value exploitation and property relations exploitation can be characterized as follows: Surplus value exploitation is a direct relation between individual proletarians and the capitalists for whom they work. The capitalist's appropriation of surplus value is viewed as a form of theft, or as akin to theft, in the sense that the appropriation of the surplus by the capitalist is conceived of as an "uncompensated taking" from the worker. This is suggested in Marx's own formulation of the charge of exploitation, though for various reasons he would be unhappy with the Lockean overtones of my way of phrasing it.[3]

On the other hand, property relations exploitation is not a relation between individuals but rather an attribute of certain classes. Roughly, the idea is that, for example, the proletariat is exploited because all of its members would be better off and all the capitalists would be worse off under different institutional arrangements.[4] It's as if the proletariat were running in a race with pebbles in their shoes; individual capitalists may not have put those pebbles in their shoes, and the proletariat are not giving the capitalists special shoes that permit them to run faster, but the latter benefit from the situation and would be worse off if the proletariat weren't disadvantaged in this way.

Roemer's classification of conceptions of exploitation is useful, but his two categories are not collectively exhaustive. A more comprehensive scheme can be constructed by noticing that surplus value exploitation is actually a special case of a more general category, which I shall call 'parasite exploitation'.[5] Unlike property relations exploitation, parasite exploitation relates particular individuals, for example, capitalists and proletarians. (It is a very personal thing.) The distinctive necessary condition common to all forms of parasite exploitation is a lack of reciprocity in the relation between the exploited and the exploiters. As we shall see, this lack of reciprocity can be spelled out in a variety of ways, not all of which can be understood as an uncompensated taking, that is, as surplus value exploitation.

This chapter is about parasite exploitation. I begin with its most common species, surplus value exploitation. First, I shall explicate more clearly the distinctive necessary condition for all forms of surplus value exploitation. Then I shall critically evaluate the arguments that have been offered for the claim that this condition is met under capitalism as it applies to the relation between the capitalist and the worker. I shall conclude that none of these arguments succeeds and that there is good reason to think that the workers are not in fact systematically surplus value exploited. In subsequent sections, other forms of parasite exploitation will be explicated and critically evaluated in a similar manner. In effect, I will investigate every way of understanding the charge that there is a crucial lack of reciprocity in the relation between the worker and the capitalist. The conclusions will be largely negative, at least from the perspective of substantiating a charge of parasite exploitation as part of a radical critique of capitalist society. Where the Marxian case does not fail, it will be shown to be incomplete in important ways. Property relations exploitation will be discussed in Chapter 5.

## Surplus Value Exploitation

Perhaps the root idea behind all conceptions of exploitation is that the exploiter is taking unfair advantage of the exploited. Surplus value exploitation locates this unfairness in the fact that the worker is causally responsible for exchange value that benefits the exploiter. As Marx says in *Capital* I,

> Whenever a part of society possesses the monopoly of the means of production, the labourer, free or not free, must add to the working-time necessary for his maintenance an extra working-time in order to produce the means of subsistence for the owners of the means of production whether this owner be the Athenian aristocrat, . . . Norman baron, American slave-owner, Wallachian Boyard, modern landlord or capitalist. [*Capital* I, p. 226]

One obvious reason why this situation is morally objectionable is that the laborer is doing unpaid labor for the exploiter. Doing unpaid labor is obviously not a sufficient condition for exploitation; for example, if Jones does volunteer work for his church, he is not being exploited (ordinarily). Moreover, as Marx recognizes in the *Critique of the Gotha Program*, the workers will not receive the full value of their product in the first stage of post-capitalist society because of various necessary deductions, yet presumably they will not be exploited.[6]

However, all this is consistent with the idea that the worker's doing unpaid labor for the capitalist is a *necessary* condition for exploitation, if exploitation is understood as an uncompensated taking. This is the distinctive necessary condition for surplus value exploitation. If the capitalist exploits the worker in this manner, then it is because the latter is doing unpaid labor for the former. This, then, is what the Marxist must prove. Aside from its relation to the charge of exploitation, this is an independently interesting question. I suspect that most Marxists believe that the workers *do* perform unpaid labor for the capitalists and that this is a fundamental and systematic feature of the capitalist economic system. In what follows I shall examine a number of attempts to prove this.

In her article, "Exploitation," Nancy Holmstrom gives the following concise account of exploitation under capitalism: "It is the fact that the [capitalist's] income is derived through forced, unpaid, surplus labor, the product of which the producers do not control, which makes it [capitalism] exploitative."[7] This passage states that there are four necessary and jointly sufficient conditions for exploitation. One of these is that the worker is doing unpaid labor for the capitalist. Given Marx's labor theory of value (LTV) and theory of surplus value (TSV), the argument for saying that proletarians perform unpaid labor is straightforward: For part of the working day, the worker cranks out a quantity of product whose sale value covers the value of the means of production used up and the value of the worker's labor power (for the entire day). But, in the remaining part of the work day, the worker cranks out a product whose sale value covers not only the value of the means of production consumed, but some additional value—that is, surplus value— which drops into the lap of the capitalist. Although the worker may indirectly get some of this value back in the future (to the extent that the capitalist reinvests his profits), he does not get all of it back, since the capitalist consumes some of this value or otherwise uses it in ways that do not redound to the worker's benefit. It follows that for part of the working day, the worker is creating value for which he is not compensated.[8]

This is a fair interpretation of Marx's account of exploitation under capitalism. As with much of the secondary literature on Marx, it is unclear if Holmstrom is merely trying to give an interpretation of Marx or if she is advancing a substantive claim about capitalism that she thinks is true. (Regrettably, many Marxists still conflate these separate tasks.) Although as an interpretation it is unobjectionable enough, substantively, it is highly problematical. In this article, Holmstrom does not discuss any of the objections to the LTV and the difficulties this creates for the TSV.[9] For example, the TSV depends on Marx's claim that what the capitalist buys from the worker is not the worker's labor, but his labor power. However, as I argued in the last chapter, this claim is justified solely on the grounds that there is no other way to explain the nature of profit, given both the LV and the LTV.[10] But we are no longer given either of those claims. Both are either false or so attenuated that they cannot support the charge of exploitation. So, the TSV is unsubstantiated. For this charge to go through, it is necessary to give some alternative argument for the claim that the capitalist's income is due to unpaid labor.

One might wonder if there is any way to sustain the charge of unpaid labor which bypasses the LTV. One way to do this is suggested by the very nature of a surplus reproduction model. Any given period of time (a year, a day, whatever) can be conceived of as consisting of two parts: necessary production time (necessary for reproduction) and surplus production time (the difference between total production time and necessary production time). Similarly, the worker's labor time can be divided into necessary labor time and surplus labor time. Suppose that some of the worker's surplus labor time eventually comes back to him (as an individual, not as a Spinozistic mode of the working class) as a result of future production. Still, since the capitalist consumes some of the results of surplus production time and otherwise uses unproductively other parts of the results of this time, it follows that the goods the worker can buy with his wage, together with what comes back to him personally in the future from reinvested profits, embody less labor than the worker gives on the job.

On this account, it is not necessary to assume that value is labor time. It's just that, in the wage bargain, unequal quantities of labor are being swapped—and the workers are consistently getting the short end of the stick. The capitalist is getting a quantity of expended labor for which he is not giving full compensation; this makes the relevant portion of labor unpaid.

This view is quite widespread. Some examples:

1. *Jon Elster*: "Being exploited means, fundamentally, working more hours than are needed to produce the goods one consumes."[11]

2. *Richard Miller*: "Marx certainly does argue that a capitalist economy cannot survive without the pervasive existence of exploitation in the following sense of the word. The time a typical worker spends on the job must be greater than the time required to produce the goods he acquires with his wage."[12]

3. *Jeffrey Reiman*: "Marxists hold that workers in capitalism work more hours for their bosses than the number of hours of work it takes to produce the real equivalent of the wages their bosses give them in return, and thus that they work in part without pay."[13]

The above is just a sample; others share this view.[14]

In evaluating this view, two questions are relevant: (i) Are unequal quantities of labor exchanged in the wage transaction? (ii) If so, does this support the charge that the worker does unpaid labor for the capitalist? Regarding (i), it might be objected that this is not necessarily the case. Suppose that an individual labors on his own for a time to create a capital good. He then buys other means of production with funds accumulated from previous labor and hires someone to help him turn out a consumer good. (John Roemer calls this 'clean accumulation'.) It might be thought that the profit that accrues to him is simply payment for his previous labor. The problem with this is that in the capitalist mode of production, the capitalist receives a payment large enough to cover the means of production consumed in the process of production and then some, that is, he receives an interest premium. In short, capital is the gift that keeps on giving. Since the capitalist receives an interest premium (not all of which goes even indirectly to the worker) and since that exchange value represents command over expended labor, it looks like there is an unequal exchange of embodied labor even in the case of clean accumulation. An affirmative answer to (i), then, looks pretty solid.

A comparable answer to (ii), on the other hand, is much more problematic. The argument for it goes something like this:

1. The worker gets less embodied labor from the capitalist than he gives.
2. If the worker gets less embodied labor from the capitalist than he gives, then the worker is doing unpaid labor for the capitalist.
∴  3. The worker is doing unpaid labor for the capitalist.

The problem with this argument is premise (2). Presumably, it is based on the principle that, if $A$ and $B$ swap unequal quantities of labor with $A$ giving up more than he gets, and if $A$ works for $B$, then $A$ does unpaid labor for $B$. To see the difficulty with this principle, consider what it means to say that $A$ does unpaid labor for $B$. First, $A$ is laboring at $B$'s behest; this may be because $A$ is $B$'s slave, serf, or proletarian. But none of these may be the case; $A$ may have volunteered. All that this requires is that $A$ has put himself, or has been put, at $B$'s disposal. To say that some or all of the resultant labor is unpaid is to say that some of the value contribution of $A$'s laboring at $B$'s behest does not go to $A$. Finally, to say that the unpaid labor is done *for B* implies that $B$ receives some of the value that results from $A$'s labor. If he does not, $A$ may still be doing unpaid labor, but he is not doing unpaid labor *for B*. The following explication sums it up:

$A$ does unpaid labor for $B$ if and only if
      i.   $A$ labors at $B$'s behest.
AND  ii.  The value contribution of $A$'s labor (call it $v$) is greater than the value $A$ receives (call it $v'$).
AND  iii.  $B$ realizes some of the difference between $v$ and $v'$ as a result of (i).

The value $A$ receives may, of course, be zero, as in the case of the corvée labor of the feudal serf. Conditions (ii) and (iii) capture the lack of reciprocity that is at the heart of the concept of unpaid labor and that makes this a form of parasite exploitation. Let us return to the principle behind premise (2), namely, If *A* gives up more labor to *B* than he gets, and if *A* works for *B*, then *A* does unpaid labor for *B*. It is easy to see that, if the LTV were true, this principle would also be true. On the other hand, if the LTV is false, and the value of a commodity is not the quantity of socially necessary labor contained in it, then it is possible for there to be an unequal swap of embodied labor without there being any unpaid labor,[15] that is, the general principle behind premise (2) is false. In other words, it is at least possible that the worker is getting the short end of the stick with respect to expended labor, and yet he receives his full value contribution, which entails, according to the above explication, that he is not doing unpaid labor for the capitalist. All this implies that the principle behind premise (2) of the argument is true if and only if the LTV is true. So much for that principle, and, by implication, for this argument for the claim that the worker is doing unpaid labor for the capitalist.[16]

Let us pause to summarize the state of the dialectic to this point. Surplus value exploitation is one species of parasite exploitation. I began by observing that a necessary condition for surplus value exploitation under capitalism is that the worker is doing unpaid labor for the capitalist. Two arguments for this claim were examined and found to be defective. The straightfoward Marxian argument depends on the LTV, and so it must be rejected. The second argument makes use of the undeniable truth that there is an unequal (embodied) labor exchange between the capitalist and the worker. The problem with this approach is that unequal labor exchange does not imply unpaid labor. More exactly, a key premise of the argument is true if and only if the LTV is true, so it too must be rejected. As yet we have no good reason to believe that the worker does unpaid labor for the capitalist.

The above account of what it means for one person to do unpaid labor for another implies that the defender of this charge against capitalism has to show that the value contribution of the laborer is greater than what he receives from the capitalist. The notion of value contribution here is nonmoral or purely positive. To substantiate the charge of unpaid labor, it would seem to be necessary to have a general theory that explains how the value of products and how the value of all the inputs in the production process are determined. That would permit a determination of the value contribution of the worker and a determination of the value realized by the worker in the form of wages and future investment. An identification of these two is a necessary condition for showing that the worker is doing unpaid labor for the capitalist. Let me suggest some reasons why the prospects for achieving this in a way favorable to the charge of systematic surplus value exploitation are not good.

A general theory of the value of products and inputs (outputs and inputs) is just a general theory of price. Whatever the correct price theory turns out to be, the falsity of the LTV implies that things other than labor are responsible for the value (equilibrium price) of inputs and outputs.[17] Given that value is not completely reducible to labor, the mere fact that the capitalist is getting a positive return on his investment cannot be used to show that the worker is doing unpaid labor for the capitalist. (The beauty of the LTV is that it permits exactly this inference.) At most, *part* of the capitalist's profit could be due to unpaid labor. Let us suppose that this is true.

The problem now is that the elimination of unpaid labor, or its diversion into the provision of public goods, does not obviously require the elimination of capitalism. Rates of return being what they are (the fallen rate of profit and all that), a somewhat steeper progressive profits tax (perhaps together with other meliorist measures that have been advocated by moderate reformers over the past century) would make it no longer true that the worker does unpaid labor *for the capitalist*, though of course he may still do some unpaid labor. Suppose, for example, that the capitalist pockets only the value that isn't explainable by labor's contribution.

This would mean that there is no longer any exploitation under capitalism! But for a Marxist, capitalism without exploitation is like a day without sunshine. It is true that a charge of systematic exploitation could be sustained against *existing* capitalism, but this charge cannot be part of a radical critique, since the elimination of the exploitation does not obviously require the elimination of capitalism.[18] It might be objected that the ruling class would never allow this to happen, but that is what they said about the Ten Hours bill. By contrast, if Marx's account were correct, the elimination of unpaid labor (and thus exploitation) would require the elimination of all returns to the owners of capital and thus the elimination of capitalism.

There is a larger lesson to be learned from this discussion. The LTV provides the basis for a radical and devastating critique of capitalism. Those who capture profits are quite literally doing so at the expense of the workers, since those profits represent unpaid labor. It also implies that there is an absolute, ineliminable, and systematic conflict of interests between capitalists and proletarians. For these reasons, it is hard to overstate the importance of the LTV for Marx's radical critique of capitalist society. Nevertheless, once the LTV is abandoned and it is admitted that the value of a commodity is not just so much embodied labor, the charge of *surplus value* exploitation, if it could be sustained, reduces to a distributional quibble of the sort that Marx expressed impatience with in his attack on the Lassalleans in the *Critique of the Gotha Program*. The idea that the worker is doing unpaid labor for the capitalist cannot be part of a truly radical critique of capitalist society without the LTV.

In spite of this, the intuition might persist that there is an objectionable lack of reciprocity of some sort between the capitalist and the worker, a kind of parasitism on the part of the capitalist. Let us turn to some other grounds on which that charge might be based.

## More on Unequal Labor Exchange

It might be thought that there remains something to be said for the undeniable truth that there is an unequal swap of embodied labor in the capitalist-laborer exchange. Perhaps the attempt to argue from this to the claim about unpaid labor is an unnecessary theoretical shuffle. Unequal labor exchange could, by itself, be the relevant necessary condition for some form of parasite exploitation. For example, in the above-quoted definition from Holmstrom, replace the term 'unpaid, surplus labor' with 'unequal labor exchange'. If this were true, the Marxist would not have to defend the claim that the worker does unpaid labor for the capitalist.

The problem with this proposal has to do with the nature of a parasitically exploitative exchange. The parasite metaphor can serve as a useful heuristic in this connection. All parasites benefit from their association with the host. Some hosts similarly benefit from their association with their parasites, for example, the intestinal bacteria, *e. coli*, but when we speak of social parasites, this is not what we have in mind. (See the writings of J. V. Stalin on this.) A social parasite is someone who is able to extract benefit out of proportion to his contribution. He is, to a greater or lesser extent, a "free rider." This is one way of capturing the lack of reciprocity between the exploiter and the exploited that is the hallmark of parasite exploitation.

In a parasitically exploitative exchange, the exploited person is getting taken advantage of and the exploiter is getting "something for nothing." And, in the case of the capitalist and the worker, the "something" has to be value and not labor because the former is what the advantage consists in. Consider a case in which $A$ does unpaid labor for $B$. $B$ is giving up (i.e., contributing) value for which he is not compensated. Whether or not it is parasitically exploitative, there is a lack of reciprocity between the person doing the labor and the person benefiting from it. By contrast, consider the following example of an unequal labor exchange that does not involve uncompensated labor: Suppose $A$ has to labor four hours in a fast-food restaurant to earn enough money to purchase a ticket to a concert given by $B$, in which $B$ invests one hour of labor. (Suppose that $B$'s labor in practicing is covered by the sale of albums and T-shirts.) In this case there is an unequal labor exchange, but no one is doing uncompensated labor, and there is no failure of reciprocity.

This shows that the bare fact of an unequal labor exchange is not a sufficient condition for a lack of reciprocity, and thus this bare fact cannot be a necessary condition for parasite exploitation. Notice that if the labor theory of value were true, unequal labor exchange would entail a lack of reciprocity. If value just is embodied labor, then unequal labor exchange would imply unequal value exchange, and the lack of reciprocity condition for parasite exploitation would be met.

If the Marxist is to sustain a charge of parasite exploitation of workers by capitalists, he will have to show that there is a lack of reciprocity—and not merely some inequality or other—between the exploited and the exploiters. Given that it cannot be satisfactorily established that the workers do uncompensated labor for the capitalists, what other options are there? One approach would be to argue that the value the capitalist receives as a result of his association with the worker is greater than the capitalist's value contribution—and not because the laborer does unpaid labor; that is, the capitalist is getting value from the process of production not explainable by his own contribution.

This would suffice to establish the lack of reciprocity required for parasite exploitation. In other words, instead of focusing on what the laborer does, that is, contributes, it might be more fruitful to turn our attention to what the capitalist does not do. Maybe what the capitalist contributes, in his capacity as capitalist, is . . . nothing at all. This line of argument deserves careful consideration and will be investigated in the next three sections.

### Productive Asymmetry

In director Mel Brooks's film, "The Producers," an impresario and a shady accountant cook up a scheme to bilk investors in a Broadway show out of their money. The investors are asked to put up a sum of money in exchange for a certain percentage of the show's profits. Far more money is raised than is needed to cover production costs and far more than one hundred percent of the profits are promised to the various investors. The scam is to produce a show that is so bad that it will fold after one night. The investors will not have to be paid off, and the impresario and the accountant will be able to keep the extra money. To insure failure, they hire the worst scriptwriter, director, and actors they can find. The result is a musical called "Springtime for Hitler," which turns out to be a smash success as a comedy, and the producers end up in jail.

Whether or not the impresario and the accountant should be called 'producers' (in contrast to the people who actually wrote, directed, and acted in the show), no one would say that the investors were. It seems that the same can be said of the ordinary capitalist who invests in an enterprise. Throughout most of this chapter, we have examined attempts to prove the charge that the workers are parasitically exploited by the capitalists because the former receive less than they contribute. In this section and the next two, we examine the other side of the equation: Can the charge of parasitic exploitation be sustained by showing that the capitalist receives more than he contributes?

In a penetrating article on the labor theory of value and exploitation,[19] G. A. Cohen thinks that the fact that capitalists, like the investors in the play, are not producers can be used to support the conclusion that the laborer is parasitically exploited (my term, not Cohen's) by the capitalist. Cohen argues that it is not necessary to establish that the worker and only the worker creates value, as the standard Marxian account maintains. A weaker premise will do the job: The worker and only the worker creates the product—the thing that has value, whatever value happens to be. Cohen says,

> what raises a charge of exploitation is not that the capitalist gets some of the value the worker produces, but that he gets some of the value *of what* the worker produces. Whether or not workers produce value, they produce the product, that which has value.
> And no one else does.[20]

The argument, which Cohen calls the 'Plain Argument', built on this observation goes like this:

1. The laborer is the only person who creates the product.
2. The capitalist receives some of the value of the product.
∴  3. The laborer receives less value than the value of what he or she creates.
∴  4. The capitalist receives some of the value of what the laborer creates.
5.
∴  6. The laborer is exploited by the capitalist.[21]

Premise (5) is not filled in because Cohen does not offer a complete analysis of the concept of exploitation. He does, however, have some idea of what it might look like:

> an essential normative premise is not stated. Its content, in very general terms, is that, under certain conditions it is (unjust) exploitation to obtain something from someone without giving him anything in return. . . . A rough idea of exploitation as a certain kind of lack of reciprocity, is all that we require.[22]

For Cohen, the central claim in this argument is that the worker *and only the worker* creates the product. Why is that? Part of the answer is the obvious fact that workers are producers. Now, Cohen maintains that it is true by definition that *only* producers create the product. But this only raises the question,

> who is a producer? But whatever the answer may be, only those whom it identifies can be said to produce what has value. And we know before we have the full answer that owners of capital, considered as such, cannot be said to do so.[23]

On the face of it, this passage, especially the last sentence, looks very suspicious, one might say question-begging. To forestall criticism, Cohen immediately enters some caveats: First, an individual capitalist may also be a producer, as is usually the case with the petty bourgeoisie. But, of course, *qua* capitalist, he is not. More damaging is Cohen's second admission that it may be that, in their very capacity as capitalists, capitalists act productively. This leads him to draw a distinction between productive activities and producing activities: To act productively is to make some contribution to bringing the product into existence; all producers act productively, but acting productively does not make one a producer. Cohen allows that if the capitalist is a productive nonproducer, it would, in his words, "have a bearing on the charge of exploitation."[24]

The problem now is evident; the distinction between being a producer and acting productively cannot bear the moral weight Cohen puts on it. After all, if in fact the capitalist (*qua* capitalist) *contributes* to production, by acting productively, wherein consists the failure of reciprocity required for parasite exploitation? Whether or not he is a producer does not seem to be very important. If the capitalist is a productive nonproducer, as it were, it seems that this might have a very important bearing on the charge of parasite exploitation—it might make it false!

There seem to be two options open to Cohen. The cleanest move would be to argue that the capitalist is simply not productive. Call this 'Option A'. The other option is to argue that the capitalist (*qua* capitalist) is productive, but what he gets is disproportionate to his contribution. Call this 'Option B'. This could be so for one of two reasons: Either he is getting some of the value of someone else's contribution (Option $B_1$), or some of the value he is getting is contributed by no one (Option $B_2$). Both versions of Option B allow that the capitalist makes some contribution and that some of what he receives is due to that contribution—deserved, as it were.

Option $B_1$ does not look promising. If the capitalist is getting the value of someone else's contribution, it would seem that it is the laborers'. But this just takes us back to the charge that the worker does unpaid labor for the capitalist. On the other hand, if he is getting value contributed by no one (Option $B_2$), then he is, in part, (for part of the working day, as Marxists like to say) not productive—in short, a parasite. For reasons that will become clear at the end of the next section, this option will not succeed. All things considered, the most promising strategy is Option A: Deny that the capitalist is productive *tout court*.

In this article, Cohen does not assert that the capitalist is not productive, but if that proposition can be successfully defended, then Cohen will have the necessary lack of reciprocity. In other words, suppose that the capitalist (*qua* capitalist, of course) is not productive. If we add *that* as a premise to the Plain Argument, the latter begins to look like a pretty good argument. As noted above, there need be no presumption that the worker is doing unpaid labor for the capitalist. The value that accrues to the capitalist need not be attributable to the worker. All we would have (and all that we would need) is that the capitalist is getting value for which he is making no contribution, that is, he is a parasite on the productive process.

However, at second glance, this does not look like a very promising strategy. After all, some capitalists have been great innovators and farsighted directors of society's productive resources. They assemble diverse factors of production in a new ensemble to meet new needs or to meet old needs more effectively. Surely, it is difficult to deny that this is a productive activity. However, there is a way of handling this response by slightly creative definition. Economists have found it useful to distinguish between the capitalist and the entrepreneur.[25] These terms are functionally defined, so that one and the same person can be both.

It is customary to think of the capitalist as someone who buys factors of production (raw materials, machines, labor [or labor power], etc.) and sells a finished product. If we draw a distinction between the capitalist and the entrepreneur, this is not accurate. The capitalist *qua* capitalist is a factor owner—an owner of the means of production. He is (merely) a supplier of capital. It is the entrepreneur who organizes and assembles the factors of production (Marx's Unholy Trinity of land, labor, and capital). Perhaps the easiest way to grasp the distinction between capitalist and entrepreneur is to think of the entrepreneur as owning no means of production at all; he assembles the resources owned by others into a productive structure. All that he brings to the productive process are his ideas and his organizing ability. His contribution consists in marshaling the resources of owners of other factors of production.

The capital he gets from capitalists can take the form of concrete capital goods (e.g., a machine, a building) or, as is more common nowadays, money with which such goods are purchased. The capitalist can be thought of as renting out his capital for a fixed rated of return to the entrepreneur while retaining ownership of some portion or proportion of the means of production. A third form of capital investment occurs when someone buys a bond issued by a firm. Though the bondholder does not have ownership rights in the firm, the assets of the latter serve to secure the loan. This distinction between capitalist and entrepreneur is evident in the Mel Brooks movie referred to at the beginning of this section—the investors are simply passive providers of capital; it is the impresario and the accountant who have the ideas. They are pure entrepreneurs, and, ignoring the peculiarities of their scam, they act productively.

There is a counterpoint to this distinction in the distribution of the social product; the term 'profit' used to refer indiscriminately to all returns to the owners of the firm. However, to conceive of the reference of the term 'profit' in this manner is to conceal the fact that this income is properly imputed to two different sources: entrepreneurship and capital. The term 'profit' or 'pure profit' is now used to refer exclusively to entrepreneurial gains (or losses) due to exceptional(ly good or bad) foresight or luck on the part of the entrepreneur.[26] By contrast, the term 'interest' refers to the return on investment that accrues to the owners of the means of production. Unlike pure profits, which fluctuate considerably, the rate of interest tends toward equality in a competitive environment. Pure profits (not due to monopoly, etc.) tend to get wiped out as factor prices are bid up and product prices are driven down by the competitive process.

The importance of this distinction for the issue at hand is that if all of the productive activities we're inclined to attribute to capitalists as individuals are really varieties of entrepreneurship, then the capitalist *qua* capitalist is not productive. His return on investment is a sign of parasite exploitation. Furthermore, it is not plausible to suppose that capitalism could persist if all the returns to capital were taxed away or plowed back into production for the future, and the only income accruing to capitalists as individuals would be the returns to their entrepreneurship, if any. In short, meliorist measures could not eliminate the parasitism of capitalists *qua* capitalists.

**What, if Anything, Does the Capitalist Contribute?**

The view that the capitalist (*qua* capitalist will again be suppressed) is not productive is, I suspect, widely shared by socialists of all stripes. It has been argued for recently by David Schweickart in his book, *Capitalism or Worker*

*Control?*. In what follows I restate and critically evaluate this argument. The following passage sums up his position:

> The basic problem in trying to justify capitalism by an appeal to contribution is the impossibility of identifying an activity (or set of activities) engaged in by all and only capitalists which can be called (preserving the ethical connotations of the word) 'contribution'.[27]

What is it that all and only capitalists do in virtue of being capitalists? The obvious answer: They provide capital. Schweickart's argument, starting from this obvious truth, is fairly straightforward:

> 'Providing capital' means nothing more than 'allowing it to be used.' But the act of granting permission can scarcely be considered a productive activity. . . . if owners ceased to grant permission, production would cease *only if* their ownership claims were enforced. If . . . their authority over the means of production was no longer recognized, then production need not be affected at all.[28]

The argument can be conveniently restated as follows:

1. The only thing that capitalists do is to provide capital.
2. 'Providing capital' means 'allowing capital to be used'.
∴ 3. The only thing that capitalists do is to allow capital to be used.
4. Allowing something to be used is (merely) granting permission.
5. Granting permission is not a productive activity.
∴ 6. The only thing that capitalists do is not a productive activity.

It follows from (6) (or from [6] and an uncontroversial premise) that:

∴ 7. Capitalists are not productive.

The relevance of this argument to a charge of parasite exploitation is fairly obvious: Given that the capitalist receives a positive return on his investment, at least part of which he consumes, he is a nonproductive consumer, that is, a parasite.[29] This is one way of understanding the lack of reciprocity that is the hallmark of parasite exploitation.

To see if this argument is sound, let us examine the premises. The independent premises are (1), (2), (4), and (5). Premise (4) is pretty obviously true; Schweickart has produced some reason for (5) in the passage quoted immediately prior to the formal representation of his argument. Premises (1) and (2) can be considered together. As individuals, capitalists do much more than allow capital to be used. Some labor, some take risks, some innovate, and some invent. But, and this is the important point, some capitalists do none of these things. As capitalists, they merely allow capital to be used. It is this activity that is common and peculiar to all capitalists. This looks to be a solid argument.

However, one thing that should make us wary is that nowhere in it (or, apparently, behind the premises) is there any explanation of why capitalists do in fact receive interest. An answer to that question would seem highly material to the issue at hand. Indeed, nowhere in his discussion does Schweickart provide such an account. As I shall show, Schweickart's failure to understand why capitalists (usually) do receive a positive return on their investments cripples this argument and in fact allows us to conclude that (7) is false.

Why do capitalists receive interest? A Marxian answer, which does not depend on the labor theory of value, might go something like this: A distinctive feature of capitalism is the fact that workers do not own the means of production. Since they must provide for their own and their families' immediate needs, the capitalist, through his intermediary, the entrepreneur, is able to demand and receive a premium on his investment. Interest (return on investment) becomes one of the 'costs of production', which determines price. The worker-consumer is at a permanent disadvantage vis-à-vis the capitalist-investor; the latter's surplus allows him to demand a premium on his investment, whereas the former's lack of a surplus requires that he meet that demand. Ultimately, what explains interest is a decisive fact about the relations of production in that society: The workers are separated from ownership of the means of production; this allows interest to be demanded and received.

Perhaps the most serious defect in this explanation of interest is that it applies only to the capitalist mode of production. Interest is a widespread phenomenon, manifesting itself in societies where the workers are not separated from ownership of the means of production. (Witness the medieval church's struggle with usury.) Furthermore, modern socialist states have found themselves constrained to employ an interest premium as an accounting device. Even if there is some question about whether such states are "really" socialist, it is hard to see how the need for such an accounting device arises out of the relations of production.

However, even if these objections could be met, there are other problems with locating the explanation of interest in capitalist relations of production. Under modern capitalism, ownership, if not effective control, of the means of production is widely dispersed. Institutional investment by, for example, insurance companies and pension funds is quite common. Even if effective control of the means of production is concentrated in a relatively few hands, the return on investment is spread quite differently from the way it was one hundred years ago. Finally, the image of the starving proletarian is hardly applicable to advanced industrial countries.

Despite the fact that the explanation of the phenomenon of interest in terms of relations of production is inadequate, there is an element of truth in it. To see what this is, let us consider what the capitalist is exchanging (i.e., buying and selling) when he invests in an enterprise. For vividness, consider a situation in which the capitalist lends money at a fixed rate to an entrepreneur who organizes an ensemble of factors of production. Capital in this case takes the form of money, though the entrepreneur will tie it up in concrete capital goods, labor, and so forth. We suppose that the capitalist and entrepreneur are different individuals, but nothing hangs on that, since these terms are functionally defined. What the capitalist-lender is doing is selling money now (the principal) in exchange for money later (the principal plus interest). The crucial question for a theory of interest (return on investment) is to explain why the entrepreneur is willing to pay a premium on the borrowed funds. Or, looked at from another perspective, why won't the capitalist lend unless he is promised interest?[30]

Schweickart never directly answers this question, but at one level, the answer is not far to seek: For the entrepreneur, $x$ dollars now are worth more to him than $x$ dollars later. Indeed, since he expects to turn a pure profit on the invested funds, he values the $x$ dollars now even more than the $(x + n\%$ of $x)$ dollars later; otherwise, he would not assume the loan. Similarly, for the lender, $x$ dollars now is worth more than $x$ dollars later. An interest premium is needed to match and indeed surpass his preference for present goods (money now) over future goods (money later). His preference in this case is the mirror image of the entrepreneur's: $(x + n\%$ of $x)$ dollars later is worth more to him than what he is selling—$x$ dollars now. The unifying explanatory principle is the same in both cases: People generally prefer present goods to (comparable) future goods. The rate of that preference can be called the time preference rate.

Obviously, different persons will have different time preference rates. For example, Jones might be a lender of money at 6 percent, a borrower at 4 percent, and not in the market at 5 percent. Smith, on the other hand, might be a lender at 12 percent, a borrower at 9 percent and on the sidelines at 10 percent. The rate of interest that emerges in a market economy is an amalgamation of different individuals' time preference rates, determined in the usual manner by supply and demand schedules. In essence, then, interest is the price of time, and it is time that the capitalist is selling.

The element of truth in the Marxist account of interest sketched above is this: Anyone who becomes a capitalist must be in a position to forgo present consumption, which can include leisure, in exchange for (greater) future consumption. The rate at which a rich person is willing to lend might well be much lower than the rate required to coax a less well off person to restrict present consumption. The factors that determine individual time preference rates are various; it might be thought that, in general, poorer people have a higher time preference than wealthier people, since the former have greater unmet consumption needs. However, this is neither universally nor even generally true; following Marx, it might be pointed out that if poor people have a religion that endows thrift with transcendent value, they might have a very low time preference rate (i.e., they are very future-oriented). On the other hand, a rich entrepreneur who sees a unique opportunity developing or a playboy who lives life in the fast lane might have a very high rate of preference for present goods over future goods.

The time preference theory of interest[31] has remarkably wide explanatory scope. Indeed, the best argument for it is an inference to the best explanation. It explains interest on consumer loans as well as business loans and the returns to capital. Thus it accounts for the fact that a machine that will produce $100,000 worth of goods over its ten-year life span will sell for considerably less than $100,000. Indeed it explains all forms of capital formation, even in a moneyless economy. If a primitive tribe decides to invest resources in lengthier or more roundabout methods of production, they will do so only if they believe that it will "pay." For example if some hunters stay home to make more effective weapons, they will do so only if the future yield is expected to more than offset the restriction of output in the near term. In a market economy, the rate of interest is in part an expression of the social rate of time preference of present goods to future goods. Actual rates can also reflect an inflation premium, as well as ignorance of more willing buyers (sellers) in the time market.

Schweickart does not directly consider the time preference theory of interest. However, at one point he maintains that 'waiting' is not necessary for the existence of capital, though he does admit that it is necessary for capital accumulation.

> Consider a society producing its maximally feasible output of consumer goods consistent with its own reproduction. In this society some workers are producing capital goods to replace those used up in the production process. Since there is maximal production of consumer goods, the workforce in the capital goods sector produces just enough for replacements . . . Capital goods here are not the product of "waiting."[32]

It is unclear what this last sentence means. Since capital goods are produced means of production, and since all production takes time, there is an obvious sense in which capital goods, as well as consumer goods, are the product of waiting. What he must be denying is that capital formation requires the restriction of present consumption so that capital goods can be brought into existence and mature or 'fructify' into consumer goods. If he is right about this, then the interest premium that capital providers receive would not be explained by their preference for present goods over future goods.

But Schweickart is clearly mistaken. Present consumption, which can include leisure, *is* being restricted for the sake of future consumption (i.e., production)—it is being restricted relative to what it might have been. Schweickart would object that this ignores his requirement that the economy "reproduce itself." But no economy does *that*. (By contrast, economic models do it all the time.) As Marx reminds us, it is human beings who produce things and "reproduce" themselves, and economic theory must, at some point, explain that and not simply assume it. The time preference theory says that capital formation requires that someone restrict present consumption in exchange for future consumption. This is the case in any real world economy.

The general Marxist reaction to talk about abstinence on the part of the capitalist is to hoot with derision. Perhaps the best hooter is Marx himself. In some of the funniest passages in all of *Capital*, he ridicules the notion of capitalist abstinence. Marx puts the following words in the mouth of the capitalist:

> "Consider my abstinence. I might have played ducks and drakes with the 15 shillings; but instead of that I consumed it productively, and made yarn with

it." Very well [says Marx], and by way of reward he is now in possession of good yarn instead of a bad conscience. [*Capital* I, p. 186]

Fortunately, Marx does more than hoot. He offers an argument. A couple of sentences later he says,

> Besides, where nothing is, the king has lost his rights; whatever may be the merit of abstinence, there is nothing wherewith specially to remunerate it, *because the value of the product is merely the sum of the values of the commodities that were thrown into the process of production.* [*Capital* I, p. 186, emphasis added]

If the time preference theory of interest is true, then the emphasized passage is false. This passage is written on the assumption of the labor theory of value. If that were right, Marx's point would be well-taken. After all, if value just is embodied labor, then the abstinence of the capitalist cannot make a value contribution, whatever the "merit" of it may be. In a revealing passage in the paragraph immediately preceding the paragraph from which the above quotations are taken, Marx plainly reveals his lack of appreciation for the economic significance of time:

> It is clear that, whether a man buys his house ready built, or gets it built for him, in neither case will the mode of acquisition increase the amount of money laid out on the house. [*Capital* I, p. 186]

No one who knows anything about the construction industry, even in nineteenth century England, would say something like this. Marx has been caught *flagrante delicto*.

Talk of abstinence brings to mind the concept of sacrifice. It is certainly odd to talk about the sacrifice of the capitalist, especially in contrast to what the worker undergoes. But even outside of this contrast, talk about sacrifice on the part of the capitalist is highly misleading. He does sacrifice present consumption, but he is more than compensated for that by the return on his investment. In net terms, there is no sacrifice at all. But there is abstinence. And, what the capitalist contributes is neither sacrifice nor abstinence; it is time. That's what is valuable.

What ultimately explains time preference is a matter of some controversy; maybe we care less about our future selves than our present selves, as some contemporary metaphysicians might put it. On a less arcane level, a partial explanation virtually suggests itself: Since the production of consumption goods takes time (and for some consumption goods considerable time) and since humans are not immortal, time is both valuable and scarce; that is why it has exchange value. However, whatever the full explanation, time preference is a fact of life, and it can be used to explain why there is (always) a positive rate of interest.

To return to our representation of Schweickart's argument, what the capitalist does, then, in providing capital is not merely 'allowing it to be used', as Schweickart maintains. When a person invests in an enterprise, he is selling present goods (usually money now) in exchange for future goods (usually money later). Capitalists and entrepreneurs are, respectively, sellers and buyers in the time market. Schweickart's definition of providing capital in premise (3) is false, if 'allowing capital to be used' is understood in the way he suggests in premise (5). An essential element required for bringing consumer goods into existence is time. As a supplier of time in the form of command over present goods, the capitalist *qua* capitalist makes an essential contribution to production. He provides the "front money" that allows other factor providers (suppliers, laborers) to be paid before the sale of the final product.

There is a further problem with Schweickart's argument, a brief discussion of which will allow a deeper insight into the role of the capitalist in a capitalist economy. Capitalists are not only providers of time; in making their investment decisions, they are also, of necessity, entrepreneurial speculators about what consumers and intermediate producers will want. Although it is useful for analytical purposes to separate the entrepreneur from the capitalist, in the real world, the two functions are usually combined in the same person. Consider the following two cases: Capitalist *A* invests his capital in building a new steel mill. Capitalist *B* invests in a project that brings a previously submarginal coal seam into production. Either they make the decisions themselves or they hire someone else to do it. Either way, the decision is entrepreneurial. Suppose that, due to the vagaries of the marketplace, *A*'s enterprise goes broke and *B*'s enterprise prospers. In retrospect, it is clear that *A* has squandered social wealth, whereas *B* has helped to create wealth (assuming that the market at least sometimes determines which projects are socially beneficial). Suppose that we grant Schweickart's claim that *A* and *B* were simply granting permission for their capitals to be used. There is an obvious sense in which *B*'s granting permission for his capital to be used was productive while *A*'s granting permission was not. Indeed, to call an investment decision a (mere) "granting of permission" is to seriously underdescribe the situation. In a world of uncertainty, capitalists must speculate about future conditions of supply and demand. Successful speculators are productive.

The point here is not that the uncertainty that investment entails, which after all may be quite low, "justifies" a premium as a "reward" (for sleepless nights?!?). Rather, the point is that the activity of granting permission for capital to be used in one line of production rather than another can be (and is, when successful) socially valuable. If such socially valuable activities can be said to be productive, then premise (5) of Schweickart's argument is also false. It seems wholly arbitrary to restrict the term 'productive' to those activities that are coextensive with laboring.

It might be objected that the above account assumes that all of what the capitalist *qua* capitalist gets is due to his contribution of time. Marxists would insist that there is more to it than this. The capitalist's position as an owner of the means of production in a system in which most do not own means of production is in part responsible for his income as capitalist. Something about the bargaining power of the capitalist, vis-à-vis the worker, is usually invoked in this connection.

There are some complicated controversies in economic theory in this connection,[33] but for present purposes they can be avoided. Suppose that the above account of the capitalist's contribution, as capitalist and as entrepreneur, is a large part of the story about the capitalist's income. That much seems hard to deny. Now the radical critic faces a problem similar to one that arose earlier in this chapter, namely, it is no longer clear that meliorist measures cannot eliminate parasite exploitation by eliminating that part of capitalists' income not due to their contributions of time and entrepreneurship. To put the point informally, a little parasitism is not that serious a problem. For the radical to be taken seriously, he needs to convince us, as Marx tried to, that capitalists are really big bloodsuckers. However easy life is for capitalists, the above account shows that this just isn't true.

## Final Doubts About Capitalist Contribution

There are a number of responses a Marxist might make to the above argument about the contribution of the capitalist. One would be to call attention to the way capitalists came to own their capital. Not a heartwarming story. Marx supplies many of the gory details in his discussion of primitive accumulation in Part VIII of *Capital* I. That outright theft and injustice attended the original accumulation of capital can hardly be denied. However, there are two problems with this approach to undercutting the capitalist's claim on the contribution that he makes.

First, there is the problem of "clean accumulation." Just as it is clear that some capitalists originally stole or otherwise unjustly usurped what became "their" capital, it is equally clear that not all did. The capital holdings of many people today cannot be traced in any morally objectionable way to unjust primitive accumulation. There really have been—and continue to be—people (e.g., immigrants to America) who started with no capital and, through hard work and successful entrepreneurship, have made money, which is converted into capital.

But let us suppose that the proportion of capital that is cleanly held is vanishingly small. Nonetheless, the injustice of primitive accumulation is not a very good foundation on which to build a case for the charge of systematic exploitation (parasitic or otherwise) against capitalism, at least from a Marxist perspective. Here's why: First, Marxists would want to hold that there is exploitation under capitalism even if capital accumulation is clean. Otherwise, it is doubtful that the exploitative character of capitalism is explainable by capitalist relations of production or the other essential features of capitalism identified in Chapter 2. To suppose otherwise, it would be necessary to show that the reproduction of capitalist relations of production depends on something like the systematic theft that characterized primitive accumulation. But this takes us back to the charge of unpaid labor. Second, the theory of justice needed to sustain the charge that capital traceable to unjust appropriation is, by that fact, unjustly held would probably be an historical entitlement theory of justice. Down that path are bridges to libertarianism, under which live trolls like Robert Nozick.

Besides, there are easier paths for a Marxist to follow. Whatever the historical story, there is obviously a . . . how shall we say? . . . lopsided distribution in control over the means of production. Capitalists are able to get the returns to capital only because they have exclusive control over the means of production. If control over the means of production were spread more evenly, so too would be the returns to capital. Everyone (or nearly everyone) would be able to contribute time in the form of command over present goods. In the final analysis, I think this is the best way for Marxists to go, but this move is the essence of the property relations exploitation. This view will be discussed in detail in the next chapter. We are not yet finished with parasite exploitation, however. The real issue that remains for this chapter is whether or not there is any way to recover parasite exploitation in light of the time preference theory of interest.

Let us return to the notion of contribution. A feeling might persist that there is something objectionable—exploitative—about a system that permits some people to live quite well while giving little or nothing of themselves. Is there some other sense of 'contribution' on the basis of which it can be said that capitalists *qua* capitalists make no contribution? I think there is. To see what it is, consider the case of low-risk investments. Clearly, it takes little entrepreneurial astuteness to invest in low-risk securities or stocks. Schweickart quotes the following passage from J. K. Galbraith to illustrate just how little some capitalists have to do to get by in the world:

> No grant of feudal privilege has ever equalled, for effortless return, that of a grandparent who bought and endowed his descendants with a thousand shares of General Motors or General Electric. The beneficiaries of this foresight have become and remain rich by no exercise of intelligence beyond the decision to do nothing, embracing as it did the decision not to sell.[34]

The standard bourgeois response to such observations is to ask rhetorically about the fate of the owners of a thousand shares of General Carriage and Harness or General Candle Corporation. Such a response is pertinent, but it does not really address Galbraith's (and Schweickart's) challenge: Why should those who contribute so little receive so much in return? Or, is the challenge: Why should those who contribute so little *of themselves* receive so much in return? The difference between these two questions reveals an ambiguity in the concept of contribution that merits some attention.

Sometimes we speak of contribution in a straightforward consequentialist sense: A person's (positive) contribution is proportional to the (good) results that flow from his actions, or perhaps it is identical to the marginal difference his actions make. In either sense, capitalists make a significant contribution to production in the form of the time (i.e., the capital) they provide and their entrepreneurship, *however easy it is for them to make that contribution*. On the other hand, except for their entrepreneurship, (and even that might be fairly passive) perhaps it can be said that they give little *of themselves* in providing capital. Let us call this sense of contribution 'personal contribution'. The claim now is that the laborer, but not the capitalist, makes a personal contribution to production. This raises two crucial questions: (i) Is this in fact true?; and (ii) If so, can it serve to ground a charge of parasite exploitation?'

Regarding (i), it is far from clear that capitalist contribution is, in general, impersonal in the required sense. Suppose a capitalist originally accumulated some capital by saving part of his wages and/or by successfully exercising entrepreneurship.[35] In the case of the petty bourgeoisie, this is often the case. For a member of the latter, his business is as much an extension of himself as embodied labor is an extension of the laborer—perhaps more so because more than labor is required *of him* if the business is to be successful. Assuming he prospers and gets interest income (i.e., income imputed to capital), his capital will represent not only his embodied labor and his "embodied entrepreneurship"; it will also represent his restriction ('embodied restriction' sounds a bit too constipative) of present consumption for the sake of production (future consumption). Given that his entire contribution can be said to be personal, it follows that the capitalist's contribution of time is not, by its very nature, impersonal.

However, it seems that the radical critic must show that the capitalist's contribution of his capital is, by its very nature, impersonal. Here's why: Showing the latter is a necessary condition for establishing that capitalism, by its very nature, is systematically parasitically exploitative, on this understanding of parasite exploitation. The reason for this is that it is necessary, for the purposes of a radical critique, to show that exploitation is rooted in the essential features of capitalism. To put the point slightly differently, if eliminating parasite exploitation requires personal contribution by all, the elimination of this form of exploitation could be insured by petty bourgeois reforms. Marx can give us a rather long list of people we might read if we want to follow this road, but his name would not be on it.

I think this problem of "clean personal accumulation" is a decisive objection against making a lack of *personal* contribution a necessary condition for parasite exploitation—at least if the latter is to be part of a radical critique of capitalist society. To put the point informally, it doesn't seem that the lack of personal involvement by capitalists is a necessary condition for parasite exploitation under capitalism. But I am not completely sure that this is right. So, let us press on a bit and see how a radical critic might respond.

A radical critic might say that, although in principle petty bourgeois reforms might eliminate parasite exploitation, in fact this could not happen. Given the internally generated movement toward monopoly capitalism, the idea of capitalism without impersonal contribution is purely utopian. There is simply no way for capitalism to exist without the impersonal contribution of the big bourgeoisie. Many of the latter are people who have inherited their capital, which continues to pile up through the magic of compound interest. They do make a decision not to consume, but surely that is not sufficient to make their contribution personal. Let us suppose that the contribution of the big bourgeoisie is (usually) impersonal—perhaps because it comes from inherited wealth.

This brings us to question (ii). Why should this be a necessary condition for a form of parasite exploitation? The facts are these: Some people personally contribute to production (laborers, petty bourgeoisie, and, possibly, some big bourgeoisie); others make an impersonal contribution. There has to be something wrong with, or more generally, objectionable about, this lack of personal contribution, at least relative to the fact that others personally contribute. It won't do to say that this lack of personal contribution is objectionable because those who impersonally contribute give nothing of themselves. After all, this is just another way of saying their contribution is impersonal. Maybe what is behind this is a principle of distributive justice similar to the Ricardian socialists' principle that people should be rewarded according to their labor contribution; only now the principle is that people should be rewarded according to their personal contribution. But this only pushes the question back one step. Why does justice (or, more weakly, the absence of parasite exploitation) require that all returns to contribution should be to personal contribution only? This is a question I am unable to answer.

Indeed, I think there is some reason to believe that no satisfactory answer to this question can be given. Any charge to the effect that the exchange between the worker and the capitalist is parasitically exploitative must appeal to a disproportionality between what the exploited contributes and what he receives *or* between what the exploiter contributes and what he receives. This is what it means to say that there is a lack of reciprocity in the exchange between the capitalist and the worker.[36] For example, if Marx's claim that the worker does unpaid labor for the capitalist was right, this requirement would be fulfilled. Similarly, if Schweickart were right in maintaining that the capitalist simply allows capital to be used and thereby contributes nothing, this (together with some uncontroversial premises supplied by Cohen) would establish that the requisite disproportionality. But if this is what the disproportionalities must be, then 'contribution' must be understood in the consequentialist sense of value contribution. Therefore, whether that contribution is personal or impersonal is irrelevant. It seems that the time preference theory of interest shows that any charge that the exchange between the capitalist and the worker is systematically parasitically exploitative cannot be sustained—at least not in a way favorable to a radical critique of capitalist society.

Note that the time preference theory does not show that the workers are not exploited by the capitalists, since it has no impact on property relations exploitation. Indeed, there is one form of parasite exploitation that this theory has no bearing on. Recall that what is distinctive about parasite exploitation is a lack or failure of reciprocity. Lawrence Becker has pointed out that reciprocity involves returning "good for good," proportionately.[37] Now a failure of reciprocity can be a simple failure to return "good for good," proportionately. The charge that the worker does unpaid labor for the capitalist and the charge that the capitalist does not contribute would, if true, be instances of this kind of failure.

But there is another way a lack of reciprocity can be manifested, namely, by returning "evil for good." Once again, the parasite metaphor is helpful: Parasites are not only, or not merely, "free riders"; they also often harm their hosts. There is, then, one other way in which it might be argued that the worker is parasitically exploited by the capitalist: The capitalist gains from harming the worker. Profit is the gain, the good, that the capitalist gets. But what is the "evil" he returns? The Marxist answer: "alienated labor." In fine, the alienating nature of the worker's contribution might be the key to sustaining a charge of parasite exploitation. This way—the last way—of arguing for the charge of parasite exploitation is the topic of the final section of this chapter.

### Parasite Exploitation and Alienated Labor

The main idea behind this argument is that what makes capitalists exploiters is that they are benefiting from the alienating labor that the worker is forced to contribute. In other words, the capitalist is living high off the hog, though not off the laborer, as a result of associating with the laborer for the purpose of production, while the laborer is doing alienating labor. It's the deformed nature of the contribution of the laborer, from which the capitalist benefits, that makes the latter a parasite. Unlike the other forms of parasite exploitation discussed above, the locus of exploitation is not in the wage bargain per se. On this view, exploitation takes place at the point of production, behind the closed doors of the factory as it were, as opposed to the point of exchange.

There are two necessary conditions for this form of parasite exploitation: (i) The exploited is doing (is forced to do) alienated labor; and (ii) the exploiter benefits from his or her association with the exploited. The lack of reciprocity inherent in the worker-capitalist relationship consists in the conjunction of these two facts. Since the capitalist benefits and the worker is harmed as a result of their association, evil is being returned for good. Of course, it does not follow that the individual capitalist ought to be blamed for this; relations of production being what they are, this form of parasite exploitation is pervasive and unavoidable.

This charge of parasite exploitation gains credibility if we consider the situation of those few in a capitalist system who do unalienated labor, those for whom labor is life's chief want. Consider the case of a scientist in the R and D department of a large corporation. Some corporations encourage their best people to do basic research whose practical payoff is uncertain at best. Assuming that these people do no unpaid labor, it is hard to see on what grounds they may be said to be parasitically exploited.

Or, consider the case of well-paid athletes. Dogmatic Marxists look pretty silly when they earnestly assert that these individuals are exploited. And this is not just a matter of high pay. These people are doing what they most want to do in life by exercising their unusual gifts, largely as an end in itself. What is enviable about their position is less a matter of money than it is a matter of how they spend their time making a living. Which is not to say that their lives could not be better. The human condition holds a wide variety of pains and disappointments, not the least of which for athletes is early retirement.

But exploitation? On the above definition, they are not parasitically exploited, since they are not doing alienated labor—and this is as it should be. The only circumstance under which we are inclined to call people with these kind of jobs exploited is when they are clearly doing unpaid labor, as was probably the case with many early figures in professional sports. By contrast, a highly paid industrial worker can be exploited, if he is forced to do alienating labor.

If what explains the fact that people with intrinsically rewarding jobs are not being exploited is that their labor is unalienating, then alienated labor is a necessary condition for this form of exploitation, at least as it relates to the process of production. This account of exploitation implies that the exploitation and alienation are much more closely related than might have been thought, since the former is in part defined in terms of the latter. For our purposes, this connection has the advantage of making it possible to bring to bear some of the material on alienation discussed in Chapter 2. Specifically, the section, "Alienation from the Activity of Laboring," elaborates in great detail the alienating nature of labor under capitalism and suggests some reasons why that might be a bad thing.

That exploitation and alienation are intimately related in this way was first suggested by some of Allen Buchanan's interpretive arguments about different conceptions of exploitation in Marx. Our focus in this chapter has been on exploitation in the relation between individual proletarians and capitalists. Buchanan argues that Marx employs a number of different conceptions of exploitation, one of which is quite general in its application. The latter is not confined to the realm of production but can characterize any or all interpersonal relations in bourgeois society. Buchanan says:

> "This general conception of exploitation includes three elements: first, to exploit someone is to *utilize* him or her as one would a tool or natural resource; second, this utilization is *harmful* to the person so utilized; and third, the *end* of such utilization *is one's own benefit.*"[38]

Parasite exploitation, as it has been defined in this section, is simply a special case of this general conception of exploitation. As Buchanan points out, the theory of alienation can be rung in to detail the various harms the exploited suffer. In particular, in Chapter 2, eight objectionable features of capitalist society traceable to alienated labor were identified:

1. Labor is physically harmful and mindless for many proletarians.
2. Labor is a mere means to needs external to it and is not an expression of the original need for truly human labor.
3. The worker conceives of himself and/or his labor power as a mere commodity.
4. Previously autonomous spheres of life become distorted by being harnessed to the production of labor power.
5. People come to think that the value of a person is determined by the value of his labor power.
6. Alienated labor under capitalism involves the domination of the worker by those who control the means of production.

    7. Alienated labor, in the form of surplus value, is required to sustain or recreate capitalist relations of production.

    8. The mystification induced by the wage labor contract prevents the worker from recognizing number 7.

Numbers 7 and 8 are, of course, tainted by the difficulties with the Labor Theory of Value, but the others are not. Furthermore, recall from Chapter 2 that the explanations for these objectionable features of capitalist society appeal to essential characteristics of capitalism. So there are a number of identifiable harms the workers suffer as a result of being forced to engage in alienated labor, that is, from being parasitically exploited in this sense. Moreover, the relevant explanations are appropriately radical, which means that the Critical Explanations requirement is satisfied for this charge of exploitation.

In light of the above, how could the apologist for capitalism respond? It will not do to say that the capitalist returns "good for good" by paying the worker a wage for two reasons: First, the charge of exploitation is not against the wage bargain (as it was for the other forms of parasite exploitation discussed in this chapter). Exploitation is alleged at the point of production, not at the point of exchange. Relatedly, the Marxist can admit that the worker is better off as a result of the wage bargain—materially better off than he would have been without it. But he is also being harmed. At least in the absence of harm in the other direction, reciprocity also requires that evil not be returned for good. Both the capitalist and laborer benefit, but only the laborer is also harmed; thus there is a lack of reciprocity.

Perhaps the most that can be said on behalf of the apologist for capitalism is that alienating labor has been historically necessary for efficient production, but that it is becoming less so as capitalism moves into the late twentieth century. The assembly line no longer dominates; unions and government regulation have curbed some of the worst harms to body and mind associated with the nineteenth century factory system. If this is right, it suggests that alienating labor has been historically necessary for capitalism to fulfill its historic mission of developing the forces of production.

To be fair to the apologist for capitalism, it is hard to deny that the nature of labor under capitalism has changed for the better over the last century and a half. On this score, there has been an amelioration of the laborer's condition in many respects, at least as it relates to number 1 above. But the Marxist can grant this; after all, necessary evils are still evils. Moreover, the harms indicated in numbers 2 through 6 above remain endemic and pervasive in capitalist society. There has been little or no improvement on this score, and the prospects for beneficial change through ameliorative measures are not bright. This is as we should expect, given that the Marxian explanations of numbers 2 through 6 all appeal to essential features of capitalism. By contrast, the "boundless thirst for surplus value," which explains number 1, no longer dictates squeezing the workers quite as hard as in the good old days—at least not in advanced capitalist societies.

It seems that the only way for the apologist for capitalism to blunt this charge of parasite exploitation is to raise questions about how things would be under communism. As some Eastern European wags put it, "What is the difference between capitalism and communism? Answer: Under capitalism, man exploits man. Under communism, it's the other way around." This response on the part of the apologist for capitalism looks to be a variant of the "So's Your Old Man" reply, a reply that has never really been convincing. Nonetheless, I think there is something to be said for this point, although there are some subtle burden of proof questions that need to be sorted out first.

Recall from Chapter 1 that one of the requirements of a radical critique of a society is the Alternative Institutions requirement. The radical critic has to specify a set of alternative institutions and must argue that these institutions do not have the defects of the existing order. In the present context, this requires the Marxist to specify an alternative set of relations of production and to show that this form of parasite exploitation would be nonexistent or virtually nonexistent in such a system. Failure to do this would leave the Marxist open to the charge that this form of parasite exploitation is a permanent feature of the human condition (or at least the post-feudal human condition) and that the most that can be done is to ameliorate its worst effects. It is incumbent on the Marxist to demonstrate that this is not the case.

Certainly, these burdens of proof should be welcomed. The content of this form of exploitation is to be found in the harms associated with alienated labor identified above. It is reasonably clear that Marx himself believed that these harms would not be associated with labor in post-capitalist society. Nonetheless, there is a burden of proof here for Marxists that must be discharged. We return to this question in Part II of this book.

On the other hand, Marx's conservative opponents also have their crosses to bear. In matters relating to burdens of proof, we are all sinners; this is the epistemological version of Original Sin. To demonstrate that this charge of parasite exploitation cannot be part of a radical critique of capitalist society, the conservative would have to argue that this form of exploitation will accompany any set of relations of production that can be realized in the post-feudal era. Alienating labor, and its associated harms, can be lessened in both their incidence and intensity but cannot be eliminated, and the workers will always be forced to do it. If this is right, this part of the radical critique fails, and the debate between the defender of capitalism and the socialist critic moves in the reformists' orbit. Undoubtedly, the arguments will be messier and less conclusive, and they will not be distinctively Marxist.

Despite these equal burdens of proof, the Marxist has won an important but limited victory. The harms associated with alienating labor do exist, the capitalists benefit from their association with the proletariat, and (let us suppose) the workers are forced to deal with the capitalists. This entails that the workers *are* parasitically exploited under capitalism on this conception of exploitation. It's only that the larger significance of this fact is as yet unsecured.

In one respect, the focus of this chapter has been extremely narrow: We have concentrated our efforts on the allegedly exploitative relation between the capitalist and the worker, in their roles as capitalists and workers. For Marx's purposes, this is perhaps the most important human relationship in capitalist society. But it is clearly not the only one. The question arises, 'Are there any other exploitative relations characteristic of capitalist society that could form part of a radical critique of the latter?'. The most obvious way to substantiate a positive answer to this question would be to follow up on Marx's generalized conception of exploitation, as identified by Buchanan. According to that conception (hereafter designated as 'generic exploitation'), any harmful utilization of another person for one's own gain is an instance of generic exploitation. Following Buchanan, the theory of alienation can be used to identify harms that are traceable to capitalist relations of production. In accordance with the account developed in Chapter 2, we should look to harms associated with alienation from others for additional manifestations of generic exploitation.

Recall that Marx's account of alienation from others under capitalism has three facets: the alienation of the state from civil society, the estrangement between the worker and the capitalist, and the generalized interpersonal alienation that reflects the commercialization of life in capitalist society. The second of these has been discussed above; let us consider the first and the third.

The alienation of the state from civil society is associated with generic exploitation most obviously in the relation between the civil service and the population at large. This can be witnessed in its purest form at the Department of Motor Vehicles. The hapless citizen goes hat in hand to the bureaucrat, like the Russian peasant coming before the czar's functionary, begging to get her license or registration renewed. The so-called citizen is a mere means to keeping the Department going for the benefit of the bureaucrats, just as toll roads continue to exist to support the toll collectors. Even more rapacious is the tax man. No nineteenth century robber baron was more single-minded in his attempt to squeeze wealth out of another human being. Old time capitalists would blush, or more likely ask advice from, the modern auditor-inquisitor. This is twentieth century exploitation in its most naked form.[39]

For generalized interpersonal alienation that is traceable to capitalist relations of production, we need only recall from Chapter 2 Marx's characterization of the nature of interpersonal relations implicit in the credit relation and the mutual plundering characteristic of both exchange and production for exchange. Regarding the latter, it is worth quoting again the remarkable passage from the "Comments on Mill":

> Each of us sees in his product only the objectification of his *own* selfish need, and therefore in the product of the other the objectification of a *different*

selfish need, independent of him and alien to him. Far from being the *means* which would give you *power* over my production, [your need, your desire] are instead the *means* for giving me power over you. . . . The *social* relation in which I stand to you, my labour for your need, is therefore also a mere *semblance*, and our complementing each other is likewise a mere *semblance*, the basis of which is mutual plundering. The intention of *plundering*, of *deception*, is necessarily present in the background, for since our exchange is a selfish one, on your side as on mine, and since the self-ishness of each seeks to get the better of that of the other, we necessarily seek to deceive each other. ["Comments on Mill," MECW, vol. 3, pp. 225–226]

According to Marx, all exchange and production for exchange involve a mutual plundering that is the hallmark of generic exploitation. And, of course, production for exchange tends to make exchange relationships ubiquitous throughout society.

How are we to evaluate the various charges of generic exploitation against capitalism? Note that the charge of generic exploitation by the state bureau-cracy requires something like Marx's theory of the state as an outgrowth of capitalist relations of production, at least if the Critical Explanations require-ment is to be satisfied. Evaluating Marx's theory of the state is a complex task that cannot be pursued here. Secondly, Marx's view that exchange and production for exchange necessarily involve mutual plundering seems a bit over-drawn. However, an adequate evaluation of it would probably require a fuller appreciation of his theory of human nature and human relations, a topic that goes beyond the scope of this book.

In the final analysis, it is likely that the defender of capitalism would press Marxists hardest on the Alternative Institutions requirement. That is, for the various charges of generic exploitation to be part of a successful radical cri-tique of capitalist society, the Marxist must argue that there are realizable alternative institutions which do not foster the various forms of generic ex-ploitation sketched above. This would require, for example, showing that there would be no state in post-capitalist society, or if there were a state, its bureaucracy would not treat people like feudal peasants. In sum, the defender of capitalism would probably admit that the various forms of generic exploi-tation identified by Marx do exist, but he would claim that these forms can only be ameliorated by reformist measures; their elimination, or virtual elim-ination, is a purely utopian ideal. The evaluation of these claims and coun-terclaims must await developments in Part II of this book.

## Summary

The overall argument of this chapter is a bit complicated, so it might be useful to summarize briefly the main points. Central to Marx's radical critique of capitalist society is the claim that the proletariat are exploited. At the outset, I distinguished two forms of exploitation, what I called 'parasite exploitation' and 'property relations exploitation'. The main task in this chapter was to state and critically evaluate all the arguments (at any rate all the arguments that I could find) for the claim that the workers are parasitically exploited by the capitalists. To that end, it was necessary to identify different conceptions of parasite exploitation that might be employed in such arguments. Though nearly all conceptions of exploitation have some sort of "forcing" as a necessary condition, they differ in other necessary conditions. I sought to identify those distinctive necessary conditions for these different conceptions of parasite exploitation.

The first species of parasite exploitation to be discussed was surplus value exploitation. All charges of surplus value exploitation claim that the worker does unpaid labor for the capitalist. Marx's argument for this is based on the labor theory of value, so it must be rejected. Another argument for this claim about unpaid labor is based on the mere fact that there is an unequal exchange of embodied labor in the wage bargain. I showed that this argument must also assume the LTV. Finally, I offered some reason for thinking that, in the absence of the LTV, it is doubtful that a charge of unpaid labor could be sustained in a way that would be suitable for a radical critique of capitalism. I also argued that the bare fact of unequal labor exchange could not be the relevant necessary condition for parasite exploitation, since by itself it does not imply the requisite lack of reciprocity.

The next three sections of this chapter were devoted to the claim that the capitalist is not a contributor to production. If this were true, it would establish the lack of reciprocity necessary for parasite exploitation. In connection with this charge, I presented a positive theory to explain the capitalist's return on investment. This theory implies that what the capitalist contributes to production is time in the form of command over present goods. The capitalist is also of necessity an entrepreneur. In his role as capitalist and entrepreneur, the capitalist makes enough of a contribution to rebut a charge of (serious) parasitism, however easy it is for him to make his contribution.

It is an article of faith among most socialists, and nearly all Marxists, that the laborer and only the laborer contributes. Some Marxists (including Marx himself) argue for this claim. For them, the claim is, in Thomistic terms, a Truth of Reason as well as a Truth of Faith. But the arguments are defective and the claim isn't true.

There is, however, one way a charge of parasite exploitation can be sustained. Reciprocity requires that one not return "evil for good." The final conception of parasite exploitation identifies alienated labor and its consequences as the harms the worker suffers from his association with the capitalist—an association from which the capitalist suffers no comparable harm. Given Marx's analysis of the harms of alienated labor, this charge of parasite exploitation can be sustained. And this is true even if these harms are historically necessary. However, whether this charge of exploitation can be part of a radical critique of capitalist society remains to be seen. The outcome depends on whether or not there are realizable alternative institutions that do not suffer comparable defects.

Finally, we looked at generic (parasite) exploitation outside of the relation between the capitalist and the worker. Following Buchanan, this was defined as the harmful utilization of one person by another for the benefit of the "user." This conception of parasite exploitation is symmetrical. This kind of exploitation can be found in the relation between the bureaucracy and the citizenry of capitalist society; it also characterizes exchange relations generally and is a ubiquitous feature of societies in which production is for exchange. It is hard to see how the defender of capitalism could deny the reality of this form of exploitation, though he may well quarrel with Marx's explanation of it. However, in the final analysis, whether or not the charge of generic exploitation (outside of the relation between the capitalist and the worker) can be part of a radical critique of capitalist society depends on whether or not there are alternative institutions that would suppress the phenomenon.

Some form of parasite exploitation seems to be the preferred way of understanding the Marxian charge of systematic exploitation against capitalism. The essence of parasitism is a lack of reciprocity—a failure to give good for good or giving evil for good. However, there is one way of understanding the charge that the proletariat are systematically exploited that does not depend on conceiving of exploitation as involving a lack of reciprocity. This is property relations exploitation, which is the subject of the next chapter.

# 5

## Property Relations
## Exploitation

### Capitalistic Exploitation

Parasite exploitation is defined in terms of the exchange between individual exploiters and those whom they exploit; the salient feature of such exchanges is a lack of reciprocity. In the case of the capitalist and the proletarian, this alleged lack of reciprocity is to be explained by the nature of capitalist relations of production, but those relations do not enter into the definition of exploitation. By contrast, property relations exploitation is defined in terms of the relations of production. For this reason, if a charge of property relations exploitation against capitalism can be sustained, it will satisfy the Critical Explanations requirement of a radical critique discussed in Chapter 1.

The main exponent of property relations exploitation is the economist John Roemer. Despite the forbidding technical apparatus he deploys, the basic ideas are straightforward and easily grasped. Roemer identifies a generic conception of property relations exploitation, of which capitalist exploitation is only one species. Let us begin with this generic conception. For Roemer, the exploited and the exploiters are groups of persons, or to use his term, 'coalitions'. The root idea behind the generic conception of exploitation is that an exploited coalition is one that would be better off if it were to sever its economic relations with the rest of society; the exploiting coalition, on the other hand, would be worse off if that happened. The general idea is that the exploiters are especially advantaged by existing institutional arrangements and the exploited are likewise especially disadvantaged. As Roemer says,

A coalition $S$, in a larger society $N$, is exploited if and only if:

(1) There is an alternative, which we may conceive of as hypothetically feasible, in which $S$ would be better off than in its present situation.

(2) Under this alternative, the complement to $S$, the coalition $N - S = S'$,
     would be worse off than at present. . . .
(3) $S'$ is in a relation of dominance to $S$.[1]

Different species of exploitation are defined by the terms of the "withdrawal." For example, feudal exploitation exists only if the $S$-coalition would
be better off by pulling out with its own endowments (i.e., productive assets).
Roemer supposes (oversimplifying a bit) that feudal serfs owned their plots
of land and means of production. Clearly, they would be better off and the
lords worse off were the serfs to pull out with their endowments, since they
would not have to labor on the lords' demesne or perform corvée labor.[2]

Capitalist exploitation is defined by different withdrawal rules. The capitalistically exploited coalition (the proletariat) is the coalition that would be
better off were it to withdraw with its skills and its per capita share of the
society's alienable assets. Roemer defines capitalistic exploitation as follows:

> $S$ will be said to be capitalistically exploited if the following three conditions
> hold:
>
> (1) If $S$ were to withdraw from the society, endowed with its *per capita*
>     share of society's alienable property (that is, produced and non-
>     produced goods), and with its own labor and skills, then S would be
>     [better] off (in terms of income and leisure) than it is at the present
>     allocation;
> (2) If $S'$ were to withdraw under the same conditions, then $S'$ would be
>     worse off (in terms of income and leisure) than it is at present;
> (3) If $S$ were to withdraw from society with its *own* endowments (not its *per
>     capita* share), then $S'$ would be worse off than at present.[3]

The first two conditions capture the intuition that it is the proletariat's
lack of ownership of the means of production that is at the heart of capitalist
exploitation. The third condition articulates the idea that the exploiting coalition is also benefiting from the endowments of the exploited coalition.

Roemer defines a third form of exploitation, which he calls 'socialist exploitation'. A socialistically exploited coalition is one that would be better
off if it withdrew with its per capita share of society's alienable and inalienable
assets; the latter is constituted by the skills of the members of the total society.
In other words, the socialistically exploiting coalition is differentially benefiting from the unequal endowment of skills.

Roemer relates these definitions to Historical Materialism by maintaining
that it is the task of bourgeois (capitalist) revolution to eliminate feudal
exploitation, but not capitalist or socialist exploitation. Similarly, socialist
revolution aims at the elimination of capitalist exploitation but not socialist
exploitation. Presumably the latter disappears with the advent of communist
society.

It is not immediately obvious how to evaluate this imposing structure of
definitions and the associated substantive claims. A curious feature of Roemer's methodology is that he seems to assume at the outset that there is a
distinctive form of exploitation characteristic of capitalism; the task is to

construct an analytically useful definition that captures this assumption or intuition. This procedure has a question-begging air about it. There should be room in this structure of definitions for a discussion of the possibility that the workers are not in fact exploited in some distinctive way under capitalism.

More suspicious still in this connection is the fact that the above conditions for capitalistic exploitation are merely said to be jointly sufficient and not singly necessary. What this entails is that it is impossible to show, based on what Roemer says, that the workers are not in fact capitalistically exploited, because even if one or more of these jointly sufficient conditions fails to hold, one cannot conclude that the workers are not capitalistically exploited. There may be other sufficient conditions for capitalistic exploitation, not mentioned in his discussion, that the workers do satisfy.

On the other hand, at one point Roemer seems to conceive of these conditions as singly necessary as well, since he evaluates charges that the above conditions are not met. Such charges would be of minor interest if the conditions in question are only jointly sufficient. For these reasons, I shall construe the above conditions as singly necessary for capitalistic exploitation as well as jointly sufficient; this makes a discussion of Roemer's critical evaluation of the above-mentioned charges of much greater interest than it otherwise would be, since if one of these charges succeeds, the workers aren't exploited in some distinctive way in capitalist society.

In "Property Relations vs. Surplus Value," Roemer considers the following two arguments:

> The first is that (I) the distribution of the means of production is a consequence of inherent traits of agents, which are legitimate and worthy of respect: skills, including entrepreneurial talent; attitudes towards risk, rates of time preference. And (II) regardless of how the distribution came about, it is necessary for the good of all (or, one can say for the good even of the exploited, or worse off coalition) that there be private and differential ownership of the means of production; this is the incentive argument.[4]

For the reasons just indicated, both types of arguments will be construed as attempts to show that, under capitalism, the workers and the capitalists are not exploited and exploiting coalitions, respectively, because they fail to satisfy some necessary condition(s). If the capitalists uniquely possess valuable skills and traits not possessed by the proletariat, then the latter would not be better off if they were to withdraw with their skills, labor, and per capita share of alienable assets. Similarly, if private ownership of the means of production is to the benefit of all, including the proletariat, then once again, the latter would not be better off by pulling out; in short, if either argument works, the proletariat are simply not capitalistically exploited.

In light of the preceding chapter, it might be thought that I would press the first line of argument against Roemer. After all, it was argued that the capitalist contributes time, entrepreneurial ability, and, of course, is the one who takes risks with what he owns. However, all these points were made in the course of rebutting the charge that the capitalist makes no contribution to production, that is, is a parasite. It is open to Roemer to argue that capitalists may be both contributors and exploiters. The reason for this is that, although capitalists may be contributors in the ways specified above, it may be that members of the proletariat could contribute in these ways as well—if they had the opportunity, which they would, were they to withdraw.

What the "capitalist ideologue," to use Roemer's phrase, has to claim is that the proletarians could not contribute (at least not effectively enough to be better off) in these ways if they were to pull out with their per capita share of alienable assets and their labor and skills. This would justify the capitalist ideologue's (hereafter, 'the CI's') contention that the proletariat are not exploited on Roemer's definition, since they would not be better off if they were to pull out according to the withdrawal rules for capitalist exploitation. Let us consider the CI's arguments and Roemer's objections to them.[5]

*Risk.* The CI claims that the capitalist, unlike the worker, is the brave soul who risks his assets in anticipating market conditions. By contrast, the security-hungry (cowardly?) worker is risk-aversive and would not take any chances with his assets were he to become a capitalist; as a result, the economy would stagnate, and the workers would not be better off by pulling out. In light of this, the worker has nothing to complain about (or, more exactly, Marxists have nothing to complain about on their behalf). More precisely, the proletariat would not be better off by pulling out with their per capita shares of alienable assets, labor, and skills because they lack the willingness to take the risks necessary for material progress. However, Roemer correctly points out that the capitalist, unlike the worker, is usually gambling with discretionary income. Attitudes toward risk might be the same for both workers and capitalists; it's just that they are at different points on their risk schedules.

*Time Preference.* The CI argues that the capitalists are more future-oriented than the proletarians. They're willing to forgo present consumption for future consumption at a lower rate than the benighted proletarians. In Roemer's response to this, he fumbles the ball by claiming that the capitalists do not suffer as a result of their abstention from present consumption. However, suffering is not what is at issue; contribution is. And, what the capitalist contributes is time, not suffering. We can recover the fumble for Roemer by pointing out that, just as in the case of risk, the capitalist and worker may have the same rates of time preference (relative willingness to contribute time by postponing present consumption); it's just that they are at different points on their schedule of preference for present goods over future goods because of their wealth or general standard of living.

*Entrepreneurial Ability.* The CI argues that the paucity of worker-owned firms under capitalism is evidence for the claim that workers lack the entrepreneurial ability to organize enterprises. So, if they were to pull out with their per capita shares, they would not be better off, since they lack the ability to deploy their productive assets effectively. Roemer responds by pointing out that there are economies of scale in the ownership of capital; in other words, it is easier for one large asset-holder to make decisions and set up an organization than it is for, say, a thousand small asset-holders. In addition, much of what passes for entrepreneurial ability is simply knowledge of investment opportunities, which "comes with the territory" of being a capitalist. So, the lack of worker-owned firms under capitalism is not evidence that the workers lack the requisite entrepreneurial ability. Thus, there is no reason to think that the proletariat would be at a disadvantage with regard to entrepreneurial talent, were they to pull out with their per capita shares.

Let us summarize the state of the dialectic to this point. Roemer wants to claim that the workers are capitalistically exploited by the capitalists for the following reasons: First, if the workers were to pull out with their labor and skills and their per capita share of society's alienable assets, they would be better off and the capitalists would be worse off; second, the latter would be worse off if the workers were to withdraw only with their own endowments. The capitalist ideologue counters by saying that the workers would not be better off by withdrawing because existing capitalists uniquely possess traits (attitudes toward risk and relatively low time preference) and skills (entre-preneurial ability), which if lost to the proletariat by the latter's withdrawal, would make them worse off. Roemer's response in each case instantiates a certain general pattern: There is no good reason to believe that existing capitalists and existing proletarians differ systematically in the way that the CI believes. The distribution of the relevant traits and skills might be much the same across classes; historical accident (e.g., what class one is born into, what circumstances one finds oneself in) can explain why these traits are exemplified in the differential way that they are. Notice that this is consistent with my claims in the last chapter about the actual contributions of existing capitalists; Roemer is in effect saying that the workers could make these contributions, too. Indeed, this is what he needs to say in order to establish that the workers would be better off if they withdrew with their labor and skills and their per capita share of society's alienable assets.

However, it is important to note that Roemer has not shown this. That is, he has not shown that this contention of the CI is false; at most he has shown that the CI cannot support it by casual empiricism, that is, by pointing to observed behavioral disparities between existing capitalists and existing workers. The CI's contention may well be true, but he has not proved it, nor has Roemer proved its denial. Roemer is aware of this and rightly maintains that genuine empirical research is needed to adjudicate the issue.[6] The upshot is that the claim that the workers would be better off if they were to withdraw with their labor and skills and their per capita share of society's alienable assets (and, by implication, that the workers are capitalistically exploited by the capitalists) is as yet neither proved nor disproved.

Let us assume for the sake of discussion that Roemer can defeat the CI's claims about risk, time preference, and entrepreneurial ability and thus that the CI's first argument fails. What about the CI's second argument? As Roemer explains it, this "argument maintains that without differential and private ownership people would cease to work hard, and technological change would slow down or cease. Thus, even though differential ownership gives rise to inequality, everyone is better off than under an egalitarian regime."[7] Roemer's response is to call attention to the fact that his definitions of exploitation contain strong *ceteris paribus* clauses. Specifically, in testing for exploitation, incentives are held constant, so it does not really matter if everyone would be better off under the current system. This response looks to be *ad hoc*, but Roemer quickly moves to mollify his critics by admitting that, if the withdrawal resulted in the destruction of the incentive structure in such a way that the proletariat would be worse off, then the exploitation they currently suffer can be said to be socially necessary.[8]

This notion of socially necessary exploitation is further refined in his book. There he distinguishes between exploitation that is socially necessary in the static sense (as defined in the preceding paragraph) and exploitation that is socially necessary in the dynamic sense:

> Suppose, however, that the coalition would initially be better off after exercising its withdrawal option, even allowing for incentive effects, but then "soon" it would become worse off, due for instance to the lack of incentives to develop the forces of production. In this case, the exploitation is *socially necessary in the dynamic sense*.[9]

Notice that these two conceptions of socially necessary exploitation are mutually exclusive. If a certain form of exploitation is dynamically socially necessary, it follows that the exploited would be initially better off even with the changes in the incentive structure entailed by withdrawal, whereas with statically necessary exploitation, changes in the incentive structure make the exploited worse off more or less immediately. Both are forms of exploitation since the insertion of very powerful *ceteris paribus* clauses effectively stipulates a situation in which they would be better off.

There is another form of socially necessary exploitation not covered in Roemer's discussion. To see what it is, consider the position of the capitalist ideologue who takes a dim view of socialism. Suppose that the CI maintains that capitalist relations of production are the greatest thing since sliced bread (maybe even greater, since capitalism invented sliced bread). More exactly, the CI might maintain that, given a certain level of development of the forces of production, say, that reached at the end of the feudal era, no economic system can ever come close to matching capitalism's capacity to develop the forces; in consequence,[10] any genuinely realizable alternative economic system would make the proletariat worse off absolutely in the near term and over the long term as well, or more weakly, worse off relative to how well they would have done under capitalism over the long run.

Let us call this series of claims the Monstrous Hypothesis. Clearly, these claims are extraordinarily strong, and it is far from obvious how the CI could show that they are true. But let us suppose, just for a moment, that they are true. Note that, according to Roemer's definition, *even if the Monstrous Hypothesis is true, existing proletarians are capitalistically exploited.* The reason for this is to be found in the strong *ceteris paribus* clauses implicit in the subjunctive clauses of the definition. We are to imagine various kinds of transfers of wealth and income with everything else being held equal. Since all else is being held equal, it does not matter that no realizable economic system other than capitalism would in fact result in the proletariat being better off. This form of social necessity differs from the two forms defined by Roemer; let us call it, 'permanently socially necessary exploitation'.

It is likely that the CI would claim that what we have here is really not a form of exploitation at all. Although it does satisfy Roemer's definition, the CI would say, "So much the worse for that definition." In other words, he would say that, if the Monstrous Hypothesis is in fact true, there really is no distinctive form of capitalist exploitation. Notice that the mere possibility that this hypothesis is true creates a genuine dispute about whether or not a state of affairs that satisfies Roemer's definition of capitalistic exploitation is in actuality a form of exploitation. Or, to put the same point more forcefully, this possibility calls into question whether or not Roemer's definition of capitalistic exploitation is correct.

It might be objected that none of Roemer's definitions can be correct or incorrect since they are essentially stipulative. However, these definitions are supposed to capture or articulate certain intuitions about a concept, in this case, the concept of exploitation. If it can be argued that the definition fails to articulate the relevant intuition, one is entitled to say that such a definition is just false.[11]

Perhaps the central intuition about exploitation Roemer seeks to articulate is that it involves one group being unfairly disadvantaged relative to others, a kind of distributional unfairness, if you will. If that is true, it is possible to adapt some insights from John Rawls to support the CI's contention that the proletariat are not (capitalistically) exploited, if in fact the Monstrous Hypothesis is true; and, if this conditional statement is true, then Roemer's definition is not correct as it stands.

As is well-known, Rawls maintains that the principles of justice are those that would be chosen by rational and self-interested agents in an original position of equality; this original position also includes the famous "veil of ignorance," which screens out particular knowledge that individuals would otherwise have about themselves and their state. Rawls argues that they would adopt his Difference Principle, which says that inequalities are unjust unless they work to the benefit of the least advantaged. Now, whether or not Rawls has discovered the correct principles of justice by this method, his Difference Principle does neatly articulate an intuitively plausible conception of *fairness*. Indeed, he calls his conception of justice, "justice as fairness." If the distributional consequences of capitalist relations of production satisfy the Difference Principle for the proletariat from the time of late feudalism onward (which is entailed by the Monstrous Hypothesis), then it seems accurate to say that the pattern of distribution of wealth and income is fair. After all, under no realizable alternative economic system would the proletariat do as well. But if the proletariat suffer no distributional unfairness under capitalism, then they can suffer no distinctive form of property relations exploitation under capitalism.[12] This substantiates the CI's contention that what was called 'permanently socially necessary exploitation' is not really exploitation after all. It follows that Roemer's definition of capitalistic exploitation is defective. It is not too difficult to repair this defect, however, but first some diagnosis would be in order.

The root of the difficulty in Roemer's definition can be found in his characterization of the generic conception of exploitation quoted at the very beginning of this chapter. The first condition begins with the phrase, "There is an alternative, which we may conceive of as hypothetically feasible . . ." Why '*hypothetically* feasible', instead of just 'feasible', or to be more precise, 'currently feasible'? From a Marxian perspective, the answer is obvious: Historical Materialism teaches that social formations perish only at certain stage in the development of the forces of production. If exploitation required the existence of currently feasible alternatives, it would exist only when the level of development of the forces makes existing relations of production ripe for replacement. This would make the incidence of exploitation a rare and historically insignificant phenomenon. Moreover, the requirement of a currently feasible alternative seems much too strong, independently of Historical Materialism. For any stable social formation, it is almost never true that there are any currently feasible alternatives. Otherwise, that formation would not be stable. But there can still be exploitation. Indeed, it seems that exploiters take advantage of (the stability of) the existing social formation for their benefit and to the detriment of the exploited.

Roemer's solution to this problem is to add the qualifier, 'hypothetically' in the first (and second) conditions. Alternatives need not be currently feasible; they only need be hypothetically feasible, that is, feasible under certain suppositions. The problem is that this qualifier allows Roemer to stipulate his alternatives *in any way he chooses*, which in turn allows his economist's imagination to run amuck. That is, there are no constraints[13] on the alternatives other than those he stipulates in his models. *The very strong ceteris paribus clauses in his definitions make what could actually happen in the real world completely irrelevant to whether or not existing workers are capitalistically exploited.*

This problem also afflicts his discussion of what he calls 'socialist exploitation' under socialism. Socialistically exploited coalitions are those that would be better off by pulling out with their per capita share of society's alienable *and inalienable* assets. The latter is constituted primarily by the talents and skills of individuals. But what does this alternative really amount to? As Jon Elster has said, "even as a thought experiment, it remains unclear how the workers are to take with them their share of the managerial skill, while leaving the managers behind."[14]

In his formal development of the concept of socialist exploitation, Roemer constructs a model of the alternative arrangements.[15] The problem is that this model cannot be instantiated in the real world. Although in general a lack of realism in models is not a decisive objection against their use, in the present context this problem is pressing. So, even if some meaning can be assigned to the appropriate conception of withdrawal, it doesn't tell us whether or not socialist exploitation could ever actually be eliminated without making the unskilled worse off, which in turn calls into question whether or not the unskilled really are exploited by the skilled.

All of this suggests that there have to be some constraints on the hypothetical alternatives against which any existing society is tested for any form of property relations exploitation. Yet those constraints cannot be so tight as to rule out any system that is not currently feasible. What we need, in short, is some concept of *realizability*. Realizability is stronger than hypothetical feasibility and yet weaker than (current) feasibility. In a Marxian spirit, realizability might be understood in terms of what is historically possible now or at some time in the future.

For now, let us proceed with only an intuitive understanding of this notion. Much more will be said about this in Chapters 6 through 9 of this book. This conception of realizability seems to be what is needed to repair Roemer's definition in light of the position of the capitalist ideologue and his Monstrous Hypothesis. If capitalistic exploitation is defined in terms of realizable alternatives—as opposed to Roemer's hypothetically feasible alternatives—then whether or not the proletariat are capitalistically exploited will depend on whether or not the Monstrous Hypothesis is true—and this is as it should be.

But this means that the question of whether or not the proletariat are exploited in some distinctive way under capitalism is up for grabs. What socialist defenders of Roemer must show is that the Monstrous Hypothesis is false. Recall that the latter says that any realizable alternative institutional arrangements would make existing proletarians worse off more or less right away and that they would continue to be worse off, at least relative to what would have been their situation had capitalism continued to exist. On the basis of the revised definition, the workers are capitalistically exploited under capitalism if and only if the Monstrous Hypothesis is false.

These results fit nicely with the requirements for a successful radical critique of a social system laid out in Chapter 1. Recall that the Alternative Institutions requirement mandates that the critic specify (realizable) alternative institutional arrangements that are not afflicted by the defect in question.[16] There is some reason to think that Roemer would not find this way of phrasing the issue too disturbing. After all, the question comes down to whether or not it would ever be historically possible for socialism to match or surpass what capitalism can do for the workers in terms of an income-leisure package. No socialist entitled to sing the "Internationale" would express much scepticism about this. Moreover, for Roemer, this is primarily a question of incentives, and on his view, what incentives will be effective is historically conditioned. In his discussion of socially necessary exploitation he says, "the social necessity (or otherwise) of a form of inequality is in large part entailed by how the people involved think. It is the "level of consciousness" which determines whether certain incentives are necessary."[17]

According to Historical Materialism, the "level of consciousness" is determined by the level of development of the forces of production. On this basis, the argument against the CI and his Monstrous Hypothesis would then go something like the following: At a certain level of development of the forces, different incentive structures would become feasible—incentive structures that do not presuppose unequal control of the means of production in order to sustain income-leisure packages for the workers that dominate those achievable under capitalism. Given that there is a tendency for the forces of production to develop over time (capitalism's historic mission), sooner or later the point will in fact be reached (has been reached) when the incentive structure implicit in capitalist relations of production is no longer necessary to sustain the level of development of the forces and, by implication, the income-leisure packages of the workers. In other words, the realizability of an adequately efficient and productive socialism will eventually be, if it is not already, a genuine historical possibility. If that is true, the Monstrous Hypothesis is false, and it follows that existing proletarians are currently being capitalistically exploited, according to the revised definition of capitalist exploitation.

Whether or not Roemer would endorse this argument against the capitalist ideologue, I would like to suggest that this response is far too facile and that the settling of accounts with the CI is a vastly more complicated and difficult task than is suggested by the above argument. Moreover, this is so even if we grant the major contentions from Historical Materialism.

Let us begin by considering the incentive issue. The above argument in favor of at least the historical possibility of a tolerably efficient socialist system seems to assume that the only important determinant of the workers' income-leisure package under any such system is the latter's "consciousness." Now what might this mean? Presumably, any incentive system involves the association of desired behaviors with certain "goods" and undesired behaviors with certain "bads." Change the consciousness of the people involved and what counts as a "good" and what counts as a "bad" change as well. Under capitalism, the goods and bads are monetary, but this need not be the case in all possible systems. Under socialism, for example, it may be that there are satisfactions of a more communitarian sort, or perhaps there are status satisfactions that go with the desired behaviors;[18] similarly for the undesired behaviors.

But what are the desired and undesired behaviors? To see what they are, let us recall the first counterargument Roemer considers. In assessing the claim that the workers might not be better off by pulling out with their per capita share of alienable assets, Roemer considers the possibility that the capitalists have valuable traits and skills that the workers lack. These are a willingness to take risks, low time preference, and entrepreneurial ability. (Traits and skills are, or are associated with, behavioral dispositions.) Recall that direct empirical evidence on these questions is hard to come by and the verdict on this contention was, "Not Proved." Nonetheless, it is abundantly clear that the exemplification of these traits and skills is a prerequisite for continued economic development (development of the forces of production).

Let us suppose for the sake of discussion that these traits and skills, or the capacity to develop them, are randomly distributed among the population as a whole and so the proletariat possess them about to the extent that the capitalists do. What the socialist opponent of the CI, whom we will hereafter call the 'Defender of the Faith' or the DF, must show is that there is every reason to believe that these traits and skills will be developed and exemplified under socialism about to the extent that they are under capitalism. So the question is, 'Will socialism encourage risk-taking, a rate of savings, and entrepreneurial creativity comparable to what is found in capitalism?'.

If we think carefully about how the DF could answer this question, we can begin to get a sense of the complexities of what is involved in showing that a tolerably efficient form of socialism is a genuine historical possibility. (This is a special case of the more general problem of intelligently discussing a society that does not yet exist.) When the Defender of the Faith affirms that there is a realizable socialist alternative in which these behaviors would be associated with "goods" and "bads" of any sort (and thus the workers would be at least as well off as they are under capitalism), she is talking about a set of social institutions, the most important of which is the economic system. It is logically possible for any consistent specification of social institutions to be instantiated,[19] but whatever realizability amounts to, it is clear that it is stronger than mere logical possibility.

In specifying the economic system, we need to know at least what the relations of production are. Roemer's specification of the alternative to capitalism is equal ownership of alienable assets. But even if one abstracts from personal property not used as means of production, it is unclear what this amounts to. The difficulty here closely parallels the problem in specifying the alternative for *socialist* exploitation discussed above. A large part of the problem in the present case consists in the fact that the concept of ownership is deceptively complicated. Ownership consists of an entire package of rights, including rights of possession, income, use, rights to the capital, and so forth.[20] It may be that the DF need not specify all of these rights in complete detail, but surely some of them must be spelled out. Consider management rights. There cannot be equal management rights for everyone in every productive asset. (By contrast, in an economist's model, this is not a problem: Each worker gets to decide what shall be done with each asset for $1/n$ of the day, where $n$ is the number of workers in the production unit, or in the economy as a whole or on the planet.) Exactly what pattern of management rights in particular and ownership rights in general are necessary and sufficient for socialist relations of production?

Second, at a less fundamental but not less important level, we need to know how the social division of labor is to be organized. In other words, it is necessary to know if the envisioned alternative is a market system of some sort or whether it is a centrally planned economy or something else. This question is highly relevant to the issue at hand in light of the practical and theoretical problems facing central planning.[21]

Third, not only is it necessary to know the relations of production and the method of organizing the social division of labor, it is also necessary to know something about how they would interact. For instance, a centrally planned economy may not be compatible with genuine worker control of the means of production. On the other hand, it may be that market socialism, in which the workers own the factories but produce for the market, has a strong tendency to degenerate into capitalism.[22]

Fourth, it is necessary to know something about the other institutions of the envisioned alternative and whether or not they are compatible with the economic system. Some critics of central planning, including some socialists, have argued that central planning is incompatible with political democracy. A commitment to the latter might render the economic system infeasible or unable to deliver the goods.

These (and possibly other) burdens that the Defender of the Faith must shoulder suggest a methodological defect that pervades all of Roemer's definitions, namely, that his alternatives are far too abstract or thin. To substantiate a charge of exploitation, it is necessary to talk about how the allegedly exploited coalition would fare under full-fledged alternative social institutions, and it must be argued that these institutions can be realized. The incentive question is really only a small part of this. How far this specification must go is unclear, but it should be obvious that it must go beyond the excessively thin account Roemer has provided.[23]

Despite the litany of complaints I have registered against Roemer, I would maintain that his property relations conception of exploitation represents a fundamental advance in anti-capitalist thought, precisely because it makes the concept of alternatives central to substantiating a charge of exploitation. If in fact the workers are exploited, it is because there are better genuine historical possibilities for them. Absent such possibilities, they are not exploited, and the Marxist belief to the contrary reflects an unrealizable utopian vision of post-capitalist society. Though socialists have always believed there are better genuine alternatives for the working class, it has not been central to their critique of capitalist society. Roemer's work brings the question of alternatives into focus in an especially sharp way.

## Exploitation and Perspectives on Distributive Justice

Throughout the last four chapters, the term 'justice' and its cognates have rarely appeared. Nothing in my reconstruction of Marx's critical explanations of the evils or defects of capitalist society has been based on any charge of injustice. I have so far avoided a spirited controversy in the secondary literature in recent years about whether or not Marx thought that capitalism was a just social system.[24] Something needs to be said about that.

This controversy was inaugurated in an article by Allen Wood in which he argues that Marx believed that capitalism is a just system.[25] According to Wood, justice for Marx is defined in terms of whatever accords with or corresponds to the existing mode of production. Thus, the capitalist does no injustice to the worker by buying his labor power and using it to create and then appropriate surplus value. In short, Marx was a relativist about justice. Wood's interpretation has been sharply controverted by other commentators. Some have argued that Marx held exactly the opposite view, namely, that capitalism is inherently unjust.[26] In support of this, they cite passages where Marx refers to the appropriation of surplus value as "theft" and the surplus itself as "booty" or "plunder." The passages in which Marx seems to claim that capitalism is just are interpreted as ironic or as disguised sociological claims about what people in capitalist society believe. Others try to mediate between these contending interpretations in a variety of ways or reject the basic terms of the debate.[27] As the debate has worn on, some commentators have openly speculated that Marx himself might have been confused.

Part of what makes this issue so difficult is Marx's theory of ideology, which holds that dominant ideas about justice, rights, and so forth, in a society systematically reflect the interests of the ruling class. When Marx says or implies that capitalism is a just society, it is unclear whether he is making a disguised sociological claim or not. If he is a relativist about justice, the sociological claim about what people believe about justice might entail a substantive claim about justice itself. Indeed, if meta-ethical (or meta-juridical) relativism is true, the two claims are definitionally equivalent. This is, I think, one of the main reasons the interpretive dispute has been so hard to settle.

I should like to avoid this dispute to a large extent by taking what might be called an eliminativist position on Marx's views on distributive justice. This position should simultaneously mollify and aggravate all the contending parties. The essence of this position is that everything that Marx might have wanted to say about justice and capitalism can be said in terms of the analysis of exploitation (and, to a lesser extent, alienation) given in this chapter and the preceding three, supplemented with some facts about post-capitalist society. Let me explain this position in more detail.

If Marx believes that capitalism is unjust, it is highly likely that the putative fact that it is systematically exploitative entails or at least somehow supports this contention. Recall from the beginning of Chapter 4 that Marx's own charge of exploitation against capitalism is based on a form of surplus value exploitation: Holmstrom captured Marx's thought best when she said, "it is the fact that the [capitalist's] income is derived through forced, unpaid, surplus labor, the product of which the producers do not control, which makes it [capitalism] exploitative."[28]

It is not unreasonable to suppose that, if someone believed this about capitalism, he or she would believe that the latter is systematically unjust. But why, exactly? Recall that surplus value exploitation is a subspecies of parasite exploitation—and the heart of parasite exploitation is a lack of reciprocity between the exploiters and the exploited. Perhaps justice requires reciprocity. This means that a lack of reciprocity entails some kind of injustice. Indeed, all the attempts to establish parasite exploitation in Chapter 4 can be construed as attempts to show that capitalism is inherently unjust. If Marx believed the latter, it is a plausible hypothesis that he believed it primarily because he believed that capitalism is exploitative. What links the two is a lack of reciprocity. All that talk about theft and booty fits in nicely with this.

On the other hand, if Marx did not believe that capitalism is unjust, or if he rejected the terms of the debate, he can still maintain, indeed he wants to maintain, that capitalism is systematically exploitative. In his critique of capitalism, maybe there is no need to add insult to injury—or, more accurately, insult to insult.

When one turns from what Marx said, or intended to say, to what is true, or might be true, the problems are enormously simplified. The only charge of parasite exploitation against capitalism to survive critical scrutiny was generic parasite exploitation—the harmful utilization of one person by another for the benefit of the latter. Under capitalism, people in general and workers in particular are regularly exploited in this sense for a variety of reasons and in a variety of ways. The theory of alienation was brought in to detail the nature of the harms suffered. Alienated labor is, of course, the chief harm suffered by the workers at the hands of the capitalists, and on the generic conception, they are forced to suffer this harm for the benefit of the capitalists. Capitalist relations of production are part of the explanation for this. Are these and the other instances of harmful utilization of one person by another instances of injustice? If justice requires reciprocity, yes. If not, these cases are still instances of systemic social evils, and that is enough for the purposes of a radical critique.

What about property relations exploitation? Suppose that the Monstrous Hypothesis is false and that the workers are in fact capitalistically exploited by the capitalists. Is this an injustice? As I noted earlier in this chapter, perhaps the central intuition that Roemer seeks to explicate by his series of definitions is that the exploited are unfairly disadvantaged relative to the exploiters; the withdrawal rules for the various coalitions implicitly articulate various "modes of unfairness." For an exponent of "justice as fairness," this unfairness, if indeed it exists, would make capitalism unjust. I do not know what justice is. However, I suspect that if capitalism is systematically exploitative, it is systematically unjust.

To return to Marx, whatever his views on the justice or injustice of capitalism, it is reasonably clear that a charge of injustice does not figure prominently or explicitly in his critique of capitalism. Maybe that is because it is implicit in the charge of exploitation, but part of the explanation for this may go much deeper.

In his *Marx and Justice*, Allen Buchanan has argued that Marx's radical critique of capitalist society is not even in part based on some principle of distributive justice. Buchanan says,

There is, however, a . . . reason to reject this claim that Marx criticizes capitalism from an external perspective of communist distributive justice. . . .
Marx believed that communism will be a society in which what Hume and
Rawls call the circumstances of (distributive) justice either no longer exist
or have so diminished that they no longer play a significant role in social life.
The circumstances of distributive justice, roughly, are those conditions of
scarcity—and of conflict based on competition for scarce goods—that make
the use of principles of distributive justice necessary. Marx holds that the
new communist mode of production will so reduce the problems of scarcity
and conflict that principles of distributive justice will no longer be needed.[29]

Historical necessity to one side, the ultimate standard against which capitalist society is to be judged is communist society. This is as it should be, if
the rationale for the Alternative Institutions requirement sketched in Chapter
1 of this book is correct. Buchanan contends that, for Marx, communist society
is "beyond justice." His claims about this society are supported by what Marx
says in the *Critique of the Gotha Program* and by what Marx and Engels say
in *The Communist Manifesto*. In the former, Marx tells us that life will be
dramatically different by the time the boundary to communist society has
been crossed. Labor will have become "life's chief want," and the distribution
of use-values will be in accordance with human need (*CGP*, pp. 17–18). In
*The Communist Manifesto* Marx and Engels say, "in place of the old bourgeois
society, with its classes and class antagonisms, we shall have an association,
in which the free development of each is the condition for the free development of all" (*CM*, MECW, vol. 6, p. 506). For this reason, egoistic man
of capitalist society will be replaced by the more communitarian men and
women of communist society. Buchanan maintains that if these circumstances
hold for that society, considerations of (distributive) justice simply do not
arise.

If Buchanan is right, it in part explains why Marx would not make much
of the injustice of capitalism or, alternatively, why he would not claim that
capitalism is unjust: Justice is a remedial virtue ('the jealous virtue'), which
has no place in the good society that will ultimately replace capitalism. The
ultimate goal of the revolution is not to create a just society; it is to create a
society beyond justice.

For all of these reasons, I maintain that Marx's radical critique of capitalist
society can be accurately explicated without bringing in a charge of injustice,
or, more modestly, without deciding the issue of Marx's views on the distributive justice or injustice of capitalism.

# 6

# Post-Capitalist Society:
# Relations of Production and
# the Coordination of Production

## Social Visions

In their later years, Marx and Engels were highly critical of Utopian Socialists such as Saint Simon, Owen, and Fourier. It was easy to ridicule their highly specific blueprints for a new society. Fourier, for example, envisioned society divided into communities ("phalanstries") of exactly two thousand individuals. Unpleasant tasks, such as cleaning sewers and drains, would be left to children, who naturally liked to play in dirt and filth. Brothels would be among the most honored institutions in the new society.[1] For Marx and Engels, the main problem with theorists such as Fourier was not that they made false predictions about a new society or even that they attempted to predict what life would be like in a society that did not exist in reality.

Indeed, Marx and Engels found nothing in principle objectionable about predicting some of the details of future social life. Nor did they shy away from what might be called 'normative predictions', that is, predictions about the values people would live their lives by in post-capitalist society. That much is implicit in, for example, the Famous Passage from *The German Ideology* where Marx predicts that communist man will not subject himself to "one exclusive sphere of activity." Implicit in the prediction of all-round development is the supposition that people will naturally value a life-style of this sort.

The main problem with the Utopians lay elsewhere. Utopian Socialist thought (according to Marx and Engels, anyway[2]) was characterized by a kind of Idealism in the following sense: These theorists believed in constructing visions of a new society according to certain ethical or normative ideals. They further believed that the propagation of these visions, either verbally, or by example in the case of Robert Owen, would be a decisive causal factor in bringing about social change.

135

By contrast, Marx and Engels's Historical Materialism locates the fundamental causes of social change in the real interactions that human beings have with each other and with their material environment in the process of production. Furthermore, the values that members of post-capitalist society will live their lives by are not *sui generis*; they are a natural outgrowth of these interactions.

Marx and Engels encouraged the view that they were the first "scientific" socialists because of the elaborate theory of history first outlined in *The German Ideology*. But that's not really fair or accurate. There are more than mere suggestions of historical determinism in Fourier and especially Saint Simon.[3] It would be more accurate to say of at least some of the Utopian Socialists that they did have theories of history and social change; it's just that those theories were wrong—wrong both in their claims about the underlying dynamics of world history and in their predictions about the future course of events. Marx and Engels's claim that theirs was the first scientific theory of socialism might be best interpreted as saying that theirs was the first such theory that was *true*. It would not be the first time, nor the last, that scientists claim that the false scientific theories of their rivals were "unscientific."

The term 'vision', as it applies to a future society, a society that does not exist, carries with it some objectionable connotations. It calls to mind some *crank* who seeks to impose on people his or her picture of what social life (the good life) should be like. In this sense, it would of course be false to say that Marx had a vision of post-capitalist society (hereafter, 'PC society'). But these connotations need not be accepted, and there is, I think, a clear sense in which Marx can be said to have a vision of PC society. The details of this vision can be culled from two sources: The first is the materialist theory of history conceived of as a scientific theory of social institutions and social change. The second source is the demands of the Alternative Institutions requirement for a radical critique. Of course, Marx does not address the second of these as such, but as I argued in Chapter 1, if his radical critique of capitalist society is to be successful, he must be construed in such a way as to address this requirement. The goal of this chapter and the next is to articulate Marx's vision of PC society. The plan is as follows: In the remainder of this first section I indicate the elements of Marx's vision of PC society that should be derivable from his Historical Materialism; the section ends with a reminder about the demands of the Alternative Institutions requirement. Subsequent sections in this chapter and the next work out the details of Marx's vision of PC society in accordance with the results of this first section.

One of the most impressive features of Marx's Historical Materialism is its attempt to state the general laws of the historical development of social systems. This claim will seem controversial only to those who have a narrow, positivist conception of scientific laws as exceptionless universal generalizations from which specific events may be deduced. However, if a law is just any general statement describing some non-accidental regularity (which includes statements describing tendencies), then the claim that Marx attempted to discover the laws of historical development of social systems is very modest indeed.

For example, in the famous "Preface" to *A Contribution to the Critique of Political Economy*, Marx asserts something like the following proposition: 'When existing relations of production fetter further development of the forces of production, the relations of production change in such a way that the forces can continue to develop'.[4] This is a law-like claim about indefinitely many social systems; as such, it includes not just European feudalism and world capitalism, but would in theory apply to feudal and capitalist societies on other planets, if there were such societies; this assumes, of course, that all else is equal (that is, all else that is relevant is equal) on these other planets. And, although *ceteris paribus* clauses can be trivializing, they need not be.[5]

Laws such as these form the basis for Marx's prediction that capitalism is historically doomed. For reasons that need not be discussed here, Marx maintained that the proletariat would rise up and do away with capitalist relations of production; this act ushers in PC society. It is absolutely clear that Marx predicts the end of capitalist society and that this prediction is derived in some manner from the laws of Historical Materialism.

But in what sense does Marx have a *vision* of PC society, as opposed to a prediction that there will be such a society? The materialist theory of history provides part of an answer. Central to that theory is the idea that the relations of production of a society are a kind of master key to understanding that society. What, then, are the relations of production that will characterize PC society? This much is clear: The workers will control the means of production. In the beginning of the next section, I shall argue that the "de-commodification" of labor power is also part of, or implicit in, the relations of production. These facts, or these facts in conjunction with some plausible auxiliary assumptions, should allow some reasonable conjectures about other features of PC society.

But what other features? In the first place, the system of distribution. In the *Critique of the Gotha Program*, Marx upbraids the Lassalleans for criticizing the distribution of wealth under capitalism while virtually ignoring the underlying relations of production. Marx is very explicit that the former is a direct consequence of the latter. He says,

> if the elements of production are so distributed [i.e., as they are under capitalism], then the present-day distribution of the means of consumption results automatically. If the material conditions of production [the means of production] are the co-operative property of the workers themselves, then there likewise results a distribution of the means of consumption different from the present one. [*CGP*, p. 16]

So, if the relations of production for PC society are known, something can be inferred about the distribution of income (means of consumption).

And that's not all. Marx claims that the normative belief systems of a society, as well as their institutional embodiments, are expressions of (a use-fully vague term) that society's relations of production.[6] The former includes religion, law, political economy, and principles of justice and morality; the latter includes established churches and the criminal and civil justice system. Typically, Marx refers to the former as 'ideology', but that term may be inappropriate when PC society is under consideration; hence the more neutral term 'normative belief systems'.

Knowing the relations of production does not, by itself, permit detailed predictions about (or postdictions or explanations of) normative belief systems and the corresponding institutions, but some fairly general predictions are possible about, say, the kind of legal system that would characterize a society with a given set of relations of production. For example, whatever the legal code of a capitalist society turns out to be, rights to private property will be highly articulated and central to that legal system. At the very least, knowing what the relations of production are would permit the inference that certain kinds of normative belief systems and institutions will not be found to coexist with those relations of production.

Finally, Marx's critical explanations of the evils of capitalism presuppose the possibility of inferring undesirable aspects of a society from its essential features,[7] which include its relations of production. If Marx is to defeat his conservative opponents who claim that the evils of capitalist society are, in one form or another, endemic to any form of post-feudal society, he has to claim that PC society will eliminate, or virtually eliminate, those evils. And, indeed, Marx clearly believes that exploitation and alienation will not be characteristic of PC society. Presumably, what justifies such beliefs are some facts about the basic institutions of that society.

It should now be somewhat clearer the sense in which Marx can be said to have a vision of PC society. Marx does not merely predict the revolutionary overthrow of capitalism; he has a conception of the relations of production that will be found in PC society. In principle, knowledge of such relations should permit inferences about the distribution system, as well as inferences about the predominant normative belief systems and their institutional em-bodiments. In addition, Marx believes that the major systemic ills of capitalist society—alienation and exploitation—will not be prevalent in PC society. The absence of these defects should be explainable by reference to the basic institutions of PC society. Finally, it should be possible to say something about what social life would be like without exploitation and alienation.

All of the above goes to make up the Marxian vision of PC society. What remains to be seen is how rich and detailed this vision is. It may be that there is no perfectly general answer to this question; Marx's vision might be quite specific and detailed in some respects and very abstract or thin in other respects. The discussion that follows is driven by what Marx actually said and the main tenets of Historical Materialism, together with the demands of the Alternative Institutions requirement. A little more needs to be said about the latter.

Recall from Chapter 1 that the Alternative Institutions requirement requires a specification of the main institutions of PC society, an account of how they would function, an explanation of their stability, and an argument to show that these institutions do not produce the ills or defects of capitalist society. Let us begin with the first of these. Obviously, different radical critics will cite different institutions as the "main" ones. For example, feminists might choose the family or its analogue, and anarchists might choose the state. However, for Marx, one can be reasonably certain that the economic system would be at the top of his list of the main alternative institutions of the new society.

What, then, is the economic system of PC society? The answer is not obvious; for example, on the face of it, it is unclear whether or not the economic system just *is* the dominant set of relations of production in that society or whether it includes more than that. Though I shall later argue for the latter, it is possible to start with the uncontroversial assumption that a necessary condition for specifying the economic system of PC society is to give an account of its relations of production. Historical Materialism promises that this will be a fruitful place to begin because it assigns explanatory or predictive primacy to those relations. Looked at from another perspective, if one wants an account of PC society that will satisfy the demands of the Alternative Institutions requirement for a radical critique, Marx's theory of history directs us to begin with the relations of production that will characterize that society.

### Relations of Production in Post-Capitalist Society

The relations of production in a society specify power and authority relations between and among persons and things with respect to man's interaction with his environment for the purpose of production. Under capitalism, for example, each person owns his or her labor power. Some people sell (are forced to sell) their labor power as a commodity to others and own no means of production; they are the proletarians. Others own means of production, but do not labor; these are the capitalists.[8] In capitalist society, there may be some individuals who fit neither category (e.g., state functionaries), but capitalist relations of production predominate in that most adults occupy the role of proletarian or capitalist.

As noted above, the central fact about the relations of production that will characterize PC society (hereafter referred to as 'PC relations of production') is that the workers (and only the workers) control the means of production. As I shall argue shortly, this control can be understood by way of analogy with capitalist control of the means of production under capitalism. In any case, worker control in PC society stands in sharp contrast to capitalism, where nonworkers control the means of production.

It is a reasonable inference from the fact that the workers control the means of production that labor power will not be a commodity in PC society. If labor power is a commodity, then it is bought and sold on a labor market.

But, if the worker sells his or her labor power, he or she alienates it in the sense of giving up control over that labor power to others. Is this even possible in a system in which the workers control the means of production?

Well, one could at least imagine a system in which workers control only means of production they themselves do not labor on or with, but surely this is not Marx's conception of PC relations of production. One of the motivations for the proletariat to overthrow capitalism is to take control of their lives; it is hard to see what motive they would have for instituting a system in which other workers control their lives by controlling the means of production they work on. If they wanted that sort of system, they need only demand that their capitalist bosses get out on the factory floor a bit more often. If the workers are not alienating their labor power, then it must be that they themselves control it. The workers, then, control both the means of production and their labor power in PC society.

These two conditions delimit, in a general way, relations between and among persons and things for the purposes of production. It is fair to say, then, that for Marx, the following define the relations of production for PC society:

1. The workers and only the workers control the means of production.
2. Labor power is not a commodity and is controlled by the workers.

Notice that (1) and (2) constitute the denial of capitalist relations of production as the latter were identified in Chapter 2.[9]

How is worker control of the means of production to be understood? The most natural way is by an analogy with capitalist ownership. Capitalists, after all, control the means of production under capitalism. Full, liberal ownership, which the capitalists enjoy, is actually a rather complex concept. A. M. Honoré has shown that it consists in a "package" of rights, terms, and conditions.[10] Lawrence Becker has summarized this package as follows:

1. *The right to possess*—that is, to exclusive physical control of the thing owned. . . .
2. *The right to use*—that is, to personal enjoyment and use of the thing as distinct from (3) and (4) below.
3. *The right to manage*—that is the right to decide how and by whom a thing shall be used.
4. *The right to the income*—that is, to the benefits derived from forgoing personal use of a thing and allowing others to use it.
5. *The right to the capital*—that is, the power to alienate the thing and to consume, waste, modify, or destroy it.
6. *The right to security*—that is, immunity from expropriation.
7. *The power of transmissibility*—that is, the power to devise or bequeath the thing.
8. *The absence of term*—that is, the indeterminate length of one's ownership rights.

9. *The prohibition of harmful use*—that is, one's duty to forebear from using the thing in certain ways harmful to others.
10. *Liability to execution*—that is, liability to having the thing taken away for repayment of a debt.
11. *Residuary character*—that is, the existence of rules governing the reversion of lapsed ownership rights.[11]

Although full liberal ownership of the means of production is sufficient for control of the latter, it is surely not necessary. For present purposes, the crucial question can be focused rather narrowly, namely, 'What rights are necessary and sufficient for genuine worker control of the means of production?'.[12]

I would argue that the answer to this question is: 'The right to possess, the right to manage, and the right to the income'. If the workers have these rights over the means of production, they have the right to determine what to produce and in what quantities; they also have the right to determine how production shall be organized within the workplace and how any surplus will be divided. For these reasons, if the workers have these rights, then it seems fair to say that they control the means of production. Conversely, if the workers fail to have any one of these rights, then it seems that they do not really have control of the means of production. After all, it is in virtue of the capitalists' rights of possession and management (especially the latter) that the workers lack control of their productive lives, and the lack of control over the surplus is central to Marx's account of their exploitation.

This argument can be strengthened by a considerations of the other rights, terms, and conditions in Becker's list. It is easy to imagine that the right to (personal) use might be held by no one in PC society. Means of production have to be used for social production and not for the personal use of anyone.

The right to the capital is a little less clear. This right (or more accurately, these rights) might be vested in the community as a whole, which includes nonworkers, for example, children, the severely handicapped, and retired workers. The concept of worker control of the means of production does not seem to rule on this one way or the other. However, since most people in PC society will be workers, it may not matter much how the assignment of these rights is conceptualized.

The right of security is also somewhat unclear. If the workers have the rights of possession, management, and income for any longish period of time, they implicitly have the right of security. On the other hand, it may be that groups of workers are assigned the former set of rights subject to certain conditions; failure to fulfill those conditions might make them liable to expropriation. In this scenario, it might be accurate to say that no one enjoys the right of security, though maybe the correct description is that the workers (or the community) as a whole enjoy this right. The same might be the case for the power of transmissibility and the absence of term. Finally, the prohibition of harmful use and liability to execution would have to be specified in PC society, but neither seems to have much to do with worker control.

To summarize, worker control of the means of production consists in the workers holding the rights of possession, management, and income, with some, perhaps limited, right of security. Honoré claims that any developed society must specify all the rights, terms, and conditions in the above list. There are a variety of ways this could be done that are consistent with worker control of the means of production. On the other hand, maybe Honoré is wrong when it comes to PC society—perhaps these other rights, terms, and conditions will "wither away."

The above account of worker control, together with the assumption of the "de-commodification" of labor power, specifies, articulates, or makes clear just what PC relations of production amount to; however, it does not imply anything about how the crucial rights will be exercised in the economy at large. This is a rather important question since how it is answered will determine how the economic system will function, and a radical critique must address this concern. More specifically, *the above account of PC relations of production does not logically determine whether or not PC society will have a market economy or a centrally planned economy*. This fact is of central importance for some arguments in Part II of this book, so it worth some elaboration and defense.

Markets and central planning are alternative ways of solving a problem that arises only at a certain level of complexity in the economic life of a community: the problem of production coordination. Consumer goods in the modern age are rarely produced directly from original factors of production (natural resources and labor). Not only are there many stages involved in the production of finished goods from natural resources, but many raw materials and semifinished products go into a multitude of production lines. Any complex economic system must solve the problem of coordinating the various lines and branches of production, the final stage of which is the production of consumer goods.

Competitive markets and central planning are alternative ways of solving this problem. In a competitive market system, buyers and sellers bid factor and product prices up or down depending on their plans and resources. Surpluses indicate that selling prices are too high relative to what buyers are willing to pay, whereas shortages reveal that prices are too low. Buyers and sellers adjust their plans according to the emerging realities as the latter are revealed in the marketplace through price changes. This mutual adjustment process constitutes the coordination of production. Of course, in capitalist market systems coordination does not always obtain. Gluts and shortages occur, and a variety of factors hold up the equilibriating tendency toward coordination. Indeed, Marx makes much of this in his account of crisis under capitalism. But that does not invalidate the fact that it is the competitive process that effects production coordination, to the extent that it is achieved, in a market system. And, as Engels points out in *Anti-Dühring*,[13] there is no overall plan guiding production.

By contrast, in a centrally planned economy, production coordination is achieved by the conscious decisions embodied in the plan for the economy. The latter determines how many tons of steel will be produced, how many automobiles will be manufactured, how many acres of corn will be grown, and so forth. Just as in a market system, there will be gluts and glitches, shortages and supply bottlenecks. But the fact remains that, in a centrally planned economy, the predominant method by which production coordination is effected is the plan.

Later I shall argue, against many contemporary thinkers, that markets and central planning are mutually exclusive and collectively exhaustive methods of production coordination on a societywide scale in the post-feudal world. For now, I only want to point out that these are two ways of solving a problem that must be solved in any moderately complex society, the problem of production coordination.

Notice that, by contrast, the problem of production coordination is not significant in pre-capitalist economic formations. The production of most use-values was a fairly straightforward affair. The state of technology being what it was, the pathways from natural resources to use-values were few and relatively short. Production was guided very directly by the consumption needs and wants of the producers and their oppressors. For these reasons, production coordination did not pose a serious problem for pre-capitalist economic systems.

What is next on the agenda is to argue that PC relations of production do not logically entail *either* central planning or markets as a method of production coordination for PC society. On the face of it, this looks relatively noncontroversial, but proving it in detail will provide a more complete picture of different possible visions of PC society. Then I shall argue for two more substantive claims: (i) An adequate vision of PC society, that is, a vision adequate for the purposes of a successful radical critique of capitalist society, requires *some* solution to the problem of production coordination: and (ii) Marx and Engels are themselves committed to central planning as the solution to this problem for PC society. Claim (i) will be argued for at the end of the next section and (ii) will be addressed in the section after that. The latter will be followed by two sections detailing Marx's motivations for saying that PC society will have a centrally planned economy. Marx's commitment to central planning will in turn play a decisive role in the main critical argument advanced against Marx in Chapter 9. The account of worker control of the means of production sketched above, together with Marx's commitment to central planning, will specify two key elements of the economic system in Marx's vision of PC society.

Chapter 7 returns to the other elements that constitute Marx's vision of PC society: the system of distribution, superstructural institutions and their corresponding normative belief systems, and an account of social life in which exploitation and alienation are absent.

**Relations of Production and the Coordination of Production in Post-Capitalist Society: Market Socialism and Central Planning**

To prove that PC relations of production do not entail markets or central planning as a method of production coordination, it will be enough to construct two models of economic systems: Both will be characterized by PC relations of production with one employing central planning and the other employing markets to coordinate production. This will be sufficient to prove that PC relations of production are consistent with either method of organizing production.

Let us call a system that exemplifies PC relations of production with a market economy, 'market socialism'. Something like this system characterizes an actually existing society, namely, Yugoslavia, so it is relatively easy to provide a model of such a system.

Imagine a market economy that is dominated by worker-controlled firms. Such firms differ from capitalist firms in the following respects:[14]

    i. The firms are self-managed. That is, the workers have and exercise possession and management rights over the means of production by deciding: (1) how work relations are to be structured; (2) what pay differentials should be for different jobs (perhaps subject to state-imposed floors and ceilings); (3) what working conditions should be (e.g., coffee breaks, etc.); (4) who will exercise day-to-day managerial tasks and what the scope of their responsibilities will be.

    ii. All the workers in these firms have income rights with respect to the means of production. In short, the workers get the profits. These profits are subject to taxation by the state to pay for public goods, to provide for the needy, and perhaps to finance new investment. However, such taxes cannot be set at confiscatory levels without impoverishing the workers, effectively destroying their income rights in the firm, or both.

    iii. The workers in the firm collectively have the right to sell the products of the firm for whatever they can get, subject perhaps to state-imposed ceilings and/or floors. This right is implicit in the right to the income from the means of production.

    iv. Those who control the means of production (i.e., the workers) do not have full, liberal ownership of the latter. Full, liberal ownership includes, among other things, rights to sell, liquidate, and destroy. The workers in worker-controlled firms have none of these rights in this model.

It is not necessary to be so pedantic as to insist that all firms in the society must be worker-controlled. It would be sufficient if worker control is the *predominant* form of economic organization. This would allow for some private (i.e., capitalist) enterprises in, for example, agriculture and some state control of the means of production, provided that they do not individually or collectively dominate the economy.

The role of the state is underdetermined in this model of market socialism. Most theorists believe that the state should and would guide production in limited ways, for example, by direct planning for state-owned firms, credit allocation for new investment, subsidy, and tax policy. However, in both the theory and practice of market socialism there are substantial markets for producer and consumer goods.

Furthermore, it might be that markets are not as extensive under market socialism as they are under capitalism. Some goods and services that are distributed by the market under capitalism would be distributed by non-market mechanisms under market socialism. However, this phenomenon would not be so widespread as to make it false or even misleading to describe such a society as a market economy.

Notice that the workers have both extensive possession and management rights and income rights with respect to the means of production. This is what their control of the means of production consists in. For this reason, this model of market socialism satisfies one condition for PC relations of production. The model also satisfies the other condition: Workers do not sell their labor power to the firms, and thus labor power is not a commodity. This is so for two reasons. First, although the worker puts his labor power at the disposal of the firm, since the workers themselves control the firm, the worker cannot truly alienate his labor power. Second, the worker does not receive wages from the firm.[15] As a part-owner, he receives a share of whatever is left over after suppliers and the state have been paid; in short, he is a residual claimant.

It might be objected that a person can sell something to a corporate body of which he is a part (just as a capitalist can sell his personal auto to his own firm), but there is a fundamental difference between a worker's relation with his capitalist boss and the former's relation to a collective of fellow workers— a difference that seems at least in part captured by the claim that her labor power is not a commodity. If a potential critic wants to insist that labor power is still a commodity (perhaps because there could be pay differentials for scarce talents and abilities), the relevant criterion for PC relations of production could be reformulated to say that labor power cannot be a mere commodity, as it is surely is in capitalist society—and as it surely is not in this model of market socialism.


The preceding shows that it is possible for a market system to instantiate PC relations of production. What about central planning and PC relations of production? What is needed is a model of a system in which there is worker control of the means of production, labor power is not a commodity and (most) production is determined by an overarching economic plan. Recall that worker control of the means of production consists in the workers having possession and management rights as well as income rights over the means of production. Let us consider rights of possession and management first. Suppose that, just as under market socialism, individual production units are self-managed; the workers in each production unit collectively decide how work relations are to be structured, the scope and extent of the duties of those who make day-to-day management decisions, and so on, and so forth.

What about the right to the income from the means of production? This right is somewhat difficult to conceptualize in a model with a centrally planned economy. In a centrally planned economy, a certain quantity of inputs and a certain output target are assigned to each production unit. That's what central planning *is*.[16] Strictly speaking, production units do not *sell* their products, at least in the raw materials and producer goods industries. Selling implies bilateral exchange, and production units do not exchange their products with other production units. Because of the nature of planning, what takes place is a series of unilateral transfers of possession and management rights. For example, producers of refined oil have the right to requisition refining equipment from those who manufacture it, and they are directed to turn over a certain quantity of petroleum products to authorized outlets. Although prices might be attached to products for accounting purposes, no money changes hands. That being the case, how are income rights over the means of production to be understood?

Ultimately, what is at stake is control over the social product. It is possible to imagine a scheme in which the workers collectively decide on a general scheme of distribution, for example, one in which workers are entitled to claims on consumer goods in proportion to labor contribution, need, and/or other criteria. Or, the production units themselves might be assigned such claims, which can then be divided up in any way they see fit. The general point is that there are a number of ways in which workers can control the social product and thus effectively have income rights over the means of production in a centrally planned economy.

However, there is a potential difficulty with this model; central planning imposes real constraints—constraints not faced under market socialism—on the workers' exercise of possession and management rights, and indirectly on income rights, in virtue of the fact that the plan determines what gets produced and in what quantities. Production units cannot proceed as they wish in this regard. They have a certain measure of internal autonomy in that the workers in these units can decide among themselves how work is to be structured, and so forth, but they are not free to produce what they want in whatever quantities they choose. Might not the constraints imposed by the plan so effectively circumscribe workers' other rights that they would have the latter in name only? In other words, is central planning really compatible with worker control the means of production? This question has real bite in light of the fact that existing centrally planned economies (in, e.g., the Soviet

Union and Cuba) are worker-controlled only in the homeopathic Leninist sense.

The answer to this question depends in large measure on how the plan is arrived at and whose will it expresses. If the plan is simply imposed by fiat by an elite planning bureaucracy, it is clear that the workers would not effectively control the means of production. The reason for this is simple: If the product to be produced and production quotas are simply dictated to groups of workers who have no say in determining these matters, important aspects of workers' productive lives (specifically, some of their management rights) are effectively beyond their control. Moreover, production quotas may significantly constrain how work relations must be structured (e.g., assembly-line labor, speed-ups). If this happened, the workers might well have possession and management rights in name only. To put it another way, they may have possession and management rights without having the corresponding *powers*.

Similarly, if the planning bureaucracy decides on some fantastically high rate of investment, workers might not have any real control over the social product. Indeed, they might starve to death. In large measure, the politically imposed famine in the Soviet Union in the Thirties stands as awesome testimony to the power of an elite in a centrally planned economy to strip the workers of control over the social product in the name of "investment for the future."

But the above are not the only possible scenarios for a centrally planned economy. Imagine a society in which the level of development of the forces of production is sufficiently high so that all of people's genuine needs can be met rather easily. Let us further assume that the workers desire that people's needs be met by the economic system. Planning under these circumstances is essentially a technical task that can be left to specialists. Under the Kantian dictum, "Whoever wills the end, wills the means," the resultant plan could be said to be that of the workers. (This ties in with Marx's conception of positive freedom in a number of ways.)

Under these circumstances, it is likely that the economic sphere would subside in importance and the constraints imposed by the plan or the planning process on the production units' rights of management, and possession would be relatively weak—weaker than, say, constraints imposed by the market on market socialist and capitalist firms. Let us further suppose that the planning apparatus is democratically controlled with the leadership accountable to the rest of the workers in the manner of the Paris Commune. In this way, a subset of management rights would be exercised indirectly and collectively by the workers as a whole. The same could be said of income rights in this story.

I'm not certain that the above scenario is the only way the workers could genuinely control the means of production in a centrally planned economy, but it is one way, and that is sufficient for the purposes at hand, which is to prove the logical compatibility of central planning and worker control. Notice that I do not make any claims about the practical feasibility of this scenario, since the point at issue is solely one of the logical consistency of (part of) a social vision. The above shows that genuine possession, management, and income rights can be held by the workers in a centrally planned economy, and that is what needs to be shown.

What would be the status of labor power in this model? PC relations of production require that labor power not be a (mere) commodity. Is this the case in the model outlined above? The decisive feature of a centrally planned economy is that it is not a system of commodity production. Goods and services are produced with their ultimate use in view and not to get exchange value in the market. For labor power to be a commodity, there would have to be well-established labor markets. But, since labor power is a producer good, its deployment has to be largely determined by the plan. For this reason, labor power could not be a commodity. More generally, it is hard to conceive of an economic system in which commodities in general are not produced, and yet human beings' capacity to labor is bought and sold on a labor market. The "de-commodification" of labor power is certainly possible in a centrally planned economy and indeed probably inevitable.

One final point about this model: By analogy with the case of worker-controlled market socialism, it is not necessary to insist that there is no production for the market going on in a society with a centrally planned economy. What counts is what method of production coordination—and what relations of production—predominate.

Though PC relations of production are compatible with both central planning and the widespread use of markets, a radical critic must commit himself to one or the other, or argue for some third alternative. The reason for this has to do with the Alternative Institutions requirement for a radical critique.

As noted above, a radical critic must spell out the alternative institutions he believes will characterize the new society, and for Marx the most important of these institutions is the economic system. Clearly Marx believes that worker control of the means of production will predominate in PC society; in addition, it is a reasonable inference that labor power will not be a commodity (or a mere commodity). Does this specify the economic system of that society? In other words, should the economic system of a society be identified with its relations of production, or must it include a method of production coordination as well?

The above discussion would seem to favor the latter. On the face of it, market socialism and centrally planned socialism are very different economic systems, so the type-individuation of economic systems by relations of production alone seems inadequate. Indeed, these days, economic systems are often identified by their methods of production coordination; people often speak of a market economy or a centrally planned economy. For this reason, henceforth I shall use the term 'economic system' to include not only the relations of production but also the method of production coordination. This implies that, for Marx to specify the economic system of PC society, it would be necessary to commit to some method of production coordination.

However, and on the other hand, even if the economic system of a society just *is* the set of relations of production, the radical critic still must provide an account of the method of production coordination *for the simple reason that the latter specifies how production will be organized across society at large.*[17] Without that, there is no way of knowing how the economic system (defined in terms of relations of production) will function or whether it can persist as a stable social form. In short, there is no way of knowing if the economic system in question is *realizable*. This implies that specifying a method of production coordination for the alternative economic system is a necessary condition for a successful radical critique of capitalist society.

However the conceptual question about the nature of the economic system is answered, the above considerations may seem to provide the basis for a serious criticism of Marx's radical critique of capitalist society, namely, that he really provides no answer to the question of how production is to be organized in PC society. After all, he had little to say about the details of PC society beyond the claim that the workers will control the means of production. In consequence, it might be thought that he has not met the Alternative Institutions requirement for a radical critique.

This objection does not succeed. In the next section I shall show that both Marx and Engels did in fact believe that PC society would have a centrally planned economy. In the two sections following that, I shall show that Marx's commitment to central planning is deeply rooted in, and motivated by, some of his critical explanations of the phenomenon of alienation in capitalist society and in his theory of history and social change.

## Marx's Commitment to Central Planning

The preceding section establishes that worker control of the means of production is consistent with markets and central planning as methods of production coordination. The main purpose of this section is to establish Marx's commitment to central planning for the economic system of PC society. Before proceeding to the textual evidence to substantiate this, it would be helpful to construct an analytical framework within which alternative methods of production coordination can be located. One of Marx's taxonomies of economic systems provides a useful point of departure. As Chapter 2 indicates, for Marx there is a fundamental difference between systems of production for exchange, or commodity production, and systems of production for use. In the former, the purpose of production is to get exchange value in the market. Recall that what makes something a commodity is that it is produced for exchange in the market. The distinguishing feature of a system of production for exchange, or to use a more familiar term, a market economy, is that there are autonomous and independent producers or production units. They are autonomous in the sense that those who control these production units decide what to produce and in what quantities. Of course, in one sense they are not free to do anything they want; prices for factors of production and finished goods shape the structure of production in that goods are produced only if the price of the product covers its cost. Marx claims that the only historically significant system of production for exchange is capitalism.[18]

By contrast, in a system of production for use, products are produced with their ultimate use in view. In pre-capitalist non-exchange economies, the producer either is the consumer or is part of the same relatively small social group to which the consumer belongs (e.g., the family, the feudal manor). Production decisions are guided by the need for use-values (i.e., consumer goods), though of course not always use-values that meet the needs of the producers themselves.

All pre-capitalist societies have been systems of production for use. For Marx, this includes primitive communism, Oriental despotism, slavery, and feudalism. This is not to deny that commodity production and market exchange took place in such societies. Indeed, Marx emphasizes that commodity production was present in all pre-capitalist "societies, excepting possibly primitive communism. However, in no case was commodity production the predominant form of production. That is, market phenomena did not determine most of what was produced, nor could the operation of market forces explain the main lines of the division of labor, to the extent that the latter existed.

What about central planning? The ultimate goal of production under a regime of central planning is the production of certain quantities of consumer goods and perhaps a certain rate of growth in the quantity of means of production, though the latter makes sense only in light of the future consumer goods that will flow from them. Thus central planning can be considered a system of production for use.

The plan informs production by assigning certain quantities of inputs and certain production targets (outputs) to various production units. Both inputs and outputs are expressed in material terms and not monetary or value terms, though a record of the latter may be kept for accounting purposes. A steel plant is told to produce certain quantities of various grades of steel instead of being directed to generate a certain monetary volume of sales. Indeed, in a centrally planned economy, there are strictly speaking no sales at all, at least in raw materials and in producer goods. Production units in a centrally planned economy do not individually own the products they produce. If they were to own these products (in any meaningful sense of 'ownership'), they would have the right to set the terms under which their products would be alienated, that is, they would have the right to sell them. But then the economy would be a market economy.

Central planning is incompatible with external autonomy, that is, autonomous decision making by individual production units about what gets produced and in what quantities. However much it may be informed by democratic procedures and ultimately controlled by the producers themselves, central planning cannot allow autonomy for individual production units about what and how much to produce, as well as what shall be done with the products. These things must be determined by the plan. Indeed, the whole point of central planning is to eliminate the chaos ("the anarchy of production") that results from autonomous control of producer goods and raw materials by diverse groups and individuals. In a worker-controlled centrally planned economy, the workers may collectively, though indirectly, control these products by way of their control of those who run the planning apparatus, but that control cannot be vested with individual production units. If the latter were to determine what to produce and in what quantities, markets, not planning, would have to coordinate production.

Developments in the twentieth century may seem to have blurred this distinction between central planning and markets. Capitalist societies are not pure market economies and socialist societies have employed markets more or less extensively. But that does not imply that central planning and markets are not mutually exclusive methods of production coordination, at least on a societywide scale. To say that a society has a market economy (i.e, is a system of production for exchange) does not imply that there is no production for use; it only means that the predominant method of production coordination is exchange on the market. Under capitalism, for example, much of education and national defense have been the result of production for use. Still, capitalism is a market economy. In a similar vein, a centrally planned economy can have some production for exchange. For example, the Soviet Union, which is perhaps the historical paradigm of a centrally planned economy, has for many years allowed some markets in agricultural products and personal services. It has also produced some goods for exchange on the international market. However, these cases of production for the market do not individually or together predominate in Soviet society. It is not misleading to say that the Soviet Union does not have a market economy.

On the other hand, socialist societies that rely extensively on markets (e.g., Yugoslavia) are systems of production for exchange, that is, market economies. Although the state can control most new investment, as well as the rate of growth, production is still for the market. Production units have external autonomy; they buy their inputs from whatever firms they can come to an agreement with and, in a similar manner, sell their output to other firms, the consumers or the state. Prices are allowed to fluctuate, subject to occasional and irregular state interference. What makes these systems not centrally planned is that the production of certain quantities of products is not mandated by a comprehensive plan. The firms may be state-owned or owned by the workers themselves, but in either case, since products are *sold*, profit and loss considerations guide production (subject to selective state interference, but that is true under capitalism as well).

Although it is at least logically possible for a complex society to have no predominant method of production coordination, this does not seem to be empirically possible. It is possible for a market socialist society to have a *state* sector that is nearly the same size as, or even larger than, the private or worker-controlled sector, but in such cases the state sector will contain many firms producing for the market. In addition, a market economy (whether capitalist or market socialist) may have an extensive set of price controls, subsidies, and an incomes policy grafted onto it in an attempt to influence production in various ways. However, all such policies are predicated on the assumption that individuals and firms will react according to economic theories describing market economies. In short, these interventionist policies are always interventions into an ongoing market process.

Although it may not be possible to prove it outright, I suspect that a system that had a market sector and a planned, production-for-use sector of roughly equal size (and so the economy in question could not be called a centrally planned economy or a market economy) would be inherently unstable because of coordination problems. The reason for this is straightforward: Given the highly interdependent character of production in a complex economic system, the planners would be continually frustrated by the actions of independent production units not subject to their directives. Assuming they had the power of the state on their side, they would have an obvious incentive to bring the market sector under their control, *or* they might decide to withdraw from active direction of the entire economy, perhaps to concentrate on a few targeted sectors. By that fact, they would defer to market forces for the bulk of the economy, subject of course to selective and irregular interference, such as one finds in capitalism.

It has become fashionable to talk about a mix of plan and market as a third way between capitalism and centrally planned socialism; such a system may be viable, but my point here is that, whatever its virtues and defects, it is a market economy, since the predominant microlevel coordinating mechanism is the market. On the other hand, if planning is the primary coordinator of production among the various production units, then it is a centrally planned economy. In such a system, if markets exist, either they are insignificant, or they are merely pseudomarkets. Vague talk about combining plan and market does not specify a coordinating mechanism for an economic system. Most of those who have thought through the implications of market socialism envision the use of markets as the predominant microlevel coordinating mechanism.[19]

As an aside, it seems that the popularity of much of this talk about combining plan and market can be traced in part to contemporary economic thought, which has been dominated by static equilibrium models. The conception of economic systems that underlies the arguments of this book is much more dynamic. If one is asking how production is coordinated, one is asking for the specification of a mechanism, or dynamic *process*, instead of an account of static states of systems. That centrally planned economies, or models thereof, have certain structural similarities to market economies (e.g., both have equilibrium states) obscures the fact that they are activated by very different mechanisms. To repeat, this is not to deny that an actually existing centrally planned economy can allow for some production for exchange in the market, and, similarly, market economies can have some production for direct use, but it is hard to see how an economic system could persist if neither method of coordinating production predominated.

Given that central planning and markets are mutually exclusive methods of coordinating production on a societywide scale, are they collectively exhaustive as well? In other words, are any other possible coordinating mechanisms for a economic system? In the pre-capitalist world, the answer would seem to be yes. Before the advent of capitalism, production coordination was not a significant problem, since the structure of production was relatively simple. Not only were many fewer kinds of products produced, but the production of most use-values was a relatively simple affair. In pre-capitalist economic systems, the highly conservative forces of custom and tradition informed the production of use-values and determined whatever division of labor existed. Production coordination was a piecemeal affair. Feudalism for example, was not a market economy, though there were always markets "around the edges," so to speak. On the other hand, though feudalism can be characterized as a system of production for use, it was not a centrally planned economy either. Maybe the method of production coordination characteristic of pre-capitalist societies should be called 'the custom and tradition method'.

Matters are quite different when the scale and complexity of production reaches the levels that one finds in, say Europe around the time of the rise of capitalism. More generally, in any modern economic system, a vast number of different kinds of use-values are produced, and perhaps more importantly, the production-consumption sequence of nearly everything is astonishingly complicated. Compare, for example, the steps involved in getting from sowing wheat to eating bread on the feudal manor to the steps involved in the comparable production-consumption sequence in a contemporary (market or centrally planned) economy.

Clearly the custom and tradition method of coordinating production is inadequate at a certain level of economic complexity and technological change. Although it cannot be proved outright that there is no alternative to central planning or markets as a coordination mechanism for a complex economy, it is hard to imagine what it would look like.

If the above arguments are sound, they establish that, in the post-feudal world at least, the categories of market economy and centrally planned economy, *as they have been defined here*, are effectively mutually exclusive and collectively exhaustive. This of course is not to deny that there can be important differences between different kinds of market economies or, for that matter, centrally planned economies (in particular, whether or not they are worker-controlled).

The following table summarizes the results of the above discussion.

|  | Production for Use | Production for Exchange |
| --- | --- | --- |
| Pre-Modern (simple) Societies | primitive communism; Oriental despotism; slavery; feudalism | simple commodity production[20] |
| Modern (complex) Societies | central planning | capitalism; market socialism |

With this analytical scheme in hand, it is now possible to address the question of Marx's views about how production would be organized in the economic system of PC society. There are only four possible answers:

1. Marx believed that central planning would coordinate production in PC society.
2. Marx believed that markets would coordinate production in PC society; that is, PC society would be a form of market socialism.
3. Marx believed that some third alternative method of coordinating production would predominate or no method would predominate; however, if the above arguments are sound, this alternative is either incoherent or unworkable for a complex economic system. (Of course, this does not imply that it could not have been Marx's view!)
4. Marx is not committed to any particular method of production coordination. He simply had no views on the matter, and his vision of PC society is correspondingly incomplete.

In what is left of this section I argue for the first of these alternatives and against the other three. The remaining sections of this chapter attempt to account for or motivate this commitment to central planning.

One factor that might appear to complicate a satisfactory answer to the question of which alternative Marx subscribed to is that, in the *Critique of the Gotha Program*, he clearly envisions two separate stages of PC society, which have been anachronistically designated as 'socialism' and 'communism'.[21] The main differences between them are two:[22] (1) In the first phase, distribution of the products of labor to the workers is directly proportional to the quantity of labor expended. In the higher phase, the operative principle is, 'From each according to his abilities, to each according to his needs'. (2) Socialism, the immediate successor to capitalism, is characterized by the dictatorship of the proletariat. That is, the state survives as an instrument of the ruling class—this time the working class. It is used to keep in line class enemies, notably the former bourgeoisie. The state "withers away" after a time, however. The reason for this is that PC society is a classless society, and without classes, there is little, and eventually nothing, for the state to do. When the state has withered away, communism has arrived.

Despite these important differences, both are socialist societies in that the workers control the means of production and all able-bodied adults are workers. However, by itself this does not imply central planning, since market socialism also involves worker control of the means of production. Nonetheless, Marx rules out market socialism in the *Critique of the Gotha Program* when he states that the first stage of PC society will not have a market economy: "With the co-operative society based on common ownership of the means of production, the producers do not exchange their products" (*CGP*, p. 16). This passage seems quite strange, if one thinks of exchange as a physical phenomenon. It is, however, a social phenomenon, involving mutual alienation of ownership rights (i.e., rights to possession, management, income, and, most importantly, the right to alienate) among autonomous individuals or production units. What this passage means is that there is no exchange in the sphere of production. That is, production units will not buy and sell raw materials and producer goods from one another. And, if they do not buy and sell from one another, markets cannot coordinate their production. The relation between production units in PC society will be like the relation between different departments of a capitalist firm.

On the other hand, there will be exchange in the sphere of distribution, since the operative distributive principle is, 'To each according to his labor contribution'. Marx envisions workers receiving labor certificates for the quantity of labor expended (less various deductions for social spending, etc.); these certificates are then exchanged for consumer goods.[23] That is, consumers buy products, but Marx insists that what they use is not money, since it does not circulate. That is, it cannot be used as capital. So, it's not as if there is no exchange in the first phase of PC society, but there is no *production for* exchange because market phenomena do not guide production. Since there is no suggestion that exchange in the sphere of production will be reintroduced in the second or higher phase of PC society, it is fair to say that both stages will be systems of production for use. This implies that market socialism (the second possibility listed above) is not an option for Marx.

Further evidence on this score can be found in *The Communist Manifesto*. In a derisive comment on the bourgeois conception of freedom, Marx and Engels say,

> By freedom is meant, under the present bourgeois conditions of production, free trade, free selling and buying.
>
> But if selling and buying disappears, free selling and buying disappears also. This talk about free selling and buying ... have a meaning if any, only in contrast with the fettered traders of the Middle Ages, but have no meaning when opposed to the Communist abolition of buying and selling, of the bourgeois conditions of production, and of the bourgeoisie itself. [*CM*, MECW, vol. 6, pp. 499–500]

Here Marx and Engels predict the abolition of all buying and selling, that is, not just in producer goods and raw materials. However, there is no inconsistency if the above refers to the second phase of PC society.

There is at least one other reason why Marx would deny that PC society could have a market economy. Recall from Chapter 2 that Marx's modal characterization of the essence of capitalism consists of two features: (i) (Most) products are produced as commodities, and (ii) The purpose of production is to maximize surplus value. (See *Capital* III, pp. 879, 880.) Arguably, market socialism meets both of these conditions. Does this mean that market socialism is "really" a form of capitalism? Traditionally, this sort of question gets Marxists tied up in *knots*. But it creates a serious problem only if one is disposed to believe that everything Marx says is true. It might be that one of Marx's characterizations of the essence of capitalism implies that market socialism is a form of capitalism, to which the appropriate response is, "So much the worse for that characterization." However, this difficulty does shed some light on the interpretive question of whether Marx is committed to the view that PC society would not have a market economy: If one of Marx's characterizations of the essence of the capitalist economic system is that it is a system of commodity production in which the goal of production is profit, then it is reasonable to believe that his vision of PC society does not include the widespread use of markets.

Although the above establishes that Marx believed that PC society would not have a market economy, it does not prove that he believed that it would have a centrally planned economy; perhaps he did not think about it enough to have any definite views about how production would be organized once markets had been abolished. That is, the fourth possibility identified above has not yet been ruled out.

Though detailed discussion is hard to find, there are some passages that directly support the commitment to central planning for PC society and by implication rule out the position that Marx simply had no views at all on this question:

> The national centralization of the means of production will become the nat-
> ural base for a society which will consist of an association of free and equal
> producers acting consciously according to a general and rational plan.[24]

Although the focus of this book is on Marx, given the close collaboration he enjoyed with Engels, the latter can provide indirect corroborating evidence. In *Socialism: Utopian and Scientific*, Engels describes the results of proletarian revolution as follows:

> State interference in social relations becomes, in one domain after another,
> superfluous, and then dies out of itself; the government of persons is replaced
> by the administration of things. ... Socialised production upon a predeter-
> mined plan becomes henceforth possible. ... In proportion as anarchy in
> production vanishes, the political authority of the state dies out.[25]

Engels's *Socialism: Utopian and Scientific* is a shortened version (with minor changes) of one section of *Anti-Dühring*. Elsewhere in the latter he gives further evidence of his commitment to central planning:

> With the seizing of the means of production by society, production of com-
> modities is done away with, and, simultaneously the mastery of the product
> over the producer. Anarchy in social production is replaced by systematic
> definite organization. [*A-D*, MECW, vol. 25, p. 270]

All of these passages taken together seem to indicate Marx's and Engels's commitment to central planning for PC society. Market socialists who claim a Marxist heritage might object that at most these passages indicate a commitment to planning of some sort but not necessarily highly detailed and centralized planning. A national strategic plan that leaves room for the limited operation of market forces is not explicitly ruled out by these passages. However, this reading of Marx and Engels is strained and implausible in light of

Marx's explicit rejection of markets in the above quotations from the *Critique of the Gotha Program* and *The Communist Manifesto* and Engels's rejection of commodity production in the quotation from *Anti-Dühring* just cited.

Moreover, and now we come to the third possibility identified above, there is no suggestion in Marx's or Engels's writings on PC society of any alternative method of production coordination. Indeed, I am aware of no such alternative for a complex economic system and have argued above against the possibility of such an alternative. Marx and Engels may well not have thought through the implications of their commitment to central planning, but there cannot be any serious doubt that this is what they believed.

This commitment takes on added significance when what motivates it is made clear. The next section investigates one such motivation for that commitment in some of the critical explanations of the phenomenon of alienation in capitalist society as they have been laid out in Chapter 2.

### Motivating Marx's Commitment to Central Planning: The Critical Explanations of Alienation in Capitalist Society

It is virtually an axiom of revolutionary practice that a radical's most dangerous opponent is the moderate reformer. For example, Lenin believed that his most dangerous opponent was the czar's reformist Minister of the Interior, Stolypin. Analogously, the radical critic's most formidable opponent is not the defender of the *status quo* who believes that this is the best of all possible worlds but instead the moderate critic. The latter believes that the evils and defects of capitalist society can be ameliorated to some extent but not eliminated. By contrast, the radical critic, that is, Marx, is committed to the proposition that these evils and defects can be eliminated or at least virtually eliminated, that is, reduced to insignificance, though this can be brought about only if existing institutions are destroyed and replaced by different ones.

Since alienation in its various guises is one of the two principal defects of capitalist society, and since Marx's critique of the latter is radical, he is clearly committed to the view that alienation would be eliminated in PC society. In this section I explore the implications of this commitment in light of Marx's critical explanations of the phenomenon of alienation in capitalist society. The purpose of this section is to show that Marx's implicit belief that alienation would be absent from PC society presupposes that PC society would not be a system of widespread commodity production; in other words, it would not have a market economy. Given that central planning is the only genuine alternative method of organizing production in a complex economy and given that there is good reason to think that Marx believed that PC society would have a centrally planned economy, this will show that Marx's commitment to central planning is central to his vision of PC society and, by implication, crucial to his radical critique of capitalism.

A recurring theme in Chapter 2's account of Marx's critical explanations of alienation in capitalist society is his appeal to the fact that capitalism is a system of widespread commodity production. Sometimes, but not always, this fact is the only essential feature of capitalism to which Marx's explanation appeals. Let us say that a system of widespread commodity production is *inherently alienating* in some respect just in case Marx's explanation for a particular manifestation of alienation in capitalist society appeals to only this essential feature of the latter.[26] The implication, of course, is that any system of widespread commodity production will suffer this form of alienation. If PC society is to be nonalienating in this respect, it cannot be a system of widespread commodity production. In what follows, I briefly review the respects in which a system of widespread commodity production is inherently alienating as a way of motivating Marx's commitment to the lack of markets and, by implication, to central planning, for PC society.

One respect in which commodity production is inherently alienating is that particular, concrete labor becomes "abstract" labor when it takes the form of exchange value. This is a loss for the worker since control over (the product of) his labor belongs to another (under capitalism, the capitalist) and passes to yet another through market exchange. Even if the workers were to control the means of production in a system of commodity production, they would still lose control of the products of their labor at the point of exchange.

And this loss can come back to haunt them. As pointed out in Chapter 2, market forces regulate the societywide division of labor in a regime of commodity production. In such a system market forces—notably competition—shape the structure of production.[27] Things do not get produced for long unless there is a profit to be made, and firms that do not show a profit are driven out of business by those that do. These forces can be controlled and moderated to some extent. Capitalists are notorious for trying to get the State to protect them from the rigors of competition. However, as long as the purpose of production is to get exchange value in the market, competitive forces will operate to favor some individuals and firms over others. This virtually guarantees serious problems for some of the losers. For example, the meager skills of a workman can be rendered useless by the technological advances fostered by competition. In addition, since a regime of commodity production is guided by no overall plan, coordination among independent producers can and does break down, leading to massive unemployment. This is central to Marx's explanation of economic crisis under capitalism.[28]

Commodity fetishism allows the true nature of the problem to be obscured. Market forces are experienced as alien and impersonal; they can and do crush firms as well as individual workers. The root of the problem is, of course, commodity production itself, but the fetishism of commodities consists in the fact that a regime of commodity production produces the illusion that these problems are somehow rooted in the nature of things and are transhistorical in their incidence. That is, it is not evident that they are peculiar to a particular, historically transitory method of organizing production.

One consequence of Marx's account of the fetishism of commodities is that the mystification produced by commodity production can be dispelled only by eliminating the latter[29] and replacing it with a consciously organized system of production for use. For example, if technological advances render some skills obsolete, the associated producers can consciously decide to withhold the application of this technology; even if they do not, this will be the result of a conscious decision and not something that just "happens," as is the case in a regime of commodity production. Under a system of production for use, social life has a kind of transparency that is lacking under a regime of commodity production.

A second manifestation of the inherently alienating nature of commodity production is to be found in the very activity of production, that is, in laboring itself. As Chapter 2 shows,[30] Marx thought of unalienated labor as a form of self-creation and as inherently satisfying. Man has an original need that is satisfied by the activity of laboring. However, in commodity production man is not producing to meet this need; instead, he is producing for exchange in a double sense: The creation of exchange value is the aim of production. Secondly, for the laborer himself, the ultimate goal is not the creation of the product; rather, it is wages. Labor is a mere means, not an end in itself. In consequence, the laborer becomes a mere means to a mere means. By contrast, Marx views production in its "natural state" (i.e., production for use) as inherently cooperative and directed toward the fabrication of use-values. The fact that commodity production is not like this makes it, by its very nature, perverse or pathological. By its very nature, then, labor is alienating in a regime of commodity production.

A third respect in which commodity production is inherently alienating concerns the nature of interpersonal relations implicit in the exchange relationship. As Marx says about exchange in his "Comments on Mill,"

> far from being the *means* which would give you *power* over my production, [your need, your desire] are instead the *means* for giving me power over you .... The intention of *plundering*, of *deception*, is necessarily present in the background, for since our exchange is a selfish one, on your side as on mine, and since the selfishness of each seeks to get the better of that of the other, we necessarily seek to deceive each other. ["Comments on Mill," MECW, vol. 3, pp. 225–226]

This is a rather remarkable passage in that Marx seems to be suggesting that exchange is necessarily mutually exploitative, though not in his technical sense of the term. Others' wants and needs are levers that we manipulate in satisfying our selfish desires. The inherently conflictual nature of exchange fosters the kind of egoism and individualism for which capitalism has been condemned since its inception. In addition, where exchange value is king, everything from human bodies to good will has a price. It is Marx's insight that these attitudes are rooted in the basic principle that organizes production: exchange in the market. Furthermore, the harmony of interests achieved by competitive markets is a kind of forced harmony resulting from the exploitation of comparative advantage by market participants.

In light of these explanations, it is evident why Marx would maintain that a progressive and revolutionary transformation of society that eliminates alienation requires the virtual elimination of commodity production. It should be clear how this motivates Marx's commitment to central planning. As I have argued in the previous section, it is a short step from the wholesale rejection of markets to central planning: Recall that without commodity production (and, by implication, market prices), production must be guided and coordinated by some other mechanism. The complexity of any post-feudal society, together with the considerable interdependence this fosters, make central planning the only genuine alternative way to coordinate production. It is, I think, uncharitable to maintain that Marx was unaware of this relatively immediate implication of this central aspect of his radical critique of capitalist society.

Indeed, implicit in the above explanations of the inherently alienating character of commodity production are suggestions of how central planning, as a method of organizing production, would eliminate the various forms of alienation: (1) Central planning overthrows the reign of market forces. Crises, a significant form of collective irrationality under capitalism, disappear in a system where production coordination is planned and is not the result of an anarchic invisible-hand mechanism. (2) In a centrally planned economy, labor regains its "natural" form in that its purpose is the production of use-values, rather than exchange value. (3) In a society no longer dominated by exchange value, interpersonal relations lose their mutually exploitative character. When production becomes a truly cooperative venture, others' needs and desires are no longer levers that are manipulated to satisfy selfish needs.

Marx's explanations of the respects in which commodity production is inherently alienating, or at least the radical implications of these explanations, are undoubtedly contentious. Socialists who favor markets and yet are sympathetic to Marx's radical critique of capitalism would certainly challenge much of the above. For example, they might claim that in worker-controlled firms, individuals do not labor for wages. As residual claimants, their income comes from profits, and the difference between wages and profits is not merely semantic. In general, market socialists would likely maintain that the harms associated with the inherently alienating character of commodity production can be significantly ameliorated by other socialist institutions. And perhaps they are right. But this only indicates that these socialists are, in one sense, reformist critics of capitalism, for they propose to maintain one of the main institutions of capitalist society, namely, the market. As the above accounts make clear, Marx would maintain that the evils associated with commodity production under capitalism cannot be significantly ameliorated without eliminating markets.

This completes the interpretive argument for Marx's commitment to central planning. This commitment will prove absolutely central to some of my criticisms of Marx's radical critique of capitalist society to be developed in Chapter 9; consequently, Marx's defenders would be well-advised to scrutinize carefully the foregoing arguments. In this section and the last I have attempted to show two things: (1) There is good reason to think that Marx (and Engels) in fact believed that PC society would have a centrally planned economy; and

(2) This commitment is at least in part motivated by some of Marx's critical explanations of the phenomenon of alienation in capitalist society. For Marx, the elimination of a number of forms of alienation that are found in capitalist society requires the elimination of markets. Furthermore, it is possible to see why Marx might believe that a centrally planned economy would not face these problems.

There are, I think, some additional motivations for this commitment that can be advanced, at least within a Marxist framework. The next section investigates some of these arguments. These arguments are speculative and reconstructive in that I do not maintain that Marx actually offered them. Rather, they are arguments he could have offered, since they cohere with what he did say in that they make use of materials from his theory of history and revolutionary change.

## Further Arguments for Central Planning in Post-Capitalist Society

One way that social theorists make predictions about future social organization is to project the working out of certain tendencies that are believed to be currently operative. Part of the Marxist story of the natural history of capitalism is that the latter is headed for an era in which fewer and fewer large firms dominate the capitalist world. Monopoly capitalism, as it is called, is the final stage of capitalist development. This tendency works itself out in part by vertical integration. A firm purchases its supplier and/or the outlets for its products. Planning replaces the previously conflictual and adversarial relation between the firms that have been integrated.[31] Suppose there is a tendency toward vertical integration in late capitalism. If this scenario is brightened by proletarian revolution, seizure of state power by the workers, and nationalization of the big firms, all that is necessary to complete the process is to integrate the production of these firms by comprehensive planning. This explanation would undoubtedly appeal to those with a teleological way of looking at world history, for example, Marx.

Lenin describes the natural end of this process with great clarity in *State and Revolution*: "*All* citizens become employees and workers of a *single* country-wide state 'syndicate.' . . . The whole of society will have become a single office and a single factory."[32] I make no judgments about the plausibility of the above account of the natural history of capitalism and its revolutionary overthrow. However, central planning represents a natural outgrowth of tendencies implicit in late capitalism, at least from a Marxist perspective; as such, the above provides another motivation for maintaining that the Marxian vision of PC society includes a centrally planned economy.

Another, more indirect argument for central planning comes from Marx's account of the genesis of capitalist relations of production during the decline of feudalism. Marx's explanation of the genesis of the capitalist class structure assigns a key causal role to the increasing incidence of commodity production.[33] If law-like statements of social causation are construed as tendency claims, and if Marx believes that there really are laws of social development, he would be committed to the view that any society that made widespread use of markets would have a tendency to evolve into capitalism. Consequently, if PC society is to have a stable economic system, it cannot have a market economy, and if central planning is the only alternative to a market economy, PC society must have a centrally planned economy.

Independently of Marx's historical account of the decline of feudalism and the rise of capitalism, a plausible case can be made for there being a causal connection between commodity production and capitalist relations of production. When products are produced for the market, especially efficient production units have a competitive advantage. That is, comparative advantage (in Adam Smith's sense) yields competitive advantage, at least insofar as there are markets. The accumulation of exchange value that results from success in the marketplace gives those who control these means of production disproportionate social and economic power. In other words, production for the market tends to give some producers (or those who control the means of production) social and economic power at the expense of their less advantaged rivals. Perhaps more importantly, the losers in this process lose control of the means of production they started with and are left with nothing to sell but their labor power. If production for exchange is not widespread, capitalism may not result; indeed, it did not in pre-capitalist societies, which had some production for exchange. However, wherever commodity production is or can become widespread in its incidence, it will tend to produce capitalist relations of production.

How can this problem be dealt with? The comparative advantage that some producers enjoy over others cannot be systematically eliminated. For example, some national and ethnic groups work harder than others, the fertility of land and fisheries varies, sources of raw materials are more abundant in some places than others, and so on. On the other hand, competitive advantage *can* be eliminated, but the only systematic way to do it is to eliminate competition itself, at least as a guiding principle of economic life.[34] That, of course, requires the virtual elimination of markets. If all this is true, it means that a necessary condition for stability in PC society is a non-market economic system. If central planning is the only alternative to a market economy, PC society must have a centrally planned economy.

Now that Marx's commitment to central planning has been substantiated and motivated, let us recall the point of all this. The Alternative Institutions requirement for a successful radical critique requires that the radical critic specify the main alternative institutions that are to replace existing ones. For Marx, this includes the economic system, and in PC society, the latter is one in which the workers control the means of production and labor power is not a commodity. However, a radical Marxian critique requires more than a specification and articulation of these relations of production; it must also be shown that the economic system is realizable, and this requires at least an account of how this institution will function. The commitment to central planning serves this purpose in that it articulates how the economic system of PC society will work, that is, how it will solve the major problem any complex economic system must solve, namely, the problem of coordinating production. It is, in effect, a way of making more still more definite what worker control of the means of production will come to in PC society, since it specifies how rights and/or powers over the means of production will be exercised. It is not the only way worker control of the means of production can be exercised (market socialism is another), but it is one such way, and it is Marx's. So, the commitment to central planning meets one important necessary condition for satisfaction of the Alternative Institutions requirement.

The story is far from complete, however. Even if attention is restricted to the economic system, more has to be known about this system to show that it is realizable. Economic systems distribute not only control over the means of production and labor power; they also distribute means of consumption. In ordinary parlance such systems are systems of production and distribution. So, we need to know something about Marx's views about distribution in PC society. In addition, there are other elements of Marx's vision of PC society that are importantly connected with the economic system: (i) normative belief systems and their institutional embodiments, and (ii) an account of what life would be like in the absence of the major systemic ills of capitalist society. These topics are addressed in the next chapter. By the end of that chapter, a reasonably complete account of Marx's vision of PC society should be in hand. Let us turn, then, to these topics.

# 7

## Post-Capitalist Society:
## Distribution, the State, and
## the Good Society

### Distribution in Post-Capitalist Society: The First Phase

In the previous chapter the economic system was said to include not only the relations of production but also the method of production coordination. A case can be made for also including the mode of distribution of income as part of the economic system: It is not really possible to think through how an economic system functions without having some account of how income (the means of consumption) is distributed.

As Marx makes clear in the *Critique of the Gotha Program* (*CGP*, p. 18), the distribution of the social product is a more or less direct consequence of the predominant relations of production. Moreover, this work contains specific descriptions of how this distribution is effected in both phases of PC society. The first three sections of this chapter are about distribution in PC society. The present section lays out this account for the first phase and reconstructs how Marx might have argued—as opposed to simply speculated— that this is the way distribution will work. Distribution in the second phase of PC society proceeds according to a very different principle, namely, "To each according to his needs." This occurs despite the fact that the relations of production are the same; something has to be said about why this is so. The second section addresses this issue. The third section attempts to make sense out of this distributive principle for the second phase of PC society.

Following that will be a section on the political system, that is, the state, in the first phase of PC society and its demise in the second phase of PC society. These first four sections of this chapter, together with the material of Chapter 6, will give us the economic and political systems of PC society, that is, its main institutions.

Finally, there will be two sections on the absence of exploitation and alienation (respectively) and how these features are related to the institutions of PC society. These sections will in effect be a partial sketch of Marx's conception of the good society. The larger purpose of this chapter and the last is to reconstruct Marx's social vision of PC society so as to address as completely as possible the Alternative Institutions requirement for a successful radical critique of capitalist society laid out in Chapter 1.

The third comment in the *Critique of the Gotha Program* is undoubtedly the most fruitful passage for discovering many of the details of Marx's vision of PC society. The context is an attack on the Lassallean proposal that the workers are to receive a fair distribution of the social product, which Marx interprets to mean they are to receive the undiminished proceeds of their labor. After pointing out a number of deductions that must be made from the social product, Marx launches into his description of PC society. He begins with the passage quoted in the previous chapter about the producers not exchanging their products. Following that is a long paragraph describing distribution in the first phase of PC society. It is worth quoting in full:

> What we have to deal with here is a communist society, not as it has *developed* on its own foundations, but, on the contrary, just as it *emerges* from capitalist society; which is thus in every respect, economically, morally, and intellectually, still stamped with the birth marks of the old society from whose womb it emerges. Accordingly, the individual producer receives back from society— after the deductions have been made—exactly what he gives to it. What he has given to it is his individual quantum of labour. For example, the social working day consists of the sum of the individual hours of work; the individual labour time of the individual producer is the part of the social working day contributed by him, his share in it. He receives a certificate from society that he has furnished such and such an amount of labour (after deducting his labor for the common funds), and with this certificate he draws from the social stock of means of consumption as much as costs the same amount of labour. The same amount of labour which he has given to society in one form he receives back in another. [CGP, p. 16]

This passage is revealing in a number of respects. First, Marx's "obstetrical metaphor" (to use Cohen's felicitous phrase) makes it evident that in certain respects, the transition to PC society does not represent a radical discontinuity with capitalist society. This is probably a good thing, if Marx is to have a plausible story about this transition.[1] Furthermore, it suggests that PC society, at least in its first stage, is not so radically different from capitalism that one cannot say much of anything definite about it. Thus it partially confirms the working hypothesis of this chapter and the last, which is that Marx has a genuine social vision of PC society, as opposed to a simple prediction that there would be such a society.

More importantly, however, this passage describes in some detail the distribution system of the first phase of PC society (hereafter referred to as 'PC$_1$ society'). Each worker gets a labor certificate indicating the number of (socially necessary?) hours worked, less the appropriate deductions, which in turn entitles him to the same quantum of embodied labor in the form of consumption goods.

The argument for saying that this will be the distribution system in $PC_1$ society is at best unclear. Marx does point out that there will be certain similarities between capitalism and $PC_1$ society:

> Here obviously the same principle prevails as that which regulates the exchange of commodities, as far as this is exchange of equal values. Content and form are changed . . . But, as far as the distribution of the [means of consumption] among individual producers is concerned, the same principle prevails as in the exchange of commodity-equivalents: a given amount of labor in one form is exchanged for a given amount of labor in another form. [*CGP*, p. 16.]

In this passage, Marx tells us how distribution will work in $PC_1$ society but not why it will work this way. However, there is a functionalist explanation (prediction) that would support this conception of distribution in $PC_1$ society. This explanation is based on the idea that stable economic systems (which $PC_1$ society presumably has) must reproduce themselves.

Recall from Chapter 3 Marx's model of simple commodity production. In this model, producers own their own means of production and yet produce for the market. Reproduction presupposes that equal quantities of embodied labor are being exchanged; the reasoning behind this is straightforward: No one would regularly exchange more embodied labor for less embodied labor. (For the purposes of this chapter, I ignore the objections I raised to all this in Chapter 3.) In full-blown capitalism, the tendency for things embodying equal quantities of labor to be exchanged is permanently distorted by capitalist relations of production. Maybe what happens in $PC_1$ society is that this distortion is removed, and the tendency toward equal labor exchange comes closer to the empirical surface. Of course, in $PC_1$ society there is no exchange in the sphere of production, but there is exchange in the sphere of distribution, and that is what must be explained at this point.

This explanation for the distribution system of $PC_1$ society might be further elaborated by pointing out that such a system implies an incentive principle the operation of which is required for former proletarians to work as hard as they must to insure reproduction. Though as we shall see, Marx envisions a very different incentive principle (or perhaps the disappearance, or *aufhebung*, of incentives) for the second phase of PC society, it is at least plausible to suppose that the kind of incentive system Marx envisions for $PC_1$ society is functionally necessary for worker control of the means of production following the overthrow of capitalism: If the workers were regularly to "draw down" more embodied labor than they contributed, the system would not reproduce itself.

Furthermore, this functional necessity may be reflected in the moral beliefs of $PC_1$ society. Recall that Marx maintains that the latter has a kind of moral continuity with capitalism. As the long quotation at the beginning of this section suggests, the proletariat bring to $PC_1$ society some of the moral beliefs inculcated under capitalism. When it comes to setting up a distribution system for $PC_1$ society, they put into effect elements of bourgeois ideology that capitalism paid lip service to but consistently violated. The main such belief is that people ought to get the value of what they contribute[2] (less necessary deductions, of course). The distribution system Marx attributes to $PC_1$ society instantiates this moral principle. This is implicit in the following passage, which follows the one just quoted: "Hence, *equal right* here is still in principle—*bourgeois right*, although principle and practice are no longer at loggerheads. . . . The right of the producers is *proportional* to the labour they supply." (*CGP*, p. 17).

It is a plausible hypothesis that, for Marx, rights-claims are to be explained by their functional necessity for the social institutions, notably the economic system, of a society.[3] If there is an equal right to consumer goods according to labor contribution, it must be because honoring that right is necessary for the reproduction, and hence the stability, of the economic system of $PC_1$ society. At this point, I make no assessment of the soundness of this argument, but it does seem to be the sort of argument Marx might have had in mind.

### The Preconditions for the Second Phase of Post-Capitalist Society

Marx's account of distribution in the higher phase of PC society (hereafter, '$PC_2$ society') is given in the second part of his famous slogan, "From each according to his abilities, to each according to his needs." This slogan appears in the *Critique of the Gotha Program* just after his discussion of distribution in $PC_1$ society and some of the defects of the distributive principle operative in the latter. It is interesting to note that Marx precedes this slogan with a lengthy list of preconditions for $PC_2$ society. The entire passage reads as follows:

> In a higher phase of communist society, after the enslaving subordination of the individual to the division of labour, and therewith also the antithesis between mental and physical labour have vanished; after labour has become not only a means of life but life's prime want, after the productive forces have also increased with the all-round development of the individual and all the springs of cooperative wealth flow more abundantly—only then can the narrow horizon of bourgeois right be crossed in its entirety and society inscribes on its banners: From each according to his ability, to each according to his needs! [*CGP* pp. 17–18]

This passage supplies us with a motion picture of PC society, as it were— or at least before and after snapshots. There is a clear implication that all of the "after clauses" in the above passage describe what actually happens in the transition from the lower to the higher phase of PC society. More will be said about this shortly, but for now it is worth noticing that a *different* distributive principle is operative in $PC_2$ society, yet the relations of production presumably have remained unchanged. How is that possible? The discussion thusfar has proceeded on the supposition (clearly supported by what Marx says at *CGP*, p. 18) that the mode of distribution is more or less immediately inferable from the relations of production, but the different distributive principles for the different phases of PC society seems to contradict that supposition.

This problem may not be as serious as it appears. Inferring a principle describing the distribution of the means of consumption from a set of relations of production would seem to require some assumptions about the natural and social environment at hand. Change those assumptions and different conclusions must be drawn. As the above quotation makes clear, conditions have radically changed when the boundary to $PC_2$ society has been crossed, and it is likely that whatever assumptions would be necessary to generate the distributive system of $PC_1$ society no longer hold when that boundary has been crossed.

But what are those assumptions? In his discussion of $PC_1$ society, Marx is not completely explicit about them, but I suspect that they can be grouped under the general heading of "birthmarks." In other words, the assumptions about $PC_1$ society that must be made to infer that it has the distribution system it will have are those that Marx vaguely refers to when he says of $PC_1$ society that it is "in every respect, economically, morally and intellectually, still stamped with the birth marks of the old society from whose womb it emerges" (*CGP*, p. 16).[4] A clearer picture of these birthmarks is implicit in the "after clauses" in the quotation ending with the description of distribution in $PC_2$ society. If one assumes that $PC_1$ society, throughout most of its history anyway, is characterized by the denial of all of these "after clauses," the need for the incentives implicit in the distributive principle of $PC_1$ society is palpable. More generally, perhaps the point of these clauses is to spell out the ways in which $PC_2$ society differs not only from capitalism, but also from $PC_1$ society. As an aside, it is worth noting that the radical nature of these changes suggests that $PC_1$ society will be around for quite some time. On the basis of this passage detailing the "after clauses," it is doubtful that Marx envisioned the advent of $PC_2$ society in his lifetime.

Let us return to $PC_2$ society. Three sorts of things are known about it:

i. The relations of production (worker control of the means of production, labor power not a commodity)
ii. Some of the attending circumstances and conditions (viz., those given in the "after clauses" in the above quotation)
iii. A principle describing both contribution and distribution ("From each according to his ability, to each according to his needs.")

Now let us make a large and important, but not unreasonable, supposition: Marx would want to argue for (iii) on the basis of (i) and (ii). In other words, the principle of contribution and distribution is supposed to be a consequence of the relations of production together with a statement of attending circumstances and conditions. Indeed, Marx follows the "after clauses" in the above-quoted passage with the phrase, "—only then can the narrow horizon of bourgeois right be crossed in its entirety." This strongly suggests that the realization of this principle presupposes that these conditions have been fulfilled and that the connection between the latter and the former is not fortuitous.

This raises an obvious question: How is the argument from (i) and (ii) to (iii) supposed to go? Marx himself never makes the argument explicit, so it will be necessary to reconstruct it. To this end, it would be helpful first to clarify the nature of the attending circumstances and conditions described by the "after clauses," as well as the $PC_2$ principle of (contribution and) distribution, "From each according to his abilities, to each according to his needs." Finally, something has to be said about what justification Marx might have for saying that these attending circumstances and conditions will eventually hold in PC society. What is at stake in all of this is of considerable importance in assessing the plausibility—or realizability—of $PC_2$ society insofar as it provides more of the details of Marx's vision of that society and how it differs from both capitalism and $PC_1$ society.

According to the above quotation, the advent of $PC_2$ society is accompanied by:

1. The end of "the enslaving subordination of the individual to the division of labor"
2. The overcoming of "the antithesis between mental and physical labor"
3. A change in the character of labor such that it "has become not only a means of life but life's prime want"
4. An increase in the productive forces concomitant with the "all-round development of the individual"
5. More abundant flow from "all the springs of cooperative wealth"

Let us begin with the first of these preconditions.

Notice that Marx does not say that the division of labor will be abolished; instead, he merely says that the enslaving subordination of the individual to the division of labor will have to go. Elsewhere, however, he does seem to suggest that the division of labor will be abolished in $PC_2$ society.[5] If the nasty connotations of the term 'division of labor' are to be retained, another term is needed to describe what will persist. John McMurtry provides a useful contrast and a new term that will help to describe the organization of labor in $PC_2$ society:

> We must avoid here the confusion, traditional since Adam Smith, of division of task and division of labor. Marx regarded the former as a principal technique of advanced productive forces, the latter as a principal form of ruling-class economic oppression. In communist society, division of task (for example, assembly-line method) is to be retained, while division of labor (confinement of individuals to one 'exclusive sphere of activity') is not.[6]

What is behind this distinction is the fact that some aspects of the organization of labor are technologically determined; PC society is not going to change that. But, people will not be riveted to one repetitive task for their entire working lives.

The second feature of $PC_2$ society listed above states that the antithesis between mental and physical labor will be overcome. What does this amount to? Surely it would be implausible to maintain that all jobs in $PC_2$ society will be mentally challenging. McMurtry, for example, implicitly admits this point when he says that the assembly line will persist in $PC_2$ society. (Anyone who has done assembly line labor will confirm that, among its many benefits, overcoming the antithesis between mental and physical labor is not and never will be one of them.) Note that Marx links this overcoming to the end of the enslaving subordination of the individual to the division of labor. Perhaps what he has in mind is something like the following: People will rotate in and out of various jobs in the economy. Many of these will be mentally challenging, but some won't be. However, since no worker will be riveted to one job, no worker will conceive of himself as an occupier of a particular social role, for example, as a street sweeper or, for that matter, as a rocket scientist. Mental and physical labor will be combined in the productive life of individuals, though not on a minute-by-minute, or even day-by-day, basis. The antithesis between mental and physical labor is overcome not always or necessarily in the work itself but in the individual worker.

The third precondition, that labor is no longer merely a means to life but life's prime want, is one of the most famous aspects of $PC_2$ society. But what does this really mean? First, it does not imply, as some of Marx's critics believe, that all or most labor is *fun*. This much is clear in Marx's criticism of Adam Smith's negative, one-sided conception of labor:

> Certainly, labour obtains its measure from the outside, through the aim to be attained and the obstacles to be overcome in attaining it. But Smith has no inkling whatever that this overcoming of obstacles is in itself a liberating activity. [Labour] becomes attractive work, the individual's self-realization, which in no way means that it becomes mere fun, mere amusement, as Fourier, with *grisette*-like naivete, conceives it. Really free working, e.g., composing, is at the same time precisely the most damned seriousness, the most intense exertion.[7]

This passage not only makes it evident that Marx is not a hedonist; it also provides a picture of his conception of how labor ought to be, his normative conception of human labor if you will. Marx's statement of this precondition suggests that much labor in $PC_2$ society does not have purely instrumental value; since people will do it, it must be because it is intrinsically rewarding. Only by making this supposition is it reasonable to suppose that $PC_2$ society will get "from each according to his abilities." Indeed, each of the first three preconditions is probably a necessary condition for contribution according to ability. Otherwise, the bourgeois objection that the principle, "From each according to his abilities, to each according to his needs," encourages people to have very small abilities and very large needs has some bite, at least on the abilities side.

The fourth and fifth preconditions predict a substantial increase in material wealth for $PC_2$ society. But how much? Presumably, it will have to have increased to the point where each person can receive "according to his needs." But what is this level? The answer is not immediately obvious. Indeed, it is not at all obvious what it means to say that distribution will be according to need.

## Distribution in Post-Capitalist Society: The Second Phase

What follows is an attempt to explicate the social instantiation of the distributive principle of $PC_2$ society, that is, to explain what society would be like when distribution is in accordance with need. This turns out to be a much more difficult and complicated task than one might initially suppose. However, carrying it out is obviously crucial for outlining Marx's vision of $PC_2$ society, since the mode of distribution is part of the economic system of that society. As will become apparent, it is difficult to say anything very definite about the absolute level of material wealth the instantiation of this principle presupposes, so the fourth and fifth preconditions must remain somewhat indeterminate.

The first thing to notice is that Marx is not saying that everyone's needs, all of them, will be met. The reason for this is straightforward: Some human needs, such as love and companionship, cannot really be met by the production of use-values. An economic system cannot solve all problems, and the relevant distributive principle is a distributive principle for an economic system. What an economic system does is to produce use-values. So the distributive principle really says that distribution will be in accordance with needs for use-values.

But what are use-values? How are they to be distinguished from other "needs-satisfiers?" All commodities (i.e., things that are produced for sale) are use-values, but since PC society does not produce commodities, it will not do to define use-values as commodities. To avoid this problem, it is easy enough to define use-values as things (physical objects, capacities, powers) that (i) meet human wants or needs and (ii) could be bought or sold, whether or not they are in fact produced for sale. Use-values, then, are by definition *exchangeable*. (Notice that labor power is a use-value on this definition.) Hereafter, let us call human needs that can be satisfied by use-values, 'use-value needs'.

Distribution in $PC_2$ society, then, will be in accordance with people's use-value needs. But what does 'in accordance with' mean? The most natural way to understand this is in terms of a straight proportionality, which, in principle, could be significantly less than unity. The greater a person's needs, the more use-values he has. If this is true of everyone, then Marx is to be interpreted as an egalitarian of some sort. In support of this interpretation, it might be claimed that Marx maintains that the chief advantage of distribution in $PC_2$ society over distribution in $PC_1$ society is that the former achieves an egalitarian distribution lacking in the latter. In the relevant part of Comment 3 in the *Critique of the Gotha Program* Marx points out that the distributive principle for $PC_1$ society—payment according to labor contribution—has some "defects." It

> tacitly recognizes unequal individual endowment and thus productive capacity as natural privileges.[8] ... Further, one worker is married, another not; one has more children than another, and so on and so forth. Thus, with an equal performance of labour, and hence an equal share in the social consumption fund, one will in fact receive more than another, one will be richer than another, and so on. To avoid all these defects, right instead of being equal would have to be unequal. [*CGP*, p. 17.]

It seems that the "defects" consist in inequality in needs satisfaction in a society where distribution is in accordance with labor contribution; these defects are to be cured in $PC_2$ society (though as he makes clear at *CGP*, p. 18, not by an appropriate assignment of rights). However, there is another way to read this passage which does not implicitly stress an egalitarian contrast between the two societies. The curing of the "defects" alluded to might mean that $PC_2$ society will be a society in which everyone's needs for use-values could be met. In short, $PC_2$ society will have achieved material abundance. The worker with more children will not have unmet household needs, even though he works just as many hours as an unmarried worker. On this interpretation, Marx is not predicting a *rationing* of use-values according to need but instead a situation in which rationing has been overcome.[9]

The text does not adjudicate decisively between these two interpretations, but it is not clear that in the final analysis it matters very much. On the "egalitarian" reading, it is obvious that Marx is predicting much more than mere proportionality between use-value needs and the satisfaction of such needs; he is also predicting that there will not be much if any gap between what people need in the way of use-values and the use-values they actually have. That is, material abundance, in some sense, will have been achieved. That is what the predicted growth in the forces of production is supposed to make possible. For this reason, I shall interpret Marx's claim that distribution of use-values in $PC_2$ society will be according to need as a prediction that $PC_2$ society will have achieved material abundance.

But how is material abundance to be understood? This is perhaps more a substantive question than an interpretive one. The aim of what follows is to make clear something to which Marx himself did not seem to have given a great deal of thought. It therefore goes beyond strict interpretation to what is the most plausible reconstruction of his position. That said, I think it does capture how he might have thought in the sense that it seems to be faithful to his vision of $PC_2$ society.

To get a handle on this, let us consider an obvious objection a bourgeois critic might make to this distributive principle. When the apologist for the bourgeoisie tries to imagine abundance, his brain overheats with images of fleets of luxury cars, yachts, and so forth. This in turn leads to the hasty judgment that the vision of $PC_2$ society is unrealizable because it presupposes a superabundance of consumer goods.

The main problem with this objection is that it conflates needs and wants. When Marx says that "labor has become not only a means of life," he implies that labor is a means to meeting people's needs, what they need to live; he does not say or imply that labor is a means to meeting every *want* a person might have. Perhaps more importantly, the distributive principle says that people will get use-values according to their needs, not their wants.

At this juncture it might be useful to distinguish a state of abundance from a state of superabundance. Let us say that a society enjoys material *abundance* (hereafter let this be called 'Marxian abundance') just in case its inhabitants' use-value needs can be satisfied. Similarly, a society enjoys *superabundance* if and only if all of its inhabitants' use-value needs and wants can be satisfied. Marx is to be interpreted, then, as saying that $PC_2$ society will enjoy abundance, not superabundance. Clearly, this response to the bourgeois critic turns on a distinction between use-value needs and use-value wants.

How, then, can this distinction be drawn? As a first approximation, it might be said that use-value needs are needs that must be satisfied in order for a person to live. Although this suggests a stringent biological conception of need, it is not necessary to give it such a narrow reading. Marx certainly believes,[10] as indeed does anyone who has given the matter serious thought, that such needs have a cultural and historical overlay. That said, it is still possible to distinguish use-value needs in terms of what a person needs to live—provided one is prepared to answer the question 'To live how?'.

The answer to this question that perhaps best accords with Marx's vision of $PC_2$ society is 'To live the good life'. Clearly, this involves a normative conception of human life, that is, a conception of how human beings ought to live, or how they would want to live under the right circumstances. On the face of it, this does not appear to say very much, or very much that is clear, but it is a step in the right direction. Here's why.

Post-capitalist society, especially in its second phase, is the good society. Presumably, what makes it a good society is that most of its members are leading the good life. The economic system is an "enabling condition" of that. It is supposed to provide the material preconditions for its members to lead the good life, however the latter is to be understood. That is what Marxian abundance is *for*. To meet people's use-value needs, then, is to provide them with the material preconditions for the good life.

Marxian abundance does not guarantee that everyone will in fact lead the good life. Living the good life is not simply a matter of having a certain bundle of use-values, as some apologists for the bourgeoisie would have it. Enriching human relationships, self-realization through labor, aesthetic appreciation, an appropriate relationship with nature, to name just a few, could be components of, indeed necessary conditions for, the good life. But so too are certain use-values. As an aside, the fact that there are other components to the good life suggests ways in which people can fail to achieve the good life, even in the good society. For example, a person might experience unrequited love or come to have a deep appreciation for rock music. The good society is not paradise.

However, what can be asked of the economic system—and what it must deliver if it is to be the economic system of the good society as Marx envisioned it—is that it supplies whatever use-values are necessary for the good life for its members. Clearly there are some desires for use-values, the satisfaction of which are not necessary for the good life, however the latter is conceived. Indeed, all of us occasionally have such desires.

The above provides enough of a purchase on the concept of a use-value need to show that there is a difference between it and the concept of a use-value want. This means that the bourgeois critic cannot charge that the Marxian conception of abundance implies that every use-value want must be satisfied. More importantly, if all of the above is right, it follows that Marxian abundance has been achieved (and, by definition, distribution is in accordance with need) when and only when society meets the material preconditions for the good life for all of its members. Absent an account of the good life in $PC_2$ society, this does not seem to provide a very clear picture of what Marxian abundance is. Though this issue will receive more detailed attention in the next chapter, something definite can be said without much agonizing over what constitutes the good life in $PC_2$ society.

As the term suggests, the material preconditions of the good life (hereafter referred to as the MPCs) are those material conditions without which the achievement or creation of the good life is not possible or at least is extremely unlikely. Some of these are patently obvious: adequate food, clothing, medical care, and housing come readily to mind. Clearly, some people can achieve the good life without some of these, but just as obviously most people cannot. Moreover, the good life should not be confused with a life worth living, as the latter term is used by contemporary medical ethicists. A life worth living is a life a person would prefer having to being dead. Every society has provided the material preconditions for a life worth living for the vast majority of its members. What previously existing societies, especially capitalist societies, have not provided are the material preconditions for the good life for most of its members.

Notice that this definition of Marxian abundance does not require that each and every use-value need that a person has is met exactly when it arises. Suppose Joe Communist's fearless dog has chewed up Joe's size 14EEE shoes, and the economic plan does not call for the delivery of shoes of this size to the stores in Joe's area until next week. The need for shoes is a genuine use-value need, but if Joe has to make due for a week, he is not being denied the MPCs, and society has not slipped from a state of Marxian abundance.

This is as it should be, since one does not want a conception of abundance that is so strict that it permits no failure or inefficiency whatever. All economic systems operated by human beings are less than absolutely perfect. Furthermore, abundance should be conceived of as a perduring condition of a society, not as something that comes and goes according to fortuitous circumstances. Whereas it is completely unreasonable to expect that no member of $PC_2$ society will ever have an unmet use-value need for any period of time, it is much more plausible to maintain, and thus to interpret Marx as saying, that

$PC_2$ society will at least guarantee the MPCs for all of its members. Only then can society start inscribing banners.

We have, then, an interpretation of Marxian abundance, that is, distribution according to need: The MPCs are guaranteed for all of society's members. However, the analysis is not yet complete; what, after all, does this require? That is, what conditions must be met for a society to achieve this state? Given the fourth and fifth preconditions listed above, it is obvious that a certain level of development of the productive forces is presupposed. The state of the development of the forces of production must be such that it is feasible for the MPCs to be met for all of society's members. Let this be called 'PF feasibility'. Though this is primarily a question of what is technologically feasible, the forces of production include more than technology. Labor power, including skills, and means of production are also productive forces.[11]

Although PF feasibility might be a necessary condition for Marxian abundance, it is not sufficient. Here's why: Suppose the economic system of a society maintains a level of development of the productive forces sufficient to meet the MPCs for all of its members, but it does so, and can do so, only if it systematically prevents some members of society from having the MPCs. Under these circumstances, it seems natural to say that abundance would be possible for that society but not actual. More exactly, the society would not meet, and indeed would not really be capable of meeting, the MPCs for all of its members. It seems that a certain level of productive force development is necessary, but not sufficient, for Marxian abundance.

What else might be necessary? The above counterexample presumes the existence of institutional constraints in the economic system that prevent the possibility of abundance from becoming an actuality. To fill out the counterexample a bit, one might imagine that the economic system creates incentives for individuals to put themselves first, or at least to ignore the wants and needs of others. It is this incentive structure that blocks the actualization of Marxian abundance in the counterexample but at the same time makes the satisfaction of the MPCs PF feasible. This point touches on a larger issue that warrants a brief digression.

One way, and maybe the only way, to understand how an economic system functions is to comprehend the incentive structure it creates. All economic systems have some such structure.[12] Capitalism, for example, encourages certain kinds of actions and discourages other kinds of actions in virtue of its constitutive social roles and how it distributes wealth and income. For example, people who fill the role of capitalist are encouraged to drive a hard bargain with workers, suppliers, and so forth, and are discouraged from caring about the effects of their actions on those who cannot help or hurt them financially. In like manner, those who fill the role of proletarian are encouraged to bargain for the most pay they can get and are discouraged from working any harder than is absolutely necessary. Actions that are encouraged will tend to become widespread and actions that are discouraged will tend to become comparatively rare. Moreover, these behaviors will tend to create, or at least bring to the fore, individuals who have the character traits that produce the favored behaviors. This picture is implicit in Marx's oft-stated contention that socioeconomic systems shape human nature. It is also implicit in most of the critical explanations of alienation detailed in Chapter 2.

The above can be tied into the notion of Marxian abundance in the following manner: Marxian abundance is achieved only if the incentive structure implicit in the constitutive social roles and the distribution principle of the economic system does not prevent the society from meeting the MPCs for all. In other words, the economic system does not stand between what is PF feasible for the society to achieve and what it does achieve with regard to the MPCs. This suggests the following definition of Marxian abundance:

> A society has achieved Marxian abundance if and only if
> i. The state of the development of the forces of production is such that it is PF feasible to meet the MPCs for all of its members.
> AND ii. The incentive structure implicit in the constitutive social roles and the distribution principle of the economic system does not prevent the society from meeting the MPCs for all of its members.

A shorthand way of characterizing condition (ii) is that it is economically feasible to meet the MPCs for everyone. Together, these two conditions make explicit the intuitively plausible idea that successfully meeting the MPCs is a matter of having the means and the will to do the job. The forces provide the means and the incentive structure provides the will. Ordinary human error will prevent the timely satisfaction of each and every use-value need that a person has, but the economic system will not be systematically preventing the satisfaction of the MPCs for all.

However, this definition is not quite right. One problem with it is that it contains a hidden circularity. Suppose that $PC_2$ society achieves Marxian abundance. The difficulty is that the distribution principle, which figures in the definiens, 'To each according to his needs', is itself defined in terms of Marxian abundance. This implies that $PC_2$ has achieved Marxian abundance only if it has achieved Marxian abundance, an obvious circularity.

This problem can be solved, however, by judiciously inserting different temporal references in the definiens and the definiendum to create an inductive definition. To put the point informally, a society can be said to have achieved Marxian abundance during a certain time frame just in case the forces of production are sufficiently developed during that time frame *and* the incentive system: (a) is "right" during that time frame and (b) was "right" just prior to that time frame. More formally,

> A society has achieved Marxian abundance throughout $t$ (where $t$ ranges over time periods of indeterminate length) if and only if
> i. The state of the development of the forces of production is

such that throughout $t$ it is PF feasible to meet the MPCs for all of society's members.

AND ii. The incentive structure implicit in the constitutive social roles and the distribution principle(s) of the economic system throughout the period $t$ and throughout the period $t - n$ (respectively) does not prevent the society from meeting the MPCs for all of its members.

Assuming $PC_2$ society achieves Marxian abundance, at some point, the incentive structure referred to in (ii) is the incentive structure of $PC_1$ society, and the latter does not covertly presuppose Marxian abundance.

There is, however, a more serious problem with this definition; it is too weak for the following reason: Suppose that a society meets both of the above conditions, but it also has other institutions, for example, the state or the family, which tend to frustrate systematically the satisfaction of important use-value needs. Society might be so far from satisfying enough of these needs that it cannot truly be said to have provided the MPCs for all, that is, it has not achieved Marxian abundance. Indeed, Marx himself teaches that society's institutions are highly interdependent. In other words, it may be that the economic system is not preventing the satisfaction of the MPCs for everyone, and yet society itself is not capable of meeting these needs because of the systematic interference of other institutions.

The definition of Marxian abundance can be easily amended to rule this out by adding the following condition:

AND iii. No other social institution systematically prevents the satisfaction of the MPCs for any of society's members.

These three conditions are, I believe, necessary and sufficient for Marxian abundance. Notice that they do not jointly guarantee that all use-value needs will be met in a timely manner. Ordinary human frailty will insure that this will not occur, but the fault will not lie in the society's institutions, and the failing will not be so serious that anyone will be denied the MPCs. Indeed, it is arguable, in the spirit of Marx, that if the society's main institutions are in order, much of what is ordinarily attributed to human frailty will not occur, though it would be implausible to maintain that communist society will create a race of superbeings who never make mistakes.

The purpose of the forgoing explication of the concept of Marxian abundance is to provide an understanding of what things would be like if distribution were in accordance with need. This is independently important for understanding the economic system of $PC_2$ society; it is also necessary for getting some idea of how to interpret the fourth and fifth preconditions for the transition to that society. It remains unclear just how much growth the forces must undergo and just how much unclogging the springs of social wealth require, but it is evident that Marx envisions quite a bit in both respects.

By now the thread of the argument has dropped from view, so a brief interim summary of this section and the last two might be helpful. These sections have proceeded on the tacit assumption that what is going on in the relevant passages of the *Critique of the Gotha Program* is science and not idle speculation on Marx's part. Central to Marx's vision of both phases of PC society are conceptions of the distribution of income, or more exactly, use-values. The purpose of these sections has been to articulate these conceptions and to reconstruct how Marx might have argued for them. Distribution in the first phase of PC society is in accordance with labor contribution. The argument for this is, roughly, that this mode of distribution is functionally necessary for reproduction in $PC_1$ society.

In $PC_2$ society, distribution is in accordance with need and contribution is in accordance with ability. The arguments for these claims are based on the tacit assumption that PC relations of production hold, together with some explicitly stated facts about the social and natural environment of $PC_2$ society, namely, the five preconditions discussed above.

For contribution to be in accordance with ability, Marx posits the end of the enslaving subordination of the individual to the division of labor, the overcoming of the antithesis between mental and physical labor, and a change in the character of labor such that it has become not only a means of life but life's prime want. Contribution in accordance with ability is supposed to follow from this.

For distribution to be in accordance with need, Marx posits an increase in the productive forces, concomitant with the all-round development of the individual, and a more abundant flow from all the springs of cooperative wealth. This was interpreted to mean that the level of material wealth will be so high that $PC_2$ society will have achieved abundance. The concept of (Marxian) abundance was explicated in such a way that it does not presuppose that every use-value want would be satisfied, or even that every use-value need would be satisfied exactly when it arose. This conception of Marxian abundance provides a way to understand what it means to say that a society distributes use-values according to need. A clear distinction between wants and needs was not provided, but the concept of the material preconditions for the good life was introduced to show that some such distinction can be drawn. What remains to be done in this section is to say something about why Marx might have thought the preconditions for $PC_2$ society will eventually be reached by way of $PC_1$ society. This is how the Transition requirement identified in Chapter 1 might be met for the transition from $PC_1$ to $PC_2$ society. Moreover, Marx clearly believed (correctly, if the argument of the preceding section is right) that these preconditions are necessary for the realizability of $PC_2$ society.

Unfortunately, in the *Critique of the Gotha Program* Marx says nothing about this; the five preconditions are simply sprung on the reader after Marx's discussion of the defects of distribution in $PC_1$ society. It is not clear to me just how much of an argument can be given for the claim that these five preconditions will be realized by the close of $PC_1$ society. Perhaps the most that can be done is to provide a small sketch of how this might develop. Let us begin with the last two preconditions, the ones that are supposed to produce Marxian abundance.

Presumably, central planning is supposed to eliminate all of the inefficiencies of capitalism associated with the "anarchy of production" that characterizes the latter. Crises of overproduction, which result in vast quantities of unsold goods, huge layoffs, and so forth, will disappear when production proceeds on a more rational, that is, planned basis. The returns to capital, or more accurately, returns to capitalists, will be eliminated. Furthermore, when production is not guided by the "boundless thirst for surplus labor," no one will have vested interest in stimulating the demand for use-values, especially use-values that do not meet needs. In consequence, use-values will be produced only to meet genuine human needs. Couple these efficiency gains with the continued growth and development of the forces of production, and Marxian abundance, as defined above, will become a reality.

I think it is fair to say that Marx envisions the subsidence of the economic sphere as these developments unfold. Acquiring and keeping use-values becomes less and less of a concern. Most use-value needs are met more or less automatically. At this point it is reasonable to suppose that getting use-values in exchange for labor is no longer the primary *raison d'être* of the activity of laboring. Indeed, perhaps it is at this point that the whole idea of *exchanging* labor for use-values is overcome. People labor, and they get use-values, but they do not conceive of laboring as a quid pro quo for use-values. Why, then, would they continue to labor?

The short answer is that labor has become life's chief want, something people do for its own sake. This state of affairs is not unprecedented, since there are models of this in contemporary capitalist society. Though labor may be alienating and hateful for many people under capitalism, it is not that way for everyone. To idealize just a little, some university professors, for example, do what they do because they find the work intrinsically rewarding. They get use-values for it in the form of a paycheck, but they do not conceive of their labor as something that is done for exchange value. Of course, most of them would not continue to labor if they did not get paid, but that is a reflection of their alternatives more than anything else. The main point is that work bears a much more tenuous relation to pay in their lives than it does for most people in capitalist society.

Given this model, is Marx saying that this is the way it will be for everyone in PC$_2$ society? The model has two salient features: (1) The work itself is intrinsically rewarding; and (2) Labor is not conceived of as a quid pro quo for use-values. It is probably implausible that (1) always holds for everyone at all times, if only because it is possible to imagine tasks that must be performed that are not, and could never be, intrinsically rewarding. Fourier notwithstanding, very few people will find cleaning sewers and drains enjoyable or intrinsically rewarding. Marx seems aware of this fact of the human condition. In *Capital* III he says,

> just as the savage must wrestle with Nature to satisfy his wants, to maintain
> and reproduce life, so must civilized man, and he must do so in all social
> formations and under all possible modes of production. . . . and achieving this
> with the least expenditure of energy and under conditions most favourable
> to, and worthy of, their human nature. But it nonetheless still remains a
> realm of necessity. Beyond it begins that development of human energy which
> is an end in itself, the true realm of human freedom. [*Capital* III, p. 820]

This passage implies that there will remain some human labor that is not
intrinsically valuable. However, the attitude toward this labor might well be
in accordance with (2) above: People do this necessary labor but do not
conceive of it as a quid pro quo for use-values. How could this be? Maybe
the idea is that the realm of necessity is sufficiently small, the recognition of
this realm is sufficiently clear, and the concern for the well-being of all is
sufficiently great that this work gets done. It's like taking out the trash or
mowing the lawn: It's a small matter, the whole family benefits, and someone
has to do it. Despite occasional squabbles and some whining and complaining,
it usually gets done. This should not obscure the fact that most labor will not
be like this. Much of the production of use-values will require an intrinsically
satisfying exercise of talents and abilities, and that will be the *raison d'être*
of most labor.

The above paragraphs constitute a description of what labor might be like,
given a sufficient level of development of the forces of production, but they
do not count for much as an *argument* for Marx's view that the first three
preconditions will characterize the transition from $PC_1$ to $PC_2$ society. After
all, there is no guarantee that the sphere of necessary labor will shrink quite
as much as Marx seems to think. In addition, it is commonly known that
technology brings with it problems as well as benefits; the merry development
of the forces of production may not be as unalloyed a benefit as Marx seems
to have supposed. Perhaps more importantly, the argument that central planning will result in significant efficiency gains and dramatic growth in the forces
of production is really quite weak. Central planning, while preventing some
of the inefficiencies of capitalism, may have significant inefficiencies peculiar
to it. Marx gives no reason to deny this.

At most, what has been described above is more like a picture of how
things might go. And there is a genuine defect at this juncture in Marx's
vision of PC society insofar as he really does need to substantiate the claim
that the five preconditions will come to describe what life will be like at the
threshold of $PC_2$ society—he needs this to justify his claim that the latter will
be described by the contribution-distribution principle, "From each according
to his abilities, to each according to his needs." Moreover, from the perspective of the demands for a successful radical critique, he needs to substantiate the five preconditions, if $PC_2$ society is to serve as the ultimate
standard of comparison against which capitalist society is to be judged. $PC_1$
society has much to recommend it vis-à-vis capitalism, but the truly radical
nature of Marx's critique of capitalist society cannot be appreciated without
the vision of $PC_2$ society—and that vision clearly assumes that the five preconditions have been met.

But the critic makes no great discovery on this score. It is virtually inevitable that the paucity of Marx's discussion of PC society will make itself felt somewhere. Here is as likely a place as any. This brings us back to the question of where the burden of proof lies. Marx does have a burden of proof that he has not discharged and that I have been unable to discharge for him. But this may be merely an indication of the limits of my ability to reconstruct Marx rather than a sign of an irremedial problem for Marx's vision of PC society. Marx's supporters may be able to fill in the gaps.

Despite these gaps, the above is of considerable value in that it provides many of the details of Marx's vision of PC society. Most significantly, the above account of distribution in both phases of PC society provides the last piece of the puzzle in the interpretation and reconstruction of Marx's vision of the economic system of PC society. Recall that an economic system consists of (1) a (dominant) type of relations of production, (2) a societywide method of production coordination, and (3) a mode of distribution of income (consumption goods). Chapter 6 offers interpretive (arguments and supporting textual evidence with respect to (1) and (2); the first and third sections of this chapter are directed at (3).

There is more, however, to Marx's vision of PC society. The economic system is not the only institution in PC society. In the first phase, anyway, there is a state, and the theory of history maintains that the state embodies key elements of a normative belief system ("ideologies," for capitalist and pre-capitalist societies). What can be said about the state and the related normative belief systems in $PC_1$ society? Finally, Marx maintains that the state will wither away by the time the threshold to $PC_2$ society is crossed. The purpose of the next section is to reconstruct Marx's views on the state and some of the associated normative beliefs in both phases of PC society.

The final two sections of this chapter round out Marx's vision of PC society by saying something about what life would be like in a world in which exploitation and alienation have been virtually eliminated. It is clear that Marx believed that the absence of these social ills would characterize PC society and that this would be a consequence of the major institutions of that society (or the absence thereof in the case of the state in $PC_2$ society); after all, he had nothing but contempt for those reformers who believed that these defects of capitalist society can be significantly ameliorated without radical institutional change. These sections provide the beginnings of a sketch of the good life to be found in PC society and some arguments for the proposition that this will be a consequence of the institutions of PC society.

### The State in Post-Capitalist Society

The main topic of this section is the state and its demise in PC society. Something will also be said about the associated normative belief systems. This is possible because the theory of history teaches that such systems serve to support the existing order (which in this case can be interpreted to be the economic and political systems) and systematically reflect the interests of the ruling class. An account of these systems is not strictly necessary for the purposes of a radical critique, but parts of such an account can indirectly indicate features the state must have in $PC_1$ society, and it can enhance the explanation of the demise of the state for $PC_2$ society.

In contrast to his discussion of the state in capitalist society, Marx has very little to say about the former in $PC_1$ society. However, what follows in this section is not merely speculative. In the *Critique of the Gotha Program* Marx offers some advice on how to proceed:

> What transformation will the state undergo in communist society? In other words, what social functions will remain in existence there that are analogous to present state functions? This question can only be answered scientifically, and one does not get a flea-hop nearer to the problem by a thousandfold combination of the word people with the word state. [*CGP*, p. 26]

What follows is a brief comment about the revolutionary dictatorship of the proletariat and then five paragraphs of complaining about the fact that the authors of the Gotha Program do not address this question scientifically— and that is all. In short, after telling the reader how this question is to be answered, Marx fails to answer it! But, at least he offers a fruitful suggestion about how to proceed: We are to consider the functions of the bourgeois state and ask how, and indeed whether, these functions will be discharged by the proletarian state.

The theory of history provides additional resources for a reconstruction of some of the details of, and arguments for, Marx's view of the state in PC society. I shall assume that whatever the correct interpretation of the theory of history is, Marx subscribes to the view that the economic system of a society imposes significant constraints on the nature of the state. In a more positive vein (and without deciding the issue of causation) it is possible to infer certain features that the state will have, given the economic system. Since a fairly complete picture of the economic system of both phases of PC society is in hand, there should be ample resources for a reconstruction of Marx's views on functions of the state in $PC_1$ society and his argument for its demise with the advent of $PC_2$ society.

Although different theorists operate with different conceptions of the state, there is some measure of agreement about a core concept, which, following Max Weber, can be roughly defined as follows: The state is that institution that tries, and in large measure succeeds, in being the sole authorizer and user of organized coercive power within a geographical area.[13]

But for what ends or purposes? As noted in Chapter 2, there are suggestions of two theories of the state in Marx's writings. On the first theory, the state is the executive committee of the ruling class whose purpose is to manage the common affairs of the latter.[14] As he says in *The German Ideology*, the state is "the form in which the individuals of the ruling class assert their common interests and in which the whole civil society of an epoch is epitomized" (*GI*, MECW, vol. 5, p. 90).

On Marx's other theory of the state, the latter is not merely a creature of the ruling class. Although its primary purpose is to secure the long-term interests of the ruling class, this may require sacrificing the short-term interests of the latter in particular circumstances. For example, under capitalism, the state had to pass the Ten Hours bill to curb the rapaciousness of capitalists. This was against the short-term interests of profit-hungry capitalists but served their long-term interests. However, the autonomy necessary for this sort of strategic action makes it possible for the bureaucracy to cultivate and further its own interests and agenda, at least to the extent that those interests and that agenda do not fundamentally conflict with those of the ruling class.

How do these theories fare when applied to PC society? Regarding the first theory, if the state is an instrument of the ruling class, then it will serve the interests of the proletariat in the first phase of PC society, since they are slated to be the ruling class. It is in this connection that Marx speaks of the "revolutionary dictatorship of the proletariat" (*CGP*, p. 26). In addition, it is easy to see why Marx would predict the end of the state in the second phase of PC society; if at that point there are no classes, then there is no ruling class. In consequence, there is nothing for the state to do. It "withers away."

The second theory appears to create complications for the account of $PC_1$ society. Presumably, Marx would say that this second theory does not apply to the latter. But why? Why can't the state become a semiautonomous actor with its own interests and agenda? He might claim that there are certain preconditions for the state to become partially alienated from the ruling class and that these preconditions will not be met in $PC_1$ society. For example, maybe some form of systemic alienation in the economic system is a necessary condition for this to happen. When the proletariat seize power, this alienation disappears. Another possibility is that the short-term and long-term interests of the ruling class always coincide in $PC_1$ society (an admittedly implausible assumption), and the state does not need the relative autonomy it must have under capitalism. Finally, it may be that Marx believed that the proletariat, mindful of this potential problem, would create institutional mechanisms to prevent this separation of the state from its class base in $PC_1$ society. There is a suggestion of this in his praise of some features of the Paris Commune, such as instant recall of public officials.

In any case, it is reasonably clear that Marx believed that the state would faithfully serve the interests of the proletariat in $PC_1$ society. That much is suggested in the following passage from the *Critique of the Gotha Program*: "Freedom consists in converting the state from an organ superimposed upon society into one completely subordinate to it" (*CGP*, p. 25). However, to say that the state will serve the interests of the proletariat does not tell us very much about what the state will do if only because it does not say what those interests are. But it is not difficult to spell out some of those interests.

The most fundamental interest would be the preservation of the new order, and Marx is quite clear that one task of the state is to deal with potential counterrevolutionaries who might threaten that order, notably former bourgeoisie.[15] There is an obvious parallel in this connection with the capitalist state: The latter exhibits unremitting hostility to advocates of revolutionary social change, at least insofar as their advocacy has some chance of success. (Otherwise, they engage in Marcusian "repressive tolerance.") Comparable hostility from the proletarian state can be expected to be directed at actual and potential counterrevolutionaries.

From a Marxist perspective, perhaps the main purpose of the bourgeois state is to preserve and protect the capitalist economic system. The methods by which this is achieved are various, ranging from keeping capital gains taxes low to strikebreaking. At the most basic level, however, the state enacts and enforces specific legal *rights*, which guarantee bourgeois control over the means of production and insure the continued operation of the market system.

On the basis of these considerations, it is a plausible inference that the proletarian state would act to sustain the economic system of $PC_1$ society. Recall that the economic system consists of three elements: (i) a predominant set of relations of production (worker control of the means of production, the de-commodification of labor power), (ii) a method of production coordination (central planning), and (iii) a system of distribution of income (payment according to labor contribution). In $PC_1$ society, none of these would "take care of themselves," as it were. The reasons for this are various.

It cannot be assumed that PC relations of production will be self-sustaining for at least two reasons. First, the workers themselves will lack experience in the tasks of management; the expertise of some of the bosses will be needed (as Lenin discovered in the aftermath of the Bolshevik revolution), and the atrophy of worker control of the means of production, or even the failure to institute it, will be a real possibility. Secondly, there will probably be incentives for some workers, namely, those with scarce skills, to sell their labor power as a commodity, on the sly, as it were. This will be especially true of workers who had earned high wages under capitalism.

Payment according to labor contribution cannot be assumed either. Recall that the workers enter PC society stamped with the birthmarks of capitalism. This means they are used to doing hateful, alienated labor in exchange for wages. The (Benthamite) consciousness of workers under capitalism is such that they want to do as little labor for as much money as possible. Undoubtedly, the revolution itself and the prospects of worker control of the means of production will affect proletarian consciousness about labor and pay, but it is likely that the old attitudes will not disappear immediately. The temptations to goldbrick will be quite strong for some members of the new ruling class. The inductive evidence for this in class societies generally is very impressive.

There is an obvious role for the state in dealing with these potential problems and thereby ensuring PC relations of production: It can act to guarantee worker control of the means of production by enacting and enforcing legal rights to worker control. In addition, it can prohibit or sharply curtail the buying and selling of labor power. Finally, the state can use its coercive power to insure payment according to labor contribution.[16]

Actually, not only can the state use its coercive power to insure payment according to labor contribution, it is arguable that it will have to do so. According to the argument of the first section of this chapter, this method of distribution is a necessary condition for the stability of the economic system. This stability may well have all the characteristics of a public good, in Mancur Olson's technical sense of the term.[17] The benefits (stability of the economic system) are non-excludable in that they are provided to non-contributors, doing what is necessary to provide this good (receiving payment according to labor contribution) is for most individuals a cost, at least until the advent of $PC_2$ society, and the contribution one person can make to the provision of this good will be negligible. As Olson has argued, even an altruist will find it rational not to do her share when these general conditions are met. In consequence, it is likely that the institutional structures that emerge will have to rely on coercion—the classic solution to a public goods problem—to secure the good in question, at least at the front end of PC society.

What about the method of production coordination? Will the state be the agent of central planning in $PC_1$ society? This has been a matter of some controversy. For example, Richard Miller says, "Marx sharply distinguishes the coordinating groups that modern material needs require from state power as such. State power, as against mere coordination, is in itself a denial of freedom, incompatible with a free association in which the full development of each is the condition for the free development of all."[18] The import of the first sentence is unclear; if Miller's claim is that, according to Marx, production can be coordinated (even centrally planned) without a state, then he is surely right; it is obvious that the two are conceptually distinct. Moreover, on Marx's account, $PC_2$ society is stateless and yet has a centrally planned economy. On the other hand, if Miller is saying that Marx never envisioned the state being the agent of production coordination, he needs to cite some supporting textual evidence. Not only does he fail to do this, but there is decisive evidence to the contrary in the *Communist Manifesto*. Marx and Engels say, "the proletariat will use its political supremacy to wrest, by degrees, all capital from the bourgeoisie, *to centralize all instruments of production in the hands of the state*, i.e., of the proletariat organized as the ruling class" (*CM*, MECW, vol. 6, p. 504, emphasis added).

Of course "public power will lose its political character," to use Engels's phrase, with the transition to $PC_2$ society, but it is $PC_1$ society that is under discussion. Formally, the planning apparatus need not be part of the state, but that apparatus must have the threat of force, that is, state power, to back up its assignment of inputs and output targets to individual production units. Central planning is not a matter of making suggestions about production that production units are free to ignore. Although the planning apparatus is ultimately the creature of the working class, the former must have final authority in particular planning decisions vis-à-vis production units, if it is to discharge its function. If there is an irresolvable conflict between what the planners want and what an individual production unit wants, the planners' will must carry the day, which means that they have to be ultimately backed by the threat of force.

It might be objected that this assumes more conflict between the production units and the planning apparatus than is warranted in a worker-controlled society. After all, isn't there a fundamental harmony of interests after the revolution? There are two reasons why this isn't the case: First, many key specialists whose expertise is needed to make production run smoothly will come from the ranks of the former bourgeoisie. It is not unreasonable to suppose that they will have some residual hostility toward the workers and a system of worker control of the means of production. After all, they have just had all their means of production confiscated! Secondly, the workers themselves may be skeptical of the benefits of the new method of production coordination. (This is obviously true if the revolution is to take place in the twenty first century because they will have before them a vivid historical record of how central planning can fail rather dramatically.)

More generally, one cannot assume on Marx's behalf a generalized harmony of interests throughout much of the early history of $PC_1$ society if only because the latter is instituted by a violent revolution in which many lives are lost and much property confiscated. Marx and Engels were practical men who foresaw the practical necessity for coercion in revolutionary and immediate post-revolutionary times. As long as the state exists, it will have to backstop the planning apparatus. This means that the directives of the latter will in effect have the force of law. Although this condition will cease by the time the boundary to $PC_2$ society is crossed, there can be little doubt that the plan for the economy in $PC_1$ society will have the force of law behind it. Whatever the formal arrangements, the state will effectively coordinate production in $PC_1$ society.

Further insight into the tasks the state must discharge can be gotten indirectly from a brief consideration of the dominant normative belief systems of $PC_1$ society. Something about these systems can be inferred from what is known about the requirements of the economic system of that society.

Let us begin with ideas about distributive justice. What will be the commonly accepted principle(s) of distributive justice for $PC_1$ society? Marx says that the actual pattern of distribution of income in $PC_1$ society will be proportional to labor contribution, at least for those who are able to work. This is expressed succinctly in the following passage: "Accordingly, the individual producer receives back from society—after the deductions have been made—exactly what he gives to it" (*CGP*, p. 16). Historical Materialism implies that the dominant ideas about justice will be in accordance with the actual distribution of wealth and income.[19] This permits the inference that the principle of distributive justice that will be widely endorsed in $PC_1$ society will be something like the following: 'Justice requires that an ablebodied adult's income should be proportional to labor contribution'. This is confirmed by what Marx says in the third paragraph following the above quotation: "The right of the producers is *proportional* to the labor they supply" (*CGP*, p. 16). At least since the advent of capitalism, and probably before, the state has conceived of enforcing rights as one of its main tasks.

This principle is somewhat vague, since it does not specify any particular proportion. This is as it should be, since dominant ideas tend to be vague. A further reason for this vagueness can be found in the demands of the economic system. In the third comment on the Gotha Program Marx discusses the Lassallean demand that the workers should be paid the undiminished proceeds of their labor. He immediately points out a number of deductions of unspecified size that must be made. These are summarized in the following list (*CGP*, p. 15):

1. Funds for replacing used-up means of production
2. Funds for augmenting existing means of production, that is, for economic growth
3. An insurance fund to cover losses due to natural disasters, and so forth

Before what is left can be divided among the workers, there must be further deductions, this time for consumption:

4. Funds for the costs of general administration not pertaining to production (e.g., for the criminal justice system)
5. Funds for public goods; Marx mentions schools and health care in this connection
6. Funds for those unable to work—what has been called 'poor relief' in capitalist society

Marx points out (*CGP*, p. 15) that the first three categories are not determined by considerations of equity (i.e., justice). He does not say whether or not such considerations determine the others. Is poor relief a matter of justice or charity? In any case, he does say that the funds expended on the latter three categories directly or indirectly benefit the workers as members of society, so the deductions in these categories are not as much of a net loss as they might appear to be. Actually, much the same could be said about the first three categories as well.

The proposition that income ought to be distributed in proportion to labor contribution, then, is the principle of distributive justice that will be widely subscribed to in $PC_1$ society, and for reasons discussed in the first section of this chapter, that is how distribution will actually occur. There is, then, a kind of pre-established harmony between the demands of justice (as most members of $PC_1$ society understand them) and what actually happens in $PC_1$ society. As was noted in Chapter 5, whether or not the same can be said of capitalism has been a matter of considerable dispute in the secondary literature, but this question is not directly relevant for the purposes at hand.

Corresponding to the state's enforcement of the right to worker control of the means of production and the prohibition or curtailment of the sale of labor power, it is reasonable to posit widespread acceptance of a belief in this right and perhaps negative beliefs or attitudes toward buying and selling labor power, comparable, say, to the "official position" in bourgeois society on prostitution. (One can imagine the gossip over the backyard fence: "It's shocking, I tell you! Tom has been selling his labor power on the side. If his co-workers ever found out. . . . ")

Given the economic system of $PC_1$ society, what else can be inferred about the proletarian state? The capitalist state has an elaborate apparatus to deal with noncriminal conflicts and disputes, namely, the civil justice system. This includes tort law, contract law, and property law. One can only hope that much of this will wither away as soon as most of the lawyers are shot, but it is likely that some of the problems and disputes that civil law deals with will remain. For example, although the character of contract law will change dramatically in a centrally planned economy, there will undoubtedly be disputes about the interpretation of planning directives that will have to be adjudicated. Moreover, as noted above, rights of worker control and the right to income according to labor contribution would have to be codified and enforced. So, it is fair to say that there will have to be some sort of civil justice system in $PC_1$ society.

The above account of the proletarian state is based on the assumption that a primary task of the state is to insure the survival or stability of the economic system of $PC_1$ society. What other functions or purposes must this state discharge? A key element of the capitalist state is the criminal justice system. Such a system would also have to exist in the proletarian state, if only to try counterrevolutionaries. A Marxist theory of criminal behavior under capitalism might be invoked to argue that there will not be much else for the criminal justice system to do, at least after a time. Jeffrie Murphy summarizes a Marxist theory of criminal behavior of the criminologist William Bonger as follows:

> Criminality has two primary sources:
> (1) need and deprivation on the part of disadvantaged members of society, and
> (2) motives of greed and selfishness that are generated and reinforced in competitive capitalistic societies.
>
> Thus criminality is economically based—either directly in the case of crimes from need, or indirectly in the case of crimes growing out of motives or psychological states that are encouraged and developed in capitalistic society.[20]

It is obvious how Marx's theory of alienation undergirds Bonger's theory. If Bonger's theory of criminal behavior is roughly correct, then it is reasonable to expect a dramatic decrease in ordinary crime as $PC_1$ society progresses. It is not necessary to suppose that all criminal behavior can be explained in this manner (or by unrelated psychopathology) to make the inference that the criminal justice system will be much reduced in scope and significance. A possible analogy for the socialist criminal justice system might be the fire protection system in capitalist society. The latter is often manned in part by volunteers, professionals spend much of their time practicing their skills and playing cards, and the entire system consumes relatively few resources. Bonger's theory may not true, and I am certainly unprepared to argue for it.

However, it suggests one way Marxists might think about a criminal justice system for $PC_1$ society.

Are there any other functions of the bourgeois state that would have to be discharged by the proletarian state? Since the American and French revolutions, the bourgeois state has made halting efforts to insured certain political and civil rights for all of its citizens. What of the proletarian state? Marx's only sustained discussion of these rights is in his early essay, "On the Jewish Question." In this essay, Marx offers an analysis of the "rights of man" as they are enumerated in, for example, the French Declaration and various state constitutions in the United States. These include rights of political participation ("the rights of the citizen") but also rights of liberty, security, and property. The discussion is complicated, but the main thrust of it is summarized in the following passage: "None of the so-called rights of man, therefore, go beyond man as a member of civil society, that is, an individual withdrawn into himself, into the confines of his private interests and private caprice, and separated from the community" ("On the Jewish Question," MECW, vol. 3, p. 164).

This passage indicates that Marx's dismissive attitude toward these rights is grounded in his belief that they are appropriate only for certain historically limited circumstances, namely, civil society, that is, capitalism. This implies that, with the passing of capitalist society, such rights will have no function to serve and will no longer be part of the law.

There is something to be said for Marx's views on this matter up to a point: Rights *are* boundary-markers between individuals, which protect the latter from various encroachments by others, including the state. This presupposes the potential for systematic conflicts of interests—conflicts Marx believed would eventually be overcome in PC society. However, there may be room for, and indeed a need for, some of these rights in the first phase of PC society; the latter is, after all, a class society, at least in the derivative sense of having many members with bourgeois consciousness, and class conflict may make rights both valuable and necessary. The *aufhebung* of class conflict in $PC_2$ society may well make rights nugatory, but it is not implausible to suppose that the proletariat will finds rights useful in the first stage of PC society. The right to worker control and the right to payment in accordance with labor contribution have already been mentioned, but rights of political participation and various other civil rights may also be appropriate until the last embers of class conflict have died out. If this is so, the normative belief system that predominates in $PC_1$ society will include beliefs that people have these rights.

This completes my account of the functions of the state in $PC_1$ society. Specifying these functions, however, is not sufficient to meet the demands of the Alternative Institutions requirement. To see what else is required, let us pursue a parallel with the economic system of PC society. In articulating the latter, it is not difficult to identify its main purpose or function: to meet people's use-value needs. The way, or the extent to which, these needs are met is determined by two different distributive principles, corresponding to the two phases of PC society. Clearly, saying that much does not specify the economic system of PC society. It was also necessary to spell out the constitutive social roles of the institution (the relations of production) and explain how the powers associated with these roles will be exercised (the method of production coordination). It seems that something comparable is needed for the political system of $PC_1$ society.

Unfortunately, Marx does not have much to say about either of these questions. It is clear that he is committed to democratic rule, as the following quotation from *The Communist Manifesto* makes evident: "The first step in the revolution by the working class is to raise the proletariat to the position of ruling class, to win the battle of democracy" (*CM*, MECW, vol. 6, p. 504). This commitment to democracy represents a first step in answering the "How question" (i.e., how the state will function), but it is only a first step. Democratic rule is rule by the people, but what does that amount to? This has been a matter of considerable controversy, to say the least, among Marx's followers in this century and the last. Marx himself says almost nothing about this question as it pertains to PC society.

A comparable problem arises with respect to the question of constitutive social roles for the state. Obviously, the role of voter or citizen can be identified, but the constitutive roles of the state apparatus itself remain unspecified. What is needed is a blueprint for a political constitution, that is, an abstract description of the constitutive social roles of the political system that could be instantiated by a number of different political constitutions. Marx inveighs against constructing "blueprints" on the grounds that the workers must decide these questions for themselves. But this is an unacceptable dodge for someone who claims to have a scientific understanding of society; even if the workers do decide these questions for themselves, an adequate social theory should permit some guarded or qualified predictions about what decisions they will come to—comparable, say, to what has been inferred in the last chapter and the first two sections of this chapter about the economic system. Marx's theory of the state in $PC_1$ society, then, is very much incomplete. He says almost nothing about the constitutive social roles and he offers no account of the mechanisms by which state power will be wielded.

Comparable problems have not stopped the current project before because it has been possible to reconstruct Marx's views based on other commitments he has. But it will stop us here. I see no way to reconstruct a specification of constitutive social roles of the state, nor can I see a way to articulate a theory of how the state will function in $PC_1$ society. Although this may be due to a lack of ingenuity on my part, there is some reason to think that these tasks cannot be accomplished, at least on the basis of what Marx gives us. If one knows the function or purpose an institution is to serve, one cannot, by that fact alone, infer how that function or purpose will be met, since there are in principle a multitude of ways this could be done. Moreover, none of Marx's other commitments seems to provide a basis for any but the vaguest and most general predictions about the state. This suggests that Marx's vision of PC society is significantly incomplete. The passage of time and advances in the social sciences (especially by Marxists) may make remedying this defect genuinely feasible, but on the basis of what comes above, it is fair to say that Marx has not met the demands of the Alternative Institution requirement for the political system. Needless to say, that does not show that it cannot be met.

Let us turn to the state and $PC_2$ society. The argument for saying that the former will wither away with the advent of $PC_2$ society appears to be fairly straightforward: If the function of the state is to further the interests of the ruling class by the use of organized coercive power and if all classes disappear with the advent of $PC_2$ society, then the state will cease to exist. For this to be at all plausible, it must be argued that the ends that state action served in $PC_1$ society will become nugatory or will be served without the use of organized coercive power. In $PC_1$ society the state serves the following ends:

1. It ensures worker control of the means of production.
2. It prohibits or sharply curtails the commodification of labor power.
3. It ensures that workers' income is determined by labor contribution.
4. It ensures the coordination of production by providing legal backing to the directives of the planning apparatus, even if the latter is not formally an organ of the state.
5. It runs the civil justice system.
6. It runs the criminal justice system.
7. It enforces various political and civil rights.

The main question is, 'Given what is known about $PC_2$ society, is it reasonable to believe that these purposes of state action in $PC_1$ society will become nugatory or will be served by noncoercive means in $PC_2$ society?'.

Regarding the first two, Marx might argue that once worker control and the de-commodification of labor power have been in place for a longish period of time, they will stay that way without the need for coercive enforcement. Given that the existing economic system decisively shapes people's ideas and values, and assuming that the continued operation of this system is in the short- and long-term objective interests of all, it is likely that no one will want it to be any different. Or, at least not enough people will object to make it worthwhile to maintain a state apparatus to deal with those who call into question the way the economic system is organized.

With respect to (3), it is known that $PC_2$ society operates on a different distributive principle, namely, distribution according to need. Given that the material preconditions for the good life are being met for all, and given that labor has become life's chief want, it is hard to see what role there would be for organized coercive power in the distribution of use-values. With regard to (4), the "increased flow from the springs of cooperative wealth," together with the eventual demise of the bourgeois specialists needed to manage production, renders otiose the need for coercive sanctions to back up planning decisions.[22] Planning is essentially a "scientific" question, a fact that should be evident to all at this stage of PC society. Communist society is a "transparent society" in which there are no mystifying social forces shaping people's lives.

What about the civil and criminal justice systems? Despite Lenin's fanciful idea that the armed workers can take care of criminal justice on an *ad hoc* basis, it seems highly implausible that these systems will disappear completely. Regarding the civil justice system, human beings are not omniscient and their individual interests do not and probably never will perfectly coincide in all possible cases. This guarantees misunderstandings and conflicts that will have to be adjudicated, and there are obvious advantages to having an institution of specialists to handle matters of this sort, though it would be important to keep them on a tight leash. In addition, despite the attractions of Bonger's theory of criminal behavior, it is likely that not all crime can be explained by faulty relations of production (or by unrelated psychopathology). Both the criminal and civil justice systems may well continue to shrink as $PC_2$ society gets under way, but their actual extinction is probably unlikely.

Marx's views on the nature and function of political and civil rights imply that these rights are useful only for members of class societies. It may be, as Buchanan argues,[23] that the *freedoms* these rights protect will continue to be enjoyed in $PC_2$ society, but Marx does appear to believe that the need for coercively backed rights to insure those freedoms will disappear when the fundamental conflicts of interest engendered by the existence of classes are resolved.

Assuming the existence of criminal and civil justice systems in $PC_2$ society, does this imply that the state continues to exist? Although the civil justice system may operate in an environment in which the use or threat of force is not thinkable, the criminal justice system cannot be assumed to operate without coercion. And it is organized coercion. Nevertheless, it may be that it should not be called a 'state' because it may make no effort to maintain a monopoly on coercion. If the members of $PC_2$ society are free to discharge it at will or to create an alternative, the existing criminal justice system would not really be a state. Alternatively, the worry about whether or not the state exists under these circumstances might be merely semantic; it is so far from states as they have existed historically that it might be fair to describe $PC_2$ society as stateless, even if that is not technically the case on the Weberian definition of the state given above.

The above considerations reconstruct how Marx and Engels might have argued that the purposes served by the use of state power would become nugatory or served by other methods in $PC_2$ society. To conclude from this that the state will in fact wither away, it is necessary to add a premise to the effect that if an institution serves purposes or functions that will be served in other ways or that become nugatory, then that institution will cease to exist. (This implies that anachronistic social institutions do not exist.) I do not know whether or not this is in general true. If it is, and if all the other claims about $PC_2$ society discussed above are true, then the state will indeed cease to exist.

Since the state disappears in $PC_2$ society, it is obvious that worries about constitutive social roles and institutional mechanisms simply do not arise. Perhaps something has to be said in this connection about the civil and criminal justice systems, but this should not be problematic. These systems as they exist under capitalism can provide some kind of model, though the scope and extent of these systems under the latter is assumed to be much greater than would be found in either phase of PC society.

## Post-Capitalist Society as the Good Society: The End of Exploitation

Most people have some sort of vision, however incomplete, about what constitutes a good society. Sometimes this vision is quite elaborate and detailed, as, for example, in the case of Fourier who went so far as to predict the exact size of communities. Marx and Engels' vision of the good society—PC society in either or both of its phases—is complicated. Bertell Ollman has discovered a surprising wealth of detailed speculation about specific features of PC$_2$ society in the early writings, especially the *1844 Manuscripts*.[24] But, as Ollman's discussion makes clear (perhaps unintentionally), this detailed speculation is unsupported by any argumentation. By the time Marx wrote the *Critique of the Gotha Program*, his official position was much more reticent about the details of PC society. Engels's *Anti-Dühring* has much to say about the proletarian revolution and its immediate aftermath but next to nothing about PC society beyond that. What accounts for this change?

It is hard to be sure, but I suspect that both Marx and Engels had become much more conscious of the scientific character of their theory of history (which was only in its infancy in the early years) and its practical implications. This theory is a theory of social institutions—how they function and the circumstances and conditions of radical institutional change. According to that theory, PC society was not going to be instituted according to the free-form construction of social visionaries. Instead, it would be brought into being by large-scale social forces operating within capitalist society and beyond their direct control. In addition, since the construction of PC society is to be the task of the proletariat, it was politically imprudent for them to say much about how this process would work itself out in the end, even to the extent that they could foresee what would happen. Finally, they might have become aware of the fact that even a very good social scientific theory cannot predict many things about the future. Such a theory can say something about the main institutions of the new society; indeed, that is one of the things social theory is supposed to do. However, many of the details must remain inscrutable. (This last point is probably overly charitable; Marx and Engels were not known for admitting their limitations, even, I suspect, to themselves.)

The reconstruction of Marx's vision of PC society in this book has been undertaken in the spirit of the theory of history in the sense that the focus has been squarely on the social institutions of PC society. But, from the point of view of the demands of a radical critique, there remains one important part of the story that has yet to be told: what social life would be like without exploitation and alienation. Clearly, Marx believed that PC society would do away with these social evils. The absence of these two defects of capitalist society—and what that entails—is in part, and perhaps in large measure, what makes PC society a good society. In the final analysis, that is what the revolution is *for*.

Moreover, if Marx is wrong about this aspect of PC society, that is, if PC society will be beset by exploitation and alienation, his radical critique of capitalism must be pronounced a failure. Even if the main predictions of the theory of history about social institutions are true, the evils of capitalist society would turn out to be permanent features of at least post-feudal society and possibly of the human condition. To forestall this possibility and to round out Marx's vision of PC society, it is necessary to say something about what social life would be like without exploitation and alienation and what justification might be offered for these important claims. The remainder of this section considers the end of exploitation, together with a digression on justice, in PC society. The next section is about the end of alienation.

As the reader will no doubt recall, the discussion of exploitation under capitalism is rather complicated. The elaborate critical arguments can be largely sidestepped. All that matters is that there are a number of different conceptions of exploitation in the production-distribution process that can be more or less plausibly attributed to Marx. A brief review of these conceptions would be useful for reconstructing Marx's views on what life would be like without exploitation and what the relevant arguments would look like.

As noted in Chapter 4, the conception of exploitation that most closely conforms to Marx's—surplus value exploitation—was identified by Nancy Holmstrom as follows: "It is the fact that the [capitalist's] income is derived through forced, unpaid, surplus labor, the product of which the producers do not control, which makes it [capitalism] exploitative."[25] If PC society is to be non-exploitative, it cannot be the case that anyone is getting income derived in this manner. And, indeed, this seems to be the case. It is true that children, old people, and those unable to work are getting income that results from unpaid surplus labor by the workers. However, the last condition cited by Holmstrom is not met, since the producers control the product through their control of the planning apparatus in both phases of PC society.

Surplus value exploitation is just one form of parasite exploitation. There are other forms of parasite exploitation that are possible in the production-distribution process. Would any of those be found in PC society? Recall that the root idea behind all forms of parasite exploitation is a lack of reciprocity between the exploiters and the exploited. This can be cashed out in one of three ways: (i) The exploited are "giving good" for which they are not compensated. This is surplus value exploitation. (ii) The exploiters are "getting good" without "returning good" (i.e., without contributing). (iii) The exploiters are "returning evil for good."

It seems that the second possibility could not be realized in either phase of PC society because the vast majority of people are workers. In PC$_1$ society the state ensures that all who can work are workers, and in PC$_2$ society labor has become "life's chief want." That leaves only the very young, the very old, and the severely and permanently disabled. It is true that some of them would be "getting good" without contributing. However, a mere lack of reciprocity is not a sufficient condition for parasite exploitation. If some sort of forcing is also necessary, it is arguable that the workers are not being forced to make the one-way transfers to these groups, since the former control the social product.

The third possibility—that there are some people who return evil for good in the production-distribution process—also seems unlikely. The relevant evils under capitalism are those associated with alienated labor. Assuming the absence, or virtual absence, of alienated labor in PC society (a topic to be addressed in the next section), there is no reason to believe that some people will be benefiting by forcing others to do alienated labor.

There is one other form of parasite exploitation in capitalist society that Marx would maintain would be absent in PC society—what in Chapter 4 was called 'generic parasite exploitation'. Recall that generic parasite exploitation is the harmful utilization of one person by another for the benefit of the latter. The harms relating to generic exploitation also come from the theory of alienation. These include, among other things, the generalized interpersonal alienation associated with the commercialization of life under capitalism. Assuming the disappearance of these forms of alienation, generic exploitation would also cease.

On the basis of all of the above, it is easy to see how Marx could maintain that there will be no, or virtually no, parasite exploitation in either phase of PC society.

The other basic kind of exploitation (discussed in Chapter 5) is property relations exploitation, of which capitalist exploitation is one species. Recall that Roemer maintains that the proletariat are capitalistically exploited under capitalism because, were they to pull out with their per capita share of alienable assets and their own endowments, they would be better off and the capitalists would be worse off. I argued that this definition had to be amended to require that the institutional arrangements defining the withdrawal conditions are realizable. Is any group (coalition) exploited in this manner in either phase of PC society? The answer depends on whether or not there is a group (in either phase) whose members would be better off under alternative realizable institutional arrangements (as specified above) *and* where the complement of that group would be worse off. It is hard to imagine that this would be the case in PC$_2$ society. But what about PC$_1$ society? Former capitalists may not be better off by pulling out with their per capita shares; after all, they would still have to labor in their alternative. Even if that is not true, proletarians would not be made worse off by the capitalists' withdrawal, so there would be no exploiting coalition and hence no capitalistic exploitation on this score in PC$_1$ society. Nor would the workers be capitalistically exploited in PC$_1$ society. Though for them there is a better historically realizable alternative to PC$_1$ society, to wit, PC$_2$ society, there is no one who would be worse off in that alternative as compared to their situation under $PC_1$ society, assuming, of course, that what Marx believes about $PC_2$ society is true. If all of the above is right, then there is no capitalistic exploitation in $PC_1$ society, and, as indicated above, no such exploitation in $PC_2$ society either.[26]

The above arguments make some large assumptions, assumptions that have not really been supported, about what would actually happen in PC society. These assumptions or presuppositions are highly contentious. However, my main purpose in this section thusfar has been to explain what the absence of various forms of exploitation in PC society would come to and to illustrate how Marx might have argued for that on the basis of his conception of the economic system of PC society. Before doing the same for alienation, it is necessary to say something about distributive justice in PC society.

Would Marx maintain that PC society is a distributively just society? The answer is complicated. Let us begin with $PC_1$ society. In the previous section on the state, it was argued that most members of $PC_1$ society would subscribe to a principle to the effect that 'Justice requires that an ablebodied person's (adult's) income should be proportional to labor contribution'. And, on Marx's view, there would be a legally enforceable right to this. Would Marx endorse this as the correct principle of distributive justice, or correct at least for $PC_1$ society? If Marx was a relativist about justice, the answer would be yes. On the other hand, if he really wants to reject all talk about justice, as sometimes seems to suggest,[27] then he would demur.

But, on the third hand, if he were not a relativist about justice, there are two independent grounds on which he might want to maintain that $PC_1$ society is a just society. And, given all of the assumptions about $PC_1$ society sketched above, this position would be at least defensible. Those two grounds are the following: First, there is no systematic lack of reciprocity between various members of PC society, with the exception perhaps of the severely disabled. (What to say about the expropriated bourgeoisie is unclear but, I suspect, not important.) For a "justice as reciprocity" theory, the requisite reciprocity would be present, and in this respect, $PC_1$ society would be a just society. Secondly, on a "justice as fairness" theory, he could maintain that distributive fairness in control over the means of production is present in $PC_1$ society, so the latter is just in that regard. On the basis of all of the above, it is easy to see how Marx could maintain that PC society is a just society, or at the very least, that it is not unjust. If Marx is given all the assumptions that have to be made, this too is at least a defensible position.

What about $PC_2$ society? To the best of my knowledge, nowhere does Marx describe $PC_2$ society as a just society. At the end of Chapter 5, an argument of Buchanan's was sketched which would explain that. Recall that the main idea of this argument is that $PC_2$ society is "beyond (distributive) justice" because the circumstances and conditions that give rise to concerns about justice are overcome. If this is right, $PC_2$ society is neither just nor unjust.

It might be objected that Marx clearly envisions a principle of distribution for $PC_2$ society, namely, "To each according to his needs." Isn't this a principle of distributive justice? Not according to Buchanan. He says,

> Marx is not offering the slogan as a communist principle of distributive justice, but rather as a description of the way things will in fact be in communism.
>
> Once the circumstances of distributive justice have been left behind, there will be no place for any such principle as a *prescriptive principle* that lays down demands for society to satisfy and specifies rights for it to protect.[28]

The point is that the distributive principle of $PC_2$ society is purely descriptive and not prescriptive. Buchanan maintains that it cannot be a principle of distributive justice because it lacks prescriptive force. It does seem plausible to suppose that this principle will lack prescriptive force under these circumstances, but I am not completely sure that this is sufficient for saying that it is not a principle of distributive justice.

If having prescriptive force is interpreted to imply having real social impact, it may be that *no* principle of justice has such force. According to Marx, ideas about justice, rights, and so forth, are not the levers of social change, nor in the final analysis are they even the pillars of social stability; ideas about class (or individual) interests are what count. If this is right—and it may not be—then $PC_2$ society does have a principle of distributive justice, and, presumably, it is a just society. This contradicts a number of claims made at the end of Chapter 5, so the latter would have to be revised accordingly. I see a way to pursue this dialectic through a few more iterations, but in the final analysis, it may not really matter very much. On the important points, Buchanan is surely right: $PC_2$ society has sufficiently transcended what Hume and Rawls call the circumstances of justice to the point where abundance has been achieved; whatever scarcities and conflicts remain do not require a state apparatus to resolve. In the spirit of the "eliminativist" approach to Marx's views on justice sketched in Chapter 5, it may not matter very much whether $PC_2$ society should be called a just society or whether it should be said to be beyond justice. It is a good society, and in the context of this section, it is a good society because it has eliminated all forms of exploitation.

### Post-Capitalist Society as the Good Society: The End of Alienation

On the basis of the elaborate discussion of alienation in Chapter 2, it is possible to describe what social life would be like without the various forms of alienation endemic to capitalism. Recall that the latter fell under four general headings: (i) alienation from the products of labor, (ii) alienated labor, (iii) alienation from species being, and (iv) alienation from others. Let us consider what the absence of each of these amounts to in PC society.

*Alienation from the Products of Labor.* Under capitalism, the workers do not own the products they work on and the latter pass to the control of another through the market. In a centrally planned economy, by contrast, the workers indirectly control the products of their labor through their control of the planning apparatus. Secondly, unlike under capitalism, the purpose of production is not to gain exchange value in the market but to fabricate things with use-value. Social need, and not the thirst for surplus value, determines what gets produced and in what quantities. Perhaps the most important feature of a society with a centrally planned economy is that it does not subject its members to large-scale social forces beyond the control and understanding of the society itself. The mystification of social relations brought about by the coordinating mechanism of the market is replaced by the conscious control of productive forces by the workers and their representatives in the planning apparatus.[29]

*Alienated Labor.* Presumably, Marx would maintain that alienated labor is absent from PC society. What does this mean? Does it mean the demise of all physically harmful and mindless work? Probably not. Nor does Marx seem to think so; in the discussion in the second section of this chapter on how labor could become life's chief want, it was pointed out that Marx recognizes that not all labor will be intrinsically rewarding. It is highly likely that some drudgery would have to be done. But in PC society, especially in its second phase, the amount would be minimal and presumably shared more equally than it is under capitalism. Most labor, but not all, would become "life's chief want."

An especially important change in PC society is the "de-commodification" of labor power. This has a number of important consequences. First, when labor power is no longer bought and sold, and a person's income is not determined by the exchange value of his labor power, people do not tend to think of themselves as a commodity. Second, they and others do not confuse the value of a person with the (exchange) value of his labor power. Third, when the worker is not selling his labor power to another and has a direct say in the organization of production, he is a self-determining being. Fourth, since labor power is not a commodity, activities that produce it or reproduce it (e.g., education, eating) have their instrumental character attenuated. These activities are undertaken much more for their own sakes than is the case under capitalism. This is an important way in which social life gets "de-mystified" in PC society.

In a related vein, the "de-commodification" of labor power means that the activity of laboring is no longer a mere means to other ends. Actually, the relation between labor and the benefits that flow from it is somewhat complicated. In $PC_1$ society there is a very close connection between the quantity of labor provided and income received. This is presupposed by the incentive principle operative in that society. For this reason, however satisfying labor is in $PC_1$ society, it will not lose its instrumental significance for the individual. Unlike capitalism, however, the character of the labor, that is, what kind of labor it is, bears no relation to income. All that matters is the total quantity (intensity and duration) of labor that is supplied. This means that distinctions based on the kind of labor supplied will not give rise to income differentials. In $PC_2$ society, on the other hand, this incentive principle is overcome. The connection between labor and use-values received for personal consumption becomes highly attenuated, if not completely severed. This seems implausible only if one assumes that most labor will retain an inherently distasteful character, an assumption Marx would by no means grant.

*Alienation from Species Being.* Human nature consists in the capacity to engage in an indefinitely large number of needs-meeting activities (productive labor). The alienation from species being that takes place under capitalism consists of two things: Individual proletarians lose—or never develop—their capacity for many different kinds of productive labor. Secondly, the universal character of production on a collective level is masked and mystified by the division of labor and the operation of the market. In PC society, individuals develop their talents and abilities more extensively as they take on many more kinds of labor. In addition, the collective capacity for universal production becomes manifest to all because the society-wide organization of production is the result of conscious plan or design and not the blind operation of market forces.

Under capitalism, the alienated form of man's species being is money. As Marx makes clear in the *1844 Manuscripts* (MECW, vol. 3, p. 325), what is bad about money is that it can and does distort our perceptions of ourselves and others. The uncultivated rich man can appear cultivated by buying the right kind of art. By contrast, a poor person with a thirst for knowledge is denied education because she lacks the funds. These distortions and systematic divergences between appearance and reality disappear in both phases of PC society. In the first phase, income is directly tied to the quantity of labor supplied. If all healthy adults are workers, there is no way for whatever income differentials persist to produce the discrepancies just alluded to. In the second phase of PC society, needs for use-values will, for the most part, be met. Whatever differences there are in the bundles of use-values people possess will accurately reflect different sets of needs.

Religion will also disappear because there will be no need to project humanity's perfections onto a supernatural being. God, conceived of as man's alienated species being, will be rejected in PC society, since there will be no need for Him.

*Alienation from Others.* There are a number of ways in which humans' alienation from each other is manifested under capitalism and will not be manifested in PC society. In capitalist society the state portrays itself as representing the common interests of the community when in fact it represents only the interests of the bourgeoisie. It is essentially a divisive institution. One of the primary functions of the capitalist state is the protection of the right of private property. This cements the inherent conflict of interests between capitalists and proletarians and encourages egoism. The mystification necessary for stability is achieved by the fiction of equal citizenship.

In $PC_1$ society, by contrast, the state is truly democratic, since nearly everyone is a worker and the proletariat are the ruling class. The furtherance of the common interest by the state comes closer to reality because of this fact. Moreover, with the transition to $PC_1$ society, there is no longer private property in the means of production, so a primary function of the state becomes nugatory. In addition, the egoism produced and reinforced by a regime of private property subsides.

In $PC_2$ society, the state itself ceases to exist. The organized use or threat of force diminishes because the conflicts of interests that require coercively backed adjudication are eliminated. For this reason, a major source of alienation of human beings from one another disappears. Finally, and perhaps most importantly, the elimination of an exchange economy decisively alters how people perceive each other. No longer do they view one another's needs and desires as levers to be manipulated for their own benefit. The inherently conflictual nature of interpersonal relations is replaced by the cooperative interpersonal relations required by a centrally planned economy.

All of the above constitutes at least a part of Marx's vision of the good society. It is in some respects a minimalist account. Marx, or Marxists in general, may well believe that the excellences and virtues of PC society are much more varied and extensive than the above indicates, and they may be right. However, my aim here has been to sketch only that part of the Marxian vision of a good society that is essential for a successful radical critique of capitalism. If the above correctly describes PC society, Marx is entitled to conclude that the latter does not face the same generic problems that face capitalist society. This much a radical critic must show.

But what is missing from the above account is anything other than the most cursory arguments for the various claims about the absence of the generic defects of capitalist society. Such arguments have to proceed from some facts about the institutions of PC society, notably the economic system, to these various claims. In this way, the account of the vision of PC society would exactly parallel the critical explanations of the defects of capitalist society.

Unfortunately, Marx himself does not make these arguments, and my efforts on his behalf, as the perceptive reader will have noticed, have been sketchy and fainthearted at best. But good arguments are needed. If an institutional structure or process causes some social evil, one cannot assume that the destruction of that institution will result in the elimination of that social evil. This is so for two reasons. First, the latter might be overdetermined. In other words, though the institution is causally sufficient for the social evil, it may be that permanent background conditions would have sufficed to produce the defect anyway. Secondly, social institutions are not merely demolished by revolutionary social change—new institutions arise in their place. What the radical critic must show is that these new institutions will not produce the same social evils all over again.

Marx does none of this. Moreover, in the voluminous secondary literature on Marx no one, as far as I am aware, has made a serious effort to explain (on Historical Materialist or any other principles) how the absence of alienation and exploitation is supposed to be a consequence of the main institutions of post-capitalist society. It is easy enough to *define* PC society as a society in which these evils do not occur, but definition is not argumentation. From the point of view of the demands of a radical critique, the vision of PC society remains incomplete.

## Summary

This chapter and the last constitute my reconstruction of Marx's vision of post-capitalist society in both of its phases. I have been guided in this by what can be reasonably inferred from his theory of history and by the demands of the Alternative Institutions requirement for a radical critique. A brief summary of the elements of that vision might be useful at this juncture. Marx's vision of PC society consists of:

1. *A specification of the economic system of both phases of PC society.* It was argued that an economic system consists of:
   a. a predominant type of relations of production
   b. a method of production coordination
   c. a system of distribution of the means of consumption

Letters (a) and (b) are the same for both phases of PC society. Regarding (a), the workers control the means of production and labor power is not a commodity. Regarding (b), markets are for the most part abolished and central planning coordinates production. The distributive principles for the two phases of PC society differ. In the first phase, the distribution of the means of consumption is proportional to labor contribution. In the second phase, distribution is in accordance with the need for use-values. $PC_2$ society has achieved abundance, as it was defined in the second section of this chapter. The inauguration of this second phase presupposes a number of profound and significant changes in the circumstances and conditions faced by humanity, namely,

   i. The end of "the enslaving subordination of the individual to the division of labor"
   ii. The overcoming of "the antithesis between mental and physical labor"
   iii. A change in the character of labor such that it "has become not only a means of life but life's prime want"
   iv. An increase in the productive forces concomitant with the "all-round development of the individual"
   v. More abundant flow from "all the springs of cooperative wealth"

It is worth reiterating that these enormous changes indicate that Marx did not believe, at least by the time he wrote the *Critique of the Gotha Program*, that $PC_1$ society would only be around for a few months or years. $PC_1$ society is not a mere bit player on the stage of world history.

2. *A specification of the political system of PC society.* For $PC_1$ society, the state is the "dictatorship of the proletariat." It serves the following ends or purposes:

   a. It ensures worker control of the means of production.
   b. It prohibits or sharply curtails the commodification of labor power.
   c. It ensures that workers' income is determined by labor contribution.
   d. It ensures the coordination of production by providing legal backing to the directives of the planning apparatus, even if the latter is not formally an organ of the state.
   e. It runs the civil justice system.
   f. It runs the criminal justice system.
   g. It enforces various political and civil rights.

Marx's vision of the state in $PC_1$ society is very much incomplete. He does not make any effort to spell out the constitutive social roles of that institution, nor does he provide any account of the mechanisms by which state power will be wielded.

Marx predicts the withering away of the state with the advent of $PC_2$ society. He seems to have believed that if an institution's functions are served in other ways or rendered nugatory, the institution will cease to exist. The above-listed functions served by the state in $PC_1$ society will have one or the other of these properties by the advent of $PC_2$ society, and so the state (gradually?) disappears.

3. *An account of what social life would be like in the absence of exploitation and alienation.*

Given certain assumptions about the absence of alienation, distribution of the means of consumption, and the labor contribution of all healthy adults, it is arguable that all forms of parasite exploitation would be absent from PC society. Whether or not property relations exploitation would exist basically depends on whether or not the economic system can outperform capitalism. That question was left hanging but will be addressed in Chapter 9. Unless Marx rejects all prescriptive talk about justice, he would also maintain that $PC_1$ society is a just society for about the same reasons he would believe it to be non-exploitative. Given some very large assumptions about that society, this is a defensible position, at least on certain theories of justice. There is also some reason to think that he would maintain that $PC_2$ society is beyond justice, though there is no decisive textual evidence on that score.

A description was also given of what social life would be like without all of the various forms of alienation characteristic of capitalist society. In the discussions of both alienation and exploitation, some attempt, especially with regard to the latter, was made to argue for the absence of these social ills on the basis of structural features of PC society—comparable to Marx's critical explanations of these phenomena under capitalism.

Marx's vision of PC society, especially in its second phase, is the standard of comparison against which capitalist society must in the end be judged. This vision is intended to satisfy the Alternative Institutions requirement for a successful radical critique of capitalist society. How has it fared?

Clearly, there are gaps of greater and lesser significance in the reconstruction. It is easy enough to point these out and draw the obvious conclusion that Marx's radical critique is significantly incomplete. In addition, where the vision is complete, the supporting argumentation is potentially suspect. The premises of these arguments can be attacked, or questions can be raised about the promiscuous use of functionalist explanations and predictions in the reconstruction.

On the latter point, my method has been to assume certain features of PC society as given (e.g., PC relations of production) and then to ask what else can be supposed on that basis. I have made little or no effort to give causal explanations, complete with microfoundations, for the institutions of PC society or for the absence of the various social ills of capitalism. Though this "functionalist procedure" has proved fruitful and, I think, in the spirit of Marx, it is potentially suspect.

I have not hesitated to point out gaps in Marx's vision as it has unfolded. Calling attention these gaps is important, but it is not of any ultimate philosophical interest. Questioning the supporting argumentation or the many functionalist predictions would obviously be unfair. Since Marx himself had so little to say about the institutions of PC society, it was necessary for me to work out those arguments and predictions as best I could. To raise questions in this regard would mean that my own argumentation would be under critical scrutiny! Despite my best efforts, I might have stacked the deck. Also, it wouldn't prove very much. My reconstructed arguments might be defective, but perhaps other, better arguments could be mounted on Marx's behalf.

A more powerful criticism would be to attack the vision itself. Part II of this book argues that Marx's vision of post-capitalist society, in either of its phases, is a truly utopian social vision in the sense that it cannot be realized. If this is right, the Alternative Institutions requirement for Marx's radical critique of capitalist society cannot be, and not simply has not been, met. If that is the case, Marx's radical critique must, in the final analysis, be judged a failure.

# II

## CRITIQUE

# 8

# The Unrealizability
# of the Second Phase
# of Post-Capitalist Society

## Unrealizability Arguments or A Critique of Radical Criticism

Chapter 1 identifies four necessary conditions for a successful radical critique of a society. They are:

  i. The Critical Explanations requirement
 ii. The Normative Theory requirement
iii. The Alternative Institutions requirement
 iv. The Transition requirement

Part I of this book reconstructs (i) and (iii) for Marx's radical critique of capitalist society. With regard to the Critical Explanations requirement, my reconstruction was itself highly critical at certain points, notably on exploitation. The aim (one aim, anyway) was to give Marx the best case possible. Similar intentions guided the reconstruction of Marx's vision of PC society in the preceding two chapters. This vision was summarized at the end of the last chapter. How does it stand up to the Alternative Institutions requirement?

Let us begin by recalling just what this requirement says. First, the radical critic must specify the main alternative institutions of the new society. For Marx, this would be the economic system and the political system. But what exactly does it mean to *specify* these institutions? The material of the last two chapters on the economic system of PC society permits a clarification of this that was not possible in Chapter 1. I have argued that the economic system of PC society consists of three elements:

1. a dominant set of relations of production
2. a method of production coordination

3. a distribution system for the means of consumption

Suppose these three elements specify the economic system, that is, they tell us what the latter *is*. If this is right, it is possible to determine what in general is required for a specification of any social institution. What is needed is an abstract tripartite characterization of a social institution for which each of the above is an instantiation. The following does the job:

1'. A description of the constitutive social roles
2'. A description of how the rights and powers associated with these roles will be exercised
3'. A description of the function(s) or purpose(s) of the institution and how well, or to what extent, the function(s) or purpose(s) are served

A brief comment about (3) and (3'): Institutions satisfy their function(s) or purpose(s) more or less well. For example, the function or purpose of the economic system is to meet people's wants and needs for use-values. An account of the principle(s) by which distribution actually occurs is needed to identify the extent to which this is accomplished. Similarly, for a political system, its main function might be to wield organized coercive power to further the interests of the ruling class. Some states do this better than others.

On the basis of this account of what it means to specify alternative institutions, it is clear that Marx has failed to specify the political system of $PC_1$ society. $PC_2$ society presents less of a problem in this connection, since Marx predicts the demise of the state at that stage of history. With regard to $PC_1$ society, it was possible to infer that a state would exist and that it would serve certain purposes or functions, but beyond some sort of commitment to democracy, it was not possible to define the constitutive social roles or give an account of how the relevant functions would be served. There is, then, a gap in the reconstruction in this regard. Although this point is not without interest, it is not a fatal blow to the project of a radical critique. Indeed, twentieth century Marxists have been eager to fill this gap; Lenin's "democratic centralism" is perhaps the most famous, or infamous, theory of the state in PC society.

Recall that the Alternative Institutions requirement demands more than a specification of the new institutions. It is also necessary for the radical critic to justify the proposition that these institutions will prevent or preclude the defects or social evils associated with existing institutions. Some arguments were offered to that effect in the two sections before the summary section of the last chapter, but no comprehensive critical evaluation of them was undertaken.

Finally, the radical critic has to offer some reason to believe that the new institutions can persist as stable social forms. I have made no effort at all to address this subrequirement in my reconstruction of Marx; it has proved fruitful to assume stability at each "level" and ask what else can be inferred, given that assumption. Once again, there is a gap in the reconstruction, but it is unclear whether or how it can be filled.

From the perspective of the Alternative Institutions requirement, Marx's radical critique of capitalist society must be judged to be incomplete at best. I have attempted to give him as complete and convincing a case as is possible, based on what he actually said and what can be reasonably inferred from the materialist theory of history. What remains to be done is to show that this critique is fundamentally mistaken, not merely incomplete. To this end, I shall argue that the Alternative Institutions requirement *cannot* be (and not just has not been) met because Marx's vision of PC society cannot be realized. The main purpose of this chapter and the next is to do just that.

One complicating factor that must be immediately taken into account is that Marx really has a dual vision of PC society—corresponding to the two phases. These must be evaluated independently. Here's why: Marx's natural history of PC society says that capitalism gives way to $PC_1$ society, which, after some unspecified but longish period of time, gives rise to $PC_2$ society. An account of the workings of $PC_1$ society, then, is supposed to describe how the transition from one to the other will go. However, suppose Marx is wrong about that. Suppose, for example, that Marx's social vision of $PC_1$ society is itself unrealizable. It does not follow that the same is true of $PC_2$ society. It *would* follow that his story about the transition is wrong, but that does not prove that the vision of $PC_2$ society cannot be realized. Maybe a new and different story of the transition could be told. One strategy that some contemporary socialists might find attractive is to argue that market socialism is necessary to take history to the advent of communism. The general point is that proving the unrealizability of Marx's vision of $PC_1$ society is not sufficient for proving the same about his vision of $PC_2$ society. The critic must take on each stage of the vision independently.

What, then, would it mean to say that a social vision is unrealizable? The strongest way to understand this is in terms of some kind of inconsistency in that vision. Recall that Marx's vision of PC society (in either of its phases) consists of the following elements:

1. An economic system, which was fully specified in terms of a set of relations of production, a method of production coordination, and a system of distribution of the means of consumption
2. A political system, which was incompletely specified in terms of the functions or purposes for which organized coercive power is wielded (for $PC_1$ society), or the absence of such a system (for $PC_2$ society)
3. A partial account of PC society as a good society, which was spelled out in terms of what social life would be like without the systemic ills (viz., exploitation and alienation) that afflict capitalist society

In principle, an internal inconsistency could be located in any one of these three elements, though prima facie it is unlikely to be found in either (2) or (3). This is unlikely with regard to (2) because the specification of the political system is so sketchy or thin. In addition, it is doubtful that there is any inconsistency in the account of what social life would be like without exploitation and alienation, again in part because of the relative thinness of that description. The economic system, on the other hand, has been elaborated in greater detail. As far as I can tell, the only potential inconsistency is between worker control of the means of production and central planning. In Chapter 6, I explained how it is possible for these features of the economic system to coexist.

Finding an internal inconsistency in any one of the elements of Marx's vision of PC society looks unpromising. For comparable reasons, a logical inconsistency between the different elements of the vision would also seem to be hard to come by. For the most part, there just does not seem to be enough rope for Marx to hang himself with. To my credit, I have done my best not to weave that rope in reconstructing Marx's vision of PC society.

The basic logical consistency of Marx's vision of PC society, which the reconstruction of the last two chapters bears out, has made it an enormously powerful rhetorical device for critics of capitalist society from the time of Marx down through the present day. Innumerable dark and gloomy books on the existing order, including some by first-rate thinkers, end on a hopeful note—though often only a sixteenth or thirty-second note—that a much better world is at least possible by radical institutional change. This is usually quickly followed by an admonishment to the effect that the issue can only be settled by revolutionary *praxis*; thus the call to arms is issued. Those who say it cannot be done are dismissed as socialists who have lost faith or as reactionary mouthpieces for the bourgeoisie. The basic consistency of the Marxian vision of the future seems to warrant these dismissals and the consequent call to arms, since how could one prove that an internally consistent vision cannot be realized?

Well, in principle, there is one way this could be done: It could be argued that the vision is incompatible with some facts that are not a part of the vision, "outside facts," if you will. This would make for a modified inconsistency argument, where elements of the vision would be shown to be inconsistent with these "outside facts." Unfortunately, time frame considerations make it very difficult to figure just what sort of facts could be used. This needs some explanation. At different times in his life Marx believed that the worldwide proletarian revolution was just around the corner and that he would live to see its success. Obviously, it has not happened yet, but it still could. Maybe his account of what was necessary and sufficient for the transition to $PC_1$ (or for that matter, $PC_2$) society was mistaken. This charitable view makes it hard to specify the outer limits of the time frame for the inauguration of the social institutions of PC society. What this means for the strategy sketched above is this: An argument to the effect that the vision of either phase of PC society is inconsistent with some "outside facts" must make use of factual claims that are likely to be true indefinitely far into the future.

What kind of facts would these be? What is needed are universal background conditions of the natural, biological, psychological and/or social environment that any society must face—in short, permanent features of the human condition. But what might such features be? The favorite tactic of conservatives, when faced with the kind of challenge posed by Marx, is to dredge up some sort of human nature argument. Original sin or basic human selfishness, for example, might be invoked to argue that abundance cannot be achieved and/or that the institutions Marx envisions cannot prevent the recurrence of the social evils characteristic of capitalism.

However, this strategy looks unpromising for two reasons: Increasingly sophisticated historical, sociological, and anthropological knowledge of other cultures makes clear the remarkable diversity of human beings and their social institutions. This renders suspect grand claims about human nature. Relatedly, it is likely that any universal claims of this sort that could survive a confrontation with the facts would be far too weak to generate the conclusion that Marx's vision of PC society is unrealizable. Remember, what is needed are facts that can be shown to be directly inconsistent with an element or elements of Marx's visions of PC society.

Secondly, Marx has an elaborate and imposing theory of human beings and their social condition that explains the dismal features of social life thusfar by an appeal to the fact that all hitherto existing societies (with the exception of primitive communist societies) have been class societies. This theory may be defective, but mounting a case for that appears to be a daunting task, and in and of itself provides no alternative explanation for the social evils in question. None of the above proves that a human nature argument will not succeed, but it does not look to be a very promising approach.

Despite the dubious prospects for a human nature argument, it has the right sort of "feel." It may be that better candidates for the "outside facts" can be found. Maybe the best way to find them is to approach the problem from the opposite direction. Perhaps it would be more fruitful to look for that feature of Marx's vision of either phase of PC society that looks prima facie most implausible and try to explain why. That might lead to a discovery of the facts in question.

Perhaps the most questionable part of Marx's vision is the distributive principle of $PC_2$ society: The distribution of use-values is to be in accordance with need, which was interpreted to mean that the material preconditions for the good life would be satisfied for all. This suggests a level of material wealth for $PC_2$ society that is very high, higher than anything heretofore observed; after all, it is very doubtful that contemporary capitalism has achieved it, given the current configuration of the forces of production. Could this principle really describe distribution in $PC_2$ society?

To assess this, it is necessary to review and then augment some material from the third section of Chapter 7, where an effort was made to explicate this principle. The goal of this explication was to give Marx something that is faithful to his vision of $PC_2$ society without being so extreme that no sane radical critic would propose it. The distributive principle was interpreted to mean that $PC_2$ society will have insured that the material preconditions for the good life will be satisfied for everyone. Anything more or less will not do. This point is of considerable importance for my critique of Marx in the second and fourth sections of this chapter, so some reiteration and further argumentation would be in order.

If the distributive principle of $PC_2$ society required that each and every use-value need for every person always be met in a timely manner, it would obviously never be achievable. No glitches, no hitches, no mistakes would be permitted. Every time a use-value need arose, it would be met right away. Human nature arguments do not go very far, but they go far enough to entail that human beings will occasionally slip up. The requirement that everyone's use-value needs will always be met in a timely manner is obviously too strong.

On the other hand, if the distributive principle of $PC_2$ society is interpreted to mean that this society meets the material preconditions for the good life (the MPCs) for all of its members, it will meet most people's use-value needs in a more or less timely fashion. Occasional glitches will occur, but they will not be severe, for the simple reason that one of the MPCs is not spending a third of one's life standing in line for toilet paper. (This state of affairs is an MPC because it can be insured by the productive forces and the operation of the economic system.)

By contrast, anything less than Marxian abundance will not do. If $PC_2$ society cannot meet the MPCs for all of its members, then what Marx is in effect envisioning is some sort of rationing of the MPCs, which is to be arranged so that the gap between the need for the MPCs and the satisfaction of those needs is about the same for everyone. The size of this gap would be indeterminate but potentially significant. Clearly, however, Marx had something more in mind than just this.

On a more substantive note, if Marx is predicting mere proportionality between use-value needs and their satisfaction at a lower level than what is implied by meeting the MPCs for all members of society, problems with interpersonal comparisons of utilities would probably doom the genuine realizability of this principle from the start. Some positivistic economists maintain that such comparisons are strictly impossible. The philosophy of mind needed to support this charge is probably untenable; however, even if interpersonal comparisons of utilities can be made, it is doubtful that anything more than very rough comparative judgments on particular alternatives could be made. For example, it may be possible to know that the residents of a town suffer more from pollution from a factory than consumers benefit from a 10 percent reduction in the purchase price of the factory's product this pollution makes possible. However, it is hard to see how one could know how great the difference is. Absent some invisible-hand mechanism, vastly more powerful knowledge claims would have to be substantiated for society to insure that the gap between use-value needs and needs-satisfaction is about the same for everyone, if that gap is any larger than it would be if $PC_2$ society guarantees the MPCs for all of its members.

On the interpretation/reconstruction I have offered, there is some indeterminacy in the actual extent of timely use-value needs satisfaction, despite the fact that the satisfaction of the MPCs is assured. This is as it should be for a realistic distributive principle, that is, one that could describe an actual society.

Distribution according to needs, then, is best understood in terms of the achievement of abundance, that is, meeting the material preconditions for the good life for all. The big problem in Chapter 7 was to articulate what this required. Though the final definition is complicated, it is constituted by three main ideas: (i) It is feasible, given the level of development of the forces of production, for the MPCs to be met for all; (ii) the incentive structure implicit in the economic system does not prevent the satisfaction of the MPCs for everyone; (iii) other institutions are staying out of the way.

Productive force feasibility was never really defined but this is easy enough to do. First, the productive forces are, roughly, anything that is directly and intentionally used by the producers to make use-values.[1] A shorthand way of characterizing the forces is in terms of the categories of labor power (including skills), technological knowledge, and means of production (including raw material). But how is feasibility to be understood?

When a construction boss says that doing a certain job to certain specifications is feasible, what he means is that his group can get it done. That in turn means that there is nothing in either the character of the task (the specifications) or the forces of production at his disposal that would prevent it from being done. Generalizing from this, condition (i) of the definition of the relevant sense of 'feasibility' can itself be defined as follows:

i. The state of the development of the forces of production is such that throughout $t$ it is PF feasible to meet the MPCs for all of society's members if and only if
   there is nothing in the character of the MPCs or the nature of the productive forces that would prevent the MPCs from being met for all of society's members.

This definition articulates the idea that any gap between feasibility and getting the job done is to be explained by something outside the means-needs nexus. For example, a feasible construction project that someone wants done may not get done (according to specifications) because inclement weather prevents it.

Similarly, it seems at least conceivable that capitalist society could reach the point where it is PF feasible to meet the MPCs for everyone, but the operation of the economic system may prevent what is PF feasible from actually happening. In this case, the productive forces are in place to do the job, but it cannot get done because of defects in this social institution. PF feasibility, then, is a necessary but not sufficient condition for Marxian abundance. That is why the definition requires that it is economically feasible for the MPCs to be met for all and that other institutions are not interfering; that is, all the social institutions are playing their part, even if their part is to stay out of the way. If they are not, meeting the MPCs for all is not assured.

Further support for this explication of the distributive principle of $PC_2$

society comes from a consideration of the failures of capitalism in this regard. Egalitarian concerns to one side, critics on the Left have complained bitterly that capitalism satisfies many of the (mere) wants of some while the genuine needs (MPCs, on our terminology) of others have gone unmet. Allen Wood attributes this sort of thinking to Marx: "For Marx, the problem with bourgeois society is not that scarcity is unevenly distributed, but that there is scarcity where there need be none, and that there is a surfeiting class which benefits from a system which subjects the majority to an artificial and unnecessary poverty."[2] For example, needs for health care, adequate nutrition, shelter, clothing, education, and so forth, go unmet for many people. By contrast (and not coincidentally), the rich have all manner of mere wants satisfied by luxury goods. Desires for these goods are encouraged by the capitalist economic system, again, not coincidentally.

However, two clauses in the above quotation from Wood are simultaneously misleading and revealing, at least as they apply to Marx: Wood says that under capitalism, "there is scarcity where there need be none," and capitalism is described as a "system which subjects the majority to an artificial and unnecessary poverty." I believe Marx would disagree with both of these in one sense and yet agree in another sense.

In *The Communist Manifesto* (and elsewhere) Marx and Engels stress the fact that it is the historic mission of capitalism to take the level of development of the forces of production from what existed at the end of the feudal era to the very high levels envisioned for PC society. On their account, this required the ever more intense squeezing of surplus value out of the vast majority of people for literally centuries. Moreover, it is an inevitable consequence of the workings of the capitalist economic system that this poverty results. For these reasons, the scarcity under capitalism is historically necessary, as is the poverty, which is not to say that they are not bad things.[3] But in what sense might they be said to be unnecessary?

They are unnecessary only in a very attenuated sense: If the level of development of the forces were higher, and if the economic system did not have the structural perversities that it did, this scarcity and poverty would not exist. These are very big counterfactuals. With regard to productive force failure, it is doubtful that even contemporary capitalist societies have an appropriately developed productive apparatus to meet the MPCs for everyone. Even if technology were far enough along, existing means of production and labor skills are configured in such a way that meeting the MPCs for everyone is probably beyond the production possibility frontier. At most, contemporary capitalism has the potential to construct an adequate productive apparatus. But even this claim to potentiality must be qualified, and that brings us to the second counterfactual.

Regarding the structural perversities, one could say, as anti-communists are fond of saying, 'A leopard doesn't change its spots'. Capitalism without these structural deformities is a contradiction in terms (at least a Hegelian contradiction). According to Marx anyway, there are very deep reasons why impoverishment of the masses is an inevitable accompaniment of the capitalist economic system.

Wood's choice of words is perhaps unfortunate (there is no reason to believe he is unaware of these points); nevertheless, there is an insight of first importance here into the Marxian conception of abundance. What prevents the satisfaction of the MPCs for everyone in pre-PC societies is either an insufficient level of development of the forces of production or deformities in the social system, or some combination of the two. Whatever gaps would remain between use-value needs and the satisfaction of such needs after these deformities have been cured and the forces sufficiently developed, are insignificant to the point where the satisfaction of the MPCs for all is assured. This is captured in the definition of Marxian abundance as it has been worked out in Chapter 7 and in this section. Failure to achieve Marxian abundance is a productive force failure or an institutional failure, or both.

It would seem that criticism of the Marxian vision of distribution in $PC_2$ society would require an account of the Marxian conception of the good life. Scattered through Marx's writings is perhaps enough material to reconstruct such an account, though doing this would be a formidable task.[4] A reconstruction of the Normative Theory behind Marx's radical critique of capitalism would probably include this.

But, contrary to appearances, such an account is unnecessary to show that Marx's vision of distribution in $PC_2$ society is unrealizable. The goal of the relevant parts of Chapter 7 and the last few pages has been to explicate a Marxian conception of abundance that is faithful to his vision of $PC_2$ society. If the foregoing is faithful to that vision, then enough is in hand to show that this distributive principle cannot be realized, which entails that the vision of $PC_2$ society is unrealizable. As the next section shows, some obscurity in the Marxian conception of the good life will not hinder the argument.

## Failure of the Material Preconditions of the Good Life

On the face of it, there does not seem to be anything in principle impossible about achieving the MPCs for all members of society. The provision of adequate food, clothing, and shelter for all seems eminently feasible, at least in principle, especially if what counts as adequate is determined historically. However, careful attention to what such an achievement really amounts to makes it evident that this could not really happen.

What exactly is an MPC? To say that some state of affairs is a material precondition for the good life is to say that it is something which is both a necessary condition for the good life and which could, in principle, be brought into existence by the productive forces and the economic system of a society. Economic activity does not merely produce different kinds of physical objects and human services; it issues in states of affairs that are necessary or sufficient conditions for the satisfaction of wants and needs. It is for this reason that a material precondition is conceived of as a state of affairs and not as a particular use-value or kind of use-value.[5] The materiality of such a state of affairs consists in the fact that the production and distribution of use-values insures its existence. Not all necessary conditions for the good life are material preconditions. Being loved by another human being would probably be an example of a necessary condition for the good life that is not material in this sense. The blandishments of the sex trade and the dating services notwithstanding, this is something the productive forces and the economic system can never assure. The economic system of the good society cannot guarantee the good life for its members.

There are two basic kinds of MPCs, what might be called universal MPCs and conditional MPCs. The former are material (in the sense just specified) necessary conditions for the fulfillment of any conception of the good life for humans. Having adequate shelter (in most climates and for most of the time) is an obvious example. The same could be said of an adequate diet. Conditional MPCs, on the other hand, are material preconditions for particular conceptions of the good life. Their status as MPCs is conditional on the decision to pursue a particular conception of the good life. If one's conception of the good life includes being a competent surgeon, it is obvious that proper training in surgery is a material precondition for the good life for that person. Just as obviously, it is not a material precondition for every conception of the good life. The materiality of this condition consists in the fact that training in surgery is something that could be provided by the economic system in conjunction with the productive forces of a society.

The achievement of Marxian abundance requires that the forces of production and the economic system must assure all the MPCs, universal and conditional alike. This needs a small argument. If a society that achieved Marxian abundance must only assure the universal MPCs (each person's being adequately fed, clothed, etc.), then there is no guarantee, indeed, it is not even likely based on that fact alone, that most or even many of the members of that society are leading the good life. Suppose that population is growing so fast that the society must bend all of its efforts just to assure the universal MPCs for everyone. This is hardly a picture of the good society. What the economic system of the good society must do is to make it materially possible for everyone to achieve the good life—and that requires meeting the conditional as well as the universal MPCs. Assuming that $PC_2$ society is the good society, it guarantees both the conditional and universal MPCs for all.

This would seem to involve a staggering level of material wealth, given the potential for wide and various conceptions of the good life, but that is what material abundance involves. However, it is worth pointing out that a society that has achieved this abundance need not guarantee that all desires for use-values are met, since it is highly doubtful that the satisfaction of all such desires is a material precondition for the good life for anyone, or at least any sane person. Nor does it guarantee that each and every use-value need will be met in a timely manner, since that too is not a necessary condition for the good life. Moreover, as noted above, guaranteeing all of the MPCs for everyone does not insure that all will in fact lead the good life, since there will be necessary preconditions for the good life that are not material. So, achieving Marxian abundance isn't everything, but it's quite a lot.

Now that a clearer conception of the MPCs is in hand, it is possible to show that assuring all the MPCs for everyone is something that no society can do, no matter how far into the future one cares to look. Let us begin with the universal MPCs. It seems reasonably clear that some of these (e.g., having adequate shelter and diet) could be assured for all. However, there are others that could not.

To see what they are, let us begin with the fact that there are many avoidable tragedies in contemporary capitalist society. Some babies are malnourished, and some die of illnesses and complications that could have been prevented with existing health-care technologies. Other tragedies are unavoidable now but will be avoidable in the future because of the development of better technologies or better health care delivery systems or both. What makes these incidents tragedies (whether avoidable or not) is that a universal necessary condition for achieving the good life has been denied. In this case, that condition is having a life that lasts longer than a few years. The denial of this condition for these children under capitalism is a productive force failure or a institutional failure or both. Blaming capitalism for either sort of failure is beside the point in a certain sense, as the discussion of the quotation from Wood in the last section indicates. But notice that because these tragedies are productive force failures or institutional failures or both, this universal necessary condition for the good life (having a life longer than a few years) is a material precondition for the good life. It could have been assured by a more adequate development of the productive forces or better social institutions. This implies that, once Marxian abundance has been achieved, neither kind of failure will happen.

This means that Marxian abundance has some truly astonishing implications: No young child will die of an illness that could have been cured or prevented by an appropriate development and deployment of productive forces. That is, medical technology will have advanced to the point where (and all of society's other needs will be satisfied to the point where) every fatal childhood illness for which it is physically possible to find a means of prevention or cure will have been prevented or cured.

If the technology existing at a given time cannot prevent this, but at some later time it could, then Marxian abundance would not really have been achieved. In other words, society could have—and would have, if Marxian abundance really had been achieved—poured enough talent and resources into health care to prevent every child's death that it is physically possible to prevent. In this area, medical advances will have come to an end. Similarly, children in $PC_2$ society will suffer no fatal accidents. Products (including products used by adults around children) would have been so safely designed and resources would have been so deployed (e.g., human or mechanical nannies) that any fatal accident would have been prevented.

It will not do to say that $PC_2$ society will not prevent all such illnesses and accidents but merely do the best it can with the resources it has. When it comes to resources that guarantee the MPCs, it has as much as it needs, and it has the will to deploy them. That is what Marxian abundance presupposes. What counts as adequate can be historically relativized when the MPCs are being adequately fed, clothed and housed, but this will not do when it comes to life itself or even certain aspects of physical health. We do speak of an adequate level of health care with the implication that such a level need not imply perfection, but that assumes that there are other pressing social needs (other MPCs) that would make it rational for a society to choose more of something else rather than more health care. The decisive point is that these choices do not face a regime of Marxian abundance.

Some illnesses and accidents do not end life but prevent children who suffer them from living the good life, however the latter is to be understood. These will never occur in a society which has achieved Marxian abundance. As in the case of fatal illnesses and accidents, it will not do to say that Marxian abundance only requires that resources will be available to help children cope with the effects of these conditions. Coping, even having a life worth living, is not what Marxian abundance guarantees. It guarantees the material pre-conditions of the good life, not the material preconditions of a life worth living. Capitalism, after all, does the latter for the vast majority of people.

Similar considerations apply to adults. Though a man who suffers a heart attack at age eighty or even fifty may have lived the good life, some illnesses or injuries at earlier ages are so bad that they preclude a good life, either thereafter or even on balance (e.g., terrible burns). Illnesses and injuries of this sort would never occur to anyone who has not yet, but otherwise would, achieve the good life in what time remains to her. Such injuries will never happen due to use-values that are defective in either design or manufacture. Indeed, such injuries will not occur due to negligence or stupidity on the part of the user, since the harmful effects of these human failings would have been foreseen and prevented if enough resources had been brought to bear in an appropriate way; idiot-proof use-values will line the shelves in the stores of $PC_2$ society. The lawn mowers of tomorrow will make those of today look like veritable death machines.[6]

In a regime of Marxian abundance, one can never say, "If only technology had developed more, if only the deployment of material resources and labor power had been more adequate, Jones would not have failed to achieve the good life." If all the universal material preconditions for the good life have been met for everyone, then no one will fail to achieve the good life in a way that could have been prevented by the economic system in conjunction with what is physically possible for productive forces to achieve. On the other hand, if it is always possible to do more to insure the physical—or physio-logical—preconditions of the good life for someone or other, then Marxian abundance can never be achieved. Certain kinds of knowledge can be a material precondition for the good life in cases such as these because such knowledge could in principle be produced by the economic system. In a regime of Marxian abundance, it would have been produced. In every case.

It's not that no failures of any sort can occur in such a regime. If the auto workers do not install my car radio properly, I am not denied a material precondition for the good life. I can simply get it fixed or do without. But when people's lives and health are at stake, no slipups will occur if Marxian abundance has been achieved.

The implausibility of assuring all the universal MPCs for everyone becomes still more obvious if one considers the fact that sometimes people can rationally pursue a conception of the good life that involves risking the failure of the MPCs. Suppose that Jane chooses to create the good life by becoming an astronaut. It will be rational for Jane to risk losing some universal MPCs to achieve that goal. She recognizes that she may be killed early in her training, but the risk is worth it, given her decision to pursue this path. In general, some people who pursue the good life will find it rational on occasion to risk losing the material preconditions for the good life to achieve some goals that may be part of their conception of the good life. Indeed, people sometimes find it rational to risk losing some material preconditions for the good life for mere wants.

It might be thought that in principle the economic system and the forces could develop to the point where people could be assured, not that they never experience failure, but that at least the MPCs will never be forfeited. With enough safety devices and sophisticated psychological testing, and so forth, perhaps all irrevocable life-plan disasters that an appropriate development and deployment of the forces of production could have prevented would be avoided. Unlikely as this appears to be, for this even to be possible, it must be assumed that risking the loss of the MPCs can never itself be a part of a plan for achieving or creating the good life. The problem is that sometimes the very taking of those risks is constitutive of the good life. For example, for some people a serious struggle with nature is part of the good life as they have chosen it; the development and deployment of productive forces could in principle assure that failure would not be met with, but these people rationally choose to forgo this assurance.

Of course, PC$_2$ society could simply prohibit the pursuit of life-plans that risk this sort of catastrophic failure, but then it is doubtful that PC$_2$ society would be a good society, since a blanket prohibition of this sort would very likely prevent many people from living the good life. For many people, autonomous choice of which path is worth pursuing is itself a component of the good life. The general point becomes still more obvious if one begins to think through what kind of institutional mechanisms would bring this about. (Think about the kind of people who would be attracted to, or created by, the social roles that insure this. Junior high school vice principals, dieticians, mothers with grown children, etc., would find new and challenging career opportunities.)

The prevention of an untimely death or catastrophic illness or disease is not the only kind of MPC that must be assured in a regime of Marxian abundance. As noted above, there are also conditional MPCs. These are material preconditions of particular conceptions of the good life. They create problems because some of them are scarce, that is, not everyone who needs them can have them. For example, it may be that for a number of people what they need to achieve their own conception of the good life is a certain kind of training or education, say, studying under a concert pianist of a certain caliber. By hypothesis, training by others will not suffice. Not all who need this training to achieve their conception of the good life can get it because it is and will always be scarce.

In the language of our explication of Marxian abundance, it is simply not PF feasible for all these people to get this material precondition for the good life, though it will be possible for some. This sort of situation is not unusual. If people's aspirations are high, a certain sort of training really is necessary if they are to achieve their conception of the good life; the problem is, not all can have it.

Other conditional MPCs will be scarce in a stronger sense. First, notice that there are two ways in which something can be a conditional MPC: It can be causally necessary for some state of affairs constitutive of the good life for an individual, or a conditional MPC can itself be constitutive of the good life for that individual. Necessary scarcities of the latter entail inherently conflicting conceptions of the good life in a society.

For some people, the achievement of human excellence as a producer, that is, in a chosen line of work, is part of their conception of the good life. Some people's conception of the good life includes being among the best surgeons; even if anyone who wanted the best surgical training could have it, not all who want to be among the best practitioners can be. The materiality of this precondition for the good life consists in the fact that the economic system provides not only consumption goods but also opportunities for productive excellence. However, the economic system provides only so many slots at the top of any line of work, so to speak. It is not unreasonable to suppose that the quest for productive excellence will animate some people no matter what kind of society exists. Indeed, given Marx's emphasis on self-realization through the development of productive talents and abilities, it seems that his conception of the good society involves many people striving to achieve various productive excellences as part of the good life for them. But, as Robert Nozick has pointed out, such achievements must be defined relative to the achievements of others: "We evaluate how *well* we do something by comparing our performance to others, to what others can do."[7] Nozick's point is that there is no meaningful measure of human excellence along any dimension apart from what others can do. Nozick further claims that these evaluations connect up with self-esteem in important ways. To have high self-esteem is to rank oneself high relative to others along some dimension thought to be important both by oneself and by some segment of society. High self-esteem based on excellence as a producer along a certain dimension will be a necessary condition for the good life for many people, namely, those for whom the achievement of excellence along that dimension is a component of their conception of the good life.[8] If they lead the good life, then they've done some particular kind of productive activity well, which means better than most.

It is virtually certain that not everyone can avoid failure to achieve the good life of this sort, if enough people are enough alike so that they seek the good life by achieving (productive) excellence along the same dimension. The very nature of human excellence makes its pursuit an inherently conflictual affair. Given our limited knowledge of ourselves, our potential, and the potential of others and their plans, people who strive for productive excellence will choose life plans that can and will be thwarted by others. Note that this point does not presume that failure means that one will not have a life worth living; it is only that one will have failed to achieve the good life.

It might be thought that any particular failure could have been averted by better self-knowledge, which could be delivered if enough resources were devoted to it. (And, in a regime of Marxian abundance, enough resources would have been devoted to it.) Moreover, even if there must be some losers in the struggle for productive excellence, these losses need not prevent the achievement of the good life. It is at least possible that all the losers are never banking their self-esteem—and their achievement of the good life—on these struggles. Just as a better development and deployment of more material resources could prevent some untimely death or catastrophic illness or injury, so too a better development and deployment of material resources could serve to reduce the stakes in this variety of social conflict for those who are destined not to do well. If the losers had been better advised, they would have chosen different life-plans, plans sufficiently different from those they wanted to pursue so that the achievement of these excellences would not constitutive of the good life for them. Clearly, one necessary condition for the good life is not to have its achievement locked up in a struggle with others that one will lose.

What can be said in response to this objection? Even if it were possible to foresee the results of someone's attempt to achieve excellence along some dimension (a wildly implausible assumption), it may be, and indeed likely will be, that for some people, the risk of failure to achieve productive excellence would itself be a conditional MPC.

To make this point on a more intuitive level, those whose conception of the good life involves the pursuit of productive excellence along a chosen dimension may rationally choose not to know beforehand whether or not they will fail. This parallels the case of the person for whom risk of a universal MPC is a component of the good life. Part of the reason for this is that fear concentrates the mind wonderfully. Of course, all this supposes, for the sake of argument, that it really could be feasible to know how well people will do in the pursuit of human excellence before they make irrevocable commitments. I don't know that this can be disproved, but it surely looks implausible.

If all of the above arguments succeed, there are a number of reasons why not all conditional MPCs can be satisfied for everyone. Couple that with the fact that not all universal MPCs will be satisfied for everyone, and there is more than adequate justification for concluding that Marxian abundance can never be achieved, since the latter requires guaranteeing all of the MPCs for everyone. If Marxian abundance cannot be achieved, then the distributive principle of PC$_2$ society cannot be realized, which entails that the economic system, and by implication the Marxian vision, of that society cannot be realized either.

## Some Objections

How might a Marxist respond to all this? Basically, there seem to be two options: One is to call into question the presuppositions behind the argument that Marxian abundance is in principle unrealizable. A second strategy would be to call into question the appropriateness of "Marxian" abundance, as it has been defined here, for Marx's vision of $PC_2$ society. In other words, maybe Marx had something less extreme in mind for the second phase of post-capitalist society. Let us begin with the first option.

It might be objected that the presuppositions about conceptions of the good life in the good society made in the above arguments are part of the author's bourgeois fantasy and something no Marxist would endorse. All those competitive types beating up on each other to achieve the good life is not part of the Marxian vision of the good society. One salient feature of the accounts of the good life discussed in the previous section is that they seem to be remarkably individualistic. The more cooperative virtues and excellences are nowhere in evidence. It might be argued that people's conception of the good life will be much more social in nature. Perhaps some argument from the cooperative, transparent nature of production in a non-market economy could be invoked to give some theoretical backing to this claim.

The problem with this line of criticism is that it overlooks the fact that the assumptions of the argument about the conditional MPCs are in reality very weak and fully consistent with a highly social conception of the good life for most people, if not everyone. First, it is not assumed that everyone, or nearly everyone, is engaged in the pursuit of human excellences of the sort discussed above. Most people's conception of the good life may well be much less inherently conflictual. Being in a sound state of body and mind, having good friends, a family one loves and is loved by, and time to go hunting in the morning, fishing in the afternoon, and so forth, may be all that is required for the good life for many.

However, for the above arguments to work, all that is necessary is that some people are engaged in such struggles, since Marxian abundance requires that the conditional MPCs be met for everyone, not just most people. All that is required is an assumption that some people will choose to specialize in developing their talents and abilities to the extent that their conception of the good life is tied up with achieving productive excellence along one dimension or a small cluster of related dimensions. A word needs to be said about that assumption.

Marx suggests in a number of places (e.g., the Famous Passage in *The German Ideology*) that people in communist society will develop a wide range of their talents and abilities to the fullest extent. It is not unreasonable to suggest that Marx was projecting his own conception of the good life onto others. The many-sided geniuses of human history—Aristotle, Leonardo, Leibniz, Marx, and so forth—are comparatively rare. For most people, even people of extraordinary talent and ability, outstanding achievement along one or a small number of dimensions is the most they can hope for and, not coincidentally, the most they do hope for. Kant comes to mind as an obvious example. Much of the explanation for this is, I suspect, a combination of genetic make-up and the fact of our mortality. Not everyone is multitalented and even if someone is, there just isn't enough time.[9]

Finally, even on a more social conception of the good life, the problems for assuring the conditional MPCs can simply reappear at the group level. The individualistic bias in the above discussion is purely methodological. If a person's conception of the good life is intertwined with the goals of others (either empirically or logically), and being a member of a group that achieves excellence is part of the individual's conception of the good life, then the same inherent scarcities will prevent the assurance of the conditional MPCs for all groups, and *a fortiori* all individuals. Following an example used by Marx, some Marxists have held out the symphony orchestra as a paradigm for the cooperative pursuit of the good life that will be characteristic of $PC_2$ society. This overlooks the fact that one of the driving forces for people in that line of work is to be a member of an orchestra that is among the best, that is, better than most other symphony orchestras.

Let us turn to the second objection. Upon reflection, it may seem that there is something specious about the above demonstration of the necessary failure of Marxian abundance. In the end, the unrealizability of Marx's vision of $PC_2$ society comes down to the fact that somebody will get hit by a truck and not every short guy who really wants to play professional basketball will be able to. In a related vein, the unrealizability of Marxian abundance might well raise some suspicions that this concept, as it has been developed in this chapter and the last, is not really something Marx's vision of $PC_2$ society is committed to; the requirements for "Marxian" abundance, are in reality extraordinarily stringent. Maybe something less stringent captures better Marx's vision of that society.

Both of these concerns are legitimate. My strategy for addressing them is twofold. First, I shall investigate the consequences of weakening the distributive principle for $PC_2$ society and, by implication, the demands on the productive forces and economic system of that society. I shall suggest that any weaker distributive principle calls into question the realizability of another element of $PC_2$ society—the disappearance of the state. To combat the suspicion that there is something superficial or specious about the demonstration of the necessary failure of Marxian abundance, in the next section I offer a theory that gives a deeper explanation for this necessary failure, together with a generalized and more rigorous version of the arguments of the preceding section for the unrealizability of Marxian abundance. Finally, in the last section of this chapter I will offer some suggestions about why Marx overlooked the problems this theory calls attention to together with some additional evidence that Marxian abundance is what Marx had in mind for $PC_2$ society.

The most obvious way to "climb down" from Marxian abundance is to hold the latter out as an ideal that $PC_2$ society should try to realize. Unrealistic ideals can be recognized for what they are and yet still serve as a standard society should try to approximate or come as close to realizing as possible (or so most people think). The realizability of Marx's vision of $PC_2$ society would require, on this account, not the achievement of Marxian abundance, but only an approximation to it or an ever-closer approach to it.

One line of criticism of this approach, which I shall not pursue in any detail here, suggests itself almost immediately: How does one go about measuring how far (or even whether) a society has traveled toward this ideal or how closely a society approximates it? There are formidable problems of both an empirical nature and a conceptual nature in doing either of these.[10] Interpersonal comparisons of something dubious are probably required.

A second problem with this approach becomes evident if one recalls the basic purpose of specifying a distributive principle for $PC_2$ society: Such a principle is supposed to tell us how distribution will in fact take place, not how it ought to take place. A (factual) distributive principle is an essential element in specifying the economic system (one of the alternative institutions) of a society. Holding out an unrealizable goal, an ideal, will not suffice to describe an economic system. (Is this a form of Idealism, in Marx's pejorative sense of the term?) A radical critic is not being asked for his wish list; he is being asked the hard question about what alternative social institutions he believes will replace those that constitute the existing order. The specification of this institution requires an account of how distribution will in fact take place.[11]

In response to these criticisms, suppose a defender of Marx solves the factual and conceptual problems alluded to above and offers something definite enough to be a distributive principle, which does in fact approximate or approach this ideal.[12] If this were done, it would create a serious problem for another aspect of Marx's vision of $PC_2$ society—the disappearance of the state.

To understand the nature of this problem, recall that this distributive principle, whatever its exact content, entails that not all the MPCs can be met. If this is true, there will be ample opportunity for, and reason for, systematic social conflict. If the economic system is not meeting the MPCs for all, but it is for some, Lenin's question, "*Kto kgo*?" ('Who does what to whom?') once again becomes pressing. The stakes, after all, are very high for some people. Moreover, since the failure to satisfy the MPCs is a chronic problem, there are obvious advantages for those who see themselves at risk of losing or not achieving their MPCs in organizing to press their interests. Notice that this does not suppose extreme selfishness; threats to the MPCs of one's friends, one's family, and one's co-workers can be, and historically have been, sufficient to motivate action.

This story can be further filled out by recalling what else the economic system and the productive forces are doing, so to speak. Presumably most people's use-value needs are being met in a more or less timely fashion; moreover, it is reasonable to assume that some (mere) wants are also being met. But not all the MPCs are being met for everyone. A little more of the MPCs could be assured if the satisfaction of use-value needs were met in a little less timely manner or if some people's mere wants went unmet. Under such circumstances, conflicts of interest are inevitable. Serious systematic conflicts of interest of this sort tend to be settled coercively. Indeed, what would be the motivation of those who see themselves at risk regarding the MPCs not to contemplate coercion? Or, for those who are not at perceived risk to act preemptively? After all, all these people come out of a history in which organized coercion has been widely employed to deal with just these problems—not just in pre-PC society, but also in the first phase of PC society.

If the history of all hitherto existing societies has been the history of class conflict, it is appropriate to ask what they have been fighting about. The discussion in the previous section provides an obvious and compelling answer: the material preconditions for the good life. If these conditions cannot ever be met for all, there is good reason to believe that the fight will go on, even if social classes, in Marx's sense, have disappeared.

The above considerations do not prove with certainty that the state will persist and thus the statelessness of $PC_2$ society cannot be realized. But these considerations provide enough justification to make it reasonable to believe that $PC_2$ society, as stateless society, is probably not realizable over the long term, given the fact that any society must deal with the chronic failure to achieve the MPCs for everyone. The above also hints at the possibility that the failure of Marxian abundance may be a deep and systematic feature of the human condition. In the next section, a theory of sorts is developed to explain this, and a more rigorous version of the main argument of the previous section is developed; in the last section, some suggestions are proffered to explain Marx's failure to see the problems involved.

### The Primary Evils

In about the middle of the first section of this chapter, various strategies were canvassed for proving the unrealizability of Marx's vision of post-capitalist society in either of its phases. Because of the incompleteness of that vision, showing that it is internally inconsistent looked to be unpromising. The other possibility was to argue that some element of the vision is inconsistent with "outside facts," that is, universal background conditions of the natural, biological, psychological, or social environment that any society must face—in short, permanent features of the human condition.

These facts are implicit in the arguments of the second section of this chapter. There are certain permanent features of the human condition—noninstitutional facts—that guarantee that the forces of production, together with the economic system, can never assure the MPCs for each and every person. That is, they guarantee that Marxian abundance can never be achieved. These features of the human condition I call 'primary evils'. In what follows, I give the intuitive idea of a primary evil, together with a (possibly incomplete) list thereof. Following that is a brief discussion of each of them, the purpose of which is to establish their status as primary evils. Finally, an explanation will be given of how the primary evils collectively guarantee the unrealizability of Marxian abundance.

There are some ineradicable features of the human condition that people prefer less of to more of. Susceptibility to injury and disease—biological frailty, if you will—is an obvious case in point. Indeed, mortality itself is something people would like at least marginally "less of," though it matters quite a bit where the extra time would appear, so to speak. To take another example, human beings are not omniscient. Actually, that puts it too mildly: They are profoundly ignorant about a wide variety of things, including matters of great practical importance to them. Profound ignorance, coupled with, say, susceptibility to injury or disease, can be productive of the most serious consequences.

Traditionally, when economists consider the normative evaluation of economic systems, they address the question of how well such systems deliver the goods. Perfect efficiency at meeting wants and needs, relative to a certain state of the productive forces, is the brass ring, and disputes are engaged over how close various systems come to grabbing that ring. Some of those disputes will be taken up in the next chapter.

However, it seems equally important, or at least fairly important, to ask how well economic systems (and the state of the productive forces) deal with the primary evils. I have a hunch that traditional evaluative tools in the economists' kit are ill-suited to this task. Perfect competition, for example, assumes perfect information, from which it can be inferred that no one in a perfectly competitive market society ever gets hit by a truck, unless his "taste for suicide" is such that his undergoing this experience would have made him better off without making anyone else worse off. If not, he would have hired someone to prevent it.

Primary evils need not be intrinsic evils, that is, things that are bad in and of themselves; by analogy with John Rawls' primary goods,[13] they are all-purpose instrumental "bads." For a feature of the human condition to be a primary evil for a person, two conditions must be met: (i) it must be something that person would prefer at least marginally less of to more of for himself, all else equal, and (ii) it is a permanent feature of the human condition.

The *ceteris paribus* condition in the first clause is introduced to take cognizance of the fact that these primary evils can actually serve our interests in certain unusual circumstances. Lest this clause make the concept appear to be ill-defined, the force of the "all else equal" should be understood to mean that, despite occasional cases to the contrary, there will be a strong tendency for, or it is highly likely that, the primary evils will consistently work against a person's interests at the margin.

The notion of permanence in the second condition is intended to indicate past and future presence of the feature in question, where the future extends out as far as it makes sense to talk about societies of human beings. It obviously does not mean for all eternity, since the sun will eventually expand beyond the earth, and humans might not get off in time, or they might annihilate themselves by nuclear war, or they might evolve biologically into pure energy. Though there is some indeterminacy in how far into the future "permanent" extends, it is not too much; the most important concern, of course, is not to mistake a feature of the human condition under capitalism, or class societies generally, for a permanent feature of the human condition *tout court*.

The above defines a primary evil for a particular person; to say that something is a primary evil *simpliciter* is to say that it is a primary evil for nearly every person, insofar as he is rational and desires to achieve or create the good life. These qualifying conditions deserve a brief comment: The two about rationality and the good life are intended in part to capture the kind of person who would inhabit the good society, for example, $PC_2$ society, if that were realizable. I say 'nearly' every person to avoid various walking counterexamples who provide at most a temporary embarrassment to the account that follows. Temporary, that is, until nature puts an end to their miserable existence.

What, then, are the primary evils? The following list, which may not be complete, comprises the most important ones:

1. Moderate scarcity, in Hume's sense[14]
2. Limited benevolence
3. Easy change in some of the repositories of value, that is, some measure of insecurity of possession
4. Conflicting conceptions of the good life
5. Profound ignorance
6. Intermittent stupidity
7. Mortality
8. Biological frailty

In what follows, I explicate the items on this list to show that they really are primary evils for nearly all individuals who are rational pursuers of the good life. (Hereafter this qualifying clause will be suppressed.) For this purpose, each of these has to be relativized to particular individuals. In other words, it is not the bare fact of limited benevolence that, say, Smith would prefer marginally less of to more of. Instead, it is the fact of limited benevolence of the rest of society toward Smith that she would prefer marginally less of to more of.

As noted above, the point of this discussion is to explain why no society can achieve the MPCs for all of its members, which entails that the distribution principle of $PC_2$ society cannot be realized. The main idea is that some subset (proper or not) of the primary evils will always assure some MPC failure for someone or other. It is important to notice that the claim is not that each primary evil guarantees that there will always be some MPC failure for someone (though that may happen to be true). For this reason, I make no attempt to link any of the primary evils to some particular kind of MPC failure. At the end of this section, I shall consider some potential objections to the explanation for, or prediction of, guaranteed MPC failure (of some sort for someone) that is suggested in what follows. The larger purpose of this section is to provide a deeper and more systematic explanation of what is behind the arguments of the second section of this chapter and, by implication, what ultimately explains the unrealizability of Marxian abundance for $PC_2$, or any other, society.

The first item on the above list of the primary evils has been encountered before toward the end of Chapter 5 as part of Hume's "circumstances of (distributive) justice." Actually, the first two items on the list are usually referred to as the circumstances of justice, and a case can be made for including numbers 3 and 4 as well.[15] It has been argued that these items individually or together generate concerns about justice. Though questions about justice are not directly relevant for present purposes, it would be useful to pursue in some detail Hume's thinking on these items to see what they amount to and to see whether or not they are in fact primary evils. It also makes clear the debt this entire discussion owes to Hume. In Book III of *A Treatise of Human Nature*, he says:

> I have already observ'd, that justice takes its rise from human conventions; and that these are intended as a remedy to some inconveniences, which proceed from the concurrence of certain *qualities* of the human mind with the *situation* of external objects. The qualities of the mind are *selfishness* and *limited generosity*: And the situation of external objects is their *easy change*, joined to their *scarcity* in comparison of the wants and desires of men. . . . Encrease to a sufficient degree the benevolence of men, or the bounty of nature, and you render justice useless, by supplying its place with much nobler virtues, and more valuable blessings. The selfishness of men is animated by the few possessions we have, in proportion to our wants.[16]

Traditionally, the circumstances of justice have been identified as moderate scarcity and limited benevolence, and the second part of the quotation supports that. The reasoning is that if scarcity were eliminated, there would be nothing in the way of use-values for people to fight over; they could simply have whatever they wanted. If perfect benevolence reigned, there would also be no conflicts. This argument is sometimes augmented by comparable claims about the other end of the spectrum for both of these circumstances: If scarcity were extreme, say to the point where not all could survive, there would be a war of all against all, and considerations of justice would not arise. In a similar manner, complete selfishness would also render considerations of justice otiose. These arguments can be challenged in a number of ways (indeed, the discussion of the second section of this chapter suggests ways in which a challenge might be mounted), but it provides a preliminary characterization of what the circumstances of justice are.

But notice that what Hume says in the first part of the above quotation is more complex than the second part suggests; he says that there are two "qualities of mind" and two features of external objects that raise concerns about justice. The former are selfishness and limited generosity and the latter are (moderate) scarcity and the "easy change" of external objects (use-values, in Marx's terminology). Let us begin with the latter two and leave considerations about justice behind once and for all.

*Moderate (or Humean) Scarcity.* Hume's discussion of moderate scarcity indicates that the lack of scarcity (in the direction of abundance) would imply the complete satisfaction of all use-value desires. To suppose that a society exists in an environment beyond scarcity in this sense is to suppose that it exists in an environment in which no choices have to be made, and all worthwhile projects can be done. There are no forgone alternatives and thus no costs, that is, no opportunity costs. About such a society, it is not even clear that it would make sense to say that it has an economic system. The latter is a social institution designed to deal with the fact that choices have opportunity costs, that is, to deal with Humean scarcity. Marx comes dangerously close to positing the abolition of this form of scarcity for $PC_2$ society at times, but I have interpreted him in such a way that this is not implied by the distribution principle for that society.

Humean scarcity is pretty clearly a permanent feature of the human condition for nearly every individual; it requires a stunted imagination to conceive of a situation in which every desire for use-values a person might have is satisfied. Does it meet the other condition for a primary evil? Whether the total abolition of Humean scarcity is something people in general would want for themselves is unclear; many reasonably doubt whether or not they would really like to be fantastically rich. At the margin, it seems that things are different. Nearly everyone would prefer a little less Humean scarcity for himself, if only because it expands options, including options to help others. Given the permanence of Humean scarcity, it follows that it is a primary evil.

*Insecurity of Possession.* What about the "easy change" of some of the repositories of value, that is, some measure of insecurity in one's possessions? Hume puts this in the proper perspective in the following passage:

> There are three different species of goods, which we are possess'd of; the internal satisfactions of our minds, the external advantages of our body, and the enjoyment of such possessions as we have acquir'd by our industry and good fortune. We are perfectly secure in the enjoyment of the first. The second may be ravish'd from us, but can be of no advantage to him who deprives us of them. The last only are both expos'd to the violence of others, and may be transferr'd without suffering any loss or alteration. [*Treatise*, pp. 487–488]

The point is simple and straightforward: Many things that satisfy our wants and needs can be taken from us. This fact makes us vulnerable in ways we would otherwise not be. Hume conceived of insecurity of possession as consisting of liability to confiscation of physical objects people own. But in fact insecurity of possession can take forms other than outright confiscation. As was pointed out in Chapter 6, ownership is really a package of rights and powers. It is possible to have some but not all of these taken from us. What makes this important is that the State defines legal property rights, and what it defines it can, and does, redefine. Even in the absence of a State, the community can change the rules of the game. Insecurity of possession consists in the fact that the rules governing permissible and impermissible rights, powers, and so forth, can unexpectedly change. From the point of view of the community, this may or may not be a good thing, but from the point of view of the individual it is clearly not. Insecurity of possession is clearly something a person would prefer marginally less of to more of. Moreover, it is obvious that there is no way to guarantee that the rules will never unexpectedly change, so insecurity of possession is a permanent feature of the human condition. Insecurity of possession is, therefore, a primary evil.

What of Hume's two "qualities of mind," selfishness and limited benevolence? The positing of these as ineradicable features of the human condition looks at first glance to be nothing more than the flimsiest bourgeois subterfuge. But is it really? In his discussion of selfishness, Hume rebukes "certain philosophers" who have overstated the extent of this quality and goes on to say:

> So far from thinking, that men have no affection for any thing beyond themselves, I am of opinion, that tho' it be rare to meet with one, who loves any single person better than himself; yet 'tis as rare to meet with one, in whom all the kind affections, taken together, do not over-balance all the selfish. [*Treatise*, p. 487]

Hume never says how he "adds up" affections, but that problem to one side, what does he mean when he describes men as selfish? What follows is somewhat speculative, but it might be what Hume had in mind.

Someone who is totally selfless would have no concern for his own interests, except insofar as it is required to further the interests of someone else or to further some ideal, and so forth. Like Hume's version of the Pyrrhonian skeptic, such a being is not likely to be met with in experience. A totally selfish person, on the other hand, would have no concern for anyone else's interests, except insofar as furthering those interests is instrumentally valuable. Hume suggests that this sort of individual is also rare. But, what is a selfish person without any qualifying predicates? The above quotation suggests that it is more than totally selflessness and less than total selfishness: It is someone who loves himself more than any single person. But what might that mean?

Though selfishness is a behavioral disposition, it manifests itself most clearly only when there is a conflict of interests between persons. (The same is true of benevolence.) When a selfish person feels that her own interests are powerfully at stake, positively or negatively, the interests of others do not really matter, in an all-things-considered sort of way. That others may be negatively affected is unfortunate but is no ultimate significance for action.

Indeed, calculation of how others' interests are affected may be done cursorily, if at all, under these circumstances. The selfish person may feel remorse, or at least some regret, and may even try to compensate or make amends to those who have been harmed by selfish action, but she will consistently favor her own interests over those of others when she believes her own interests to be powerfully at stake.

As the stakes for the selfish person go down, however, others' interests become systematically more important. Indeed, a selfish person may on occasion make a small to medium-sized sacrifice of her own interests for the benefit of others whose interests are powerfully at stake. The price of non-selfish behavior has gotten to affordable levels. When the stakes are at medium levels on both sides, the selfish person will consistently pursue her own interests ("Better him than me"). However, when the stakes are very small on both sides, a selfish person may behave less consistently. It is not necessary to put a Hobbesian twist on this picture to make it comprehensible, and it may be possible to elaborate this account with some formal machinery to make it more complete and perspicuous. However, as it stands, it may give us some purchase on what Hume might mean when he says that nearly all people are selfish.[17] Most people I know fit this description.

But then both Hume and I travel in bourgeois circles, so it is fitting to ask if this is really universally true. There might be some Darwinian argument for this, but given Marx's profound insights into how social institutions both shape and constrain human behavior, an argument of this sort may not succeed. Whether or not Hume is right about the universality of selfishness is hard to say; for this reason it does not appear on the list of primary evils. In addition, except in certain religious circles, it does not seem to be the sort of thing that people would prefer marginally less of to more of for themselves.

*Limited Benevolence.* Limited benevolence, however, does appear and with good reason. Benevolence can be thought of as a disposition to advance the interests of others for the sake of those interests. Like selfishness, it is most clearly manifested in cases of conflicts of interests, but there are cases where there is both harmony of interests (or even identity of interests), and yet benevolence operates. For example, a mother may act to help her child for the sake of benefiting the child, and yet doing this may also benefit her.

To decide whether or not limited benevolence is a primary evil, it has to be conceived of in a relativized way, that is, as the limited benevolence of others toward the individual for whom it is a primary evil. What, then, does limited benevolence in this sense involve? Benevolence of others toward an individual, call her Robinson, can be limited in two ways: First, it can be limited in extent; there may be some people who will do nothing to further Robinson's interests for the sake of those interests, an enemy, for example, or perhaps a perfect stranger. Second, benevolence can be limited in "intensity." A person may be disposed to act benevolently toward Robinson, but will refrain from so acting when the harm to his own interests reaches a certain level.

Conversely, for benevolence of all members of society to be unlimited toward an individual, call him Jackson, it would be necessary for both of the above-mentioned limitations to be removed: Every member of the society would desire to further Jackson's interests to the same extent as every other member of society; further, each member of society would be willing to act to further the interests of Jackson for the sake of Jackson's interests no matter what the cost to himself. Limited benevolence requires only that at least one of these conditions fails to be met.

Is limited benevolence a primary evil for nearly everyone? First, is it a permanent feature of the human condition? The ways in which benevolence can be limited that are sketched in the preceding paragraph suggest that it is limited in both respects. No one, not even Jesus Christ, is such that every other member of society would be willing to sacrifice his own interests, no matter what the costs, for the most trivial interest of that person. Second, everyone is the object of varying degrees of sympathy and antipathy, as Bentham might put it, from different people.

Limited benevolence also meets the other condition for a primary evil. It is something that, all else equal, one would prefer less of to more of, since increased benevolence of others directed toward oneself is likely to advance one's pursuit of the good life. As in the case of Humean scarcity, people might not rationally prefer the total abolition of limited benevolence directed toward themselves; it might be just too embarrassing. But a little less limited benevolence is something we all could use. There are rare counterexamples to this, namely, a certain kind of irksome pest whom one occasionally comes across, but in general this is not the case.

*Conflicting Conceptions of the Good Life.* The pursuit of human excellence along a certain dimension entails conflicting conceptions of the good life among those who are serious in this pursuit. In its strictest sense, to say that conceptions of the good life conflict is to say that they are logically incompatible. To generate such incompatibilities, it is not necessary to suppose, for example, that the conceptions of the good life held by two dozen different philosophers each includes being the world's best. All that is required is that, say, two-thirds of the world's philosophers want to be among the top one-third.[18] (I ignore complications brought on by faulty factual assessments people often make of their own standing.)

Conceptions of the good life can conflict in less direct ways. The achievement of the good life for one or a number of people may only make it unlikely that others will achieve it. For example, an education at the best colleges increases the chances of many of those who get it for achieving the good life, and correspondingly it decreases the chances of those who do not. Some of the latter may succeed anyway, but among those who did not, it is likely that there will be some for whom that education would have made the difference.

Finally, there are other, nastier ways in which conceptions of the good life can conflict. A fanatic of one sort or another, or even any nonliberal in the broadest sense, will have a conception of the good life that includes others in the society not living a life of a certain sort, irrespective of its direct effects on others. Marx might maintain that all such conflicts are ultimately rooted in class conflicts, where classes are defined relative to the means of production. The disappearance of classes would be followed by the disappearance of these conflicts. But, there is ample reason to doubt this, and there is some reason to believe that these conflicts would be found in any society.[19]

It is easy to see that these conflicts are something people would prefer marginally less of to more of in their own case. Nearly all of us know of cases in which someone did not, but probably would have, achieved the good life if some other individuals had pursued a slightly different conception of the good life. Furthermore, assuming at least that the pursuit of human excellence will be around indefinitely, this will be a permanent feature of the human condition. Conflicting conceptions of the good life is a primary evil for many people.

However, it is not clear that such conflicts would persist indefinitely for nearly everyone. It may be that someday many people's conceptions of the good life will not be in conflict; possibly, these conflicts will diminish when classes are abolished. For this reason it may be inaccurate to describe conflicting conceptions of the good life as a primary evil *simpliciter*. But, as long as there will be such conflicts for some people, conflicting conceptions of the good life will be a primary evil for some members of society.

The preceding four primary evils are all ineradicable features of the social world. This is true even of Humean scarcity, since it involves the scarcity of use-values, most of which must be cooperatively produced. The next four primary evils to be discussed are all features of individuals, though the problems they create are often social problems. Let us begin with profound ignorance.

*Profound Ignorance.* It is often said that human beings are not omniscient. That, however, is just a bit too weak. It's not just that there is some mathematical theorem that people do not know but God does; rather, human beings' lack of knowledge is both extensive and significant, even and especially for their own purposes. As it pertains to their pursuit of the good life, there are three kinds of ignorance that beset human beings. First, people often have hazy, false, and inconsistent beliefs about what the good life for them is. Call this 'personal ignorance'. It is a rare person who knows exactly what he wants out of life until that life is well along—very often too far along. Perhaps part of the explanation for this is that it takes time to know what one's talents and abilities are, as well as the exact nature and extent of the external constraints one is operating under. The utopian character of Plato's *Republic* consists at least in part in the fact that these problems do not arise there. People are assigned their respective stations and duties from birth. In a world where this doesn't happen, failure and social conflict are much more likely.

A second form of ignorance is ignorance about the beliefs and plans of others. Call this 'social ignorance'. Successful cooperation with others, a necessary condition for the good life for nearly everyone, requires a dove-tailing of plans among various individuals that cooperation presupposes. Once societies extend beyond small groups, the successful coordination of plans, which entails the successful coordination of beliefs and desires, becomes a highly problematic affair. I return to this point in great detail in the next chapter, where I argue that the method of production coordination for the economic system of both phases of PC society (central planning) is much more poorly suited to gathering and disseminating information about others' beliefs and desires than is a market economy. That point to one side, every person is ignorant about some important beliefs and desires of others.

A final form of ignorance that besets humans is ignorance of the natural world. This includes ignorance of the underlying laws of nature, but equally important is ignorance of highly particular facts about the natural world that can thwart our plans and goals. This ignorance can be particularly devastating in conjunction with, for example, our biological frailty.

There is some reason to believe that Marx saw the overcoming of ignorance of various kinds as one of the central achievements of PC society, though he does not seem to have thought of it in exactly the above terms. Some have argued that positive freedom, which entails knowledge, is a kind of all-purpose instrumental good for Marx.[20] Be that as it may, the above forms of ignorance are, to a greater or lesser extent, permanent features of the human condition and something a person would prefer less of to more of. These forms of ignorance are, therefore, primary evils.

*Intermittent Stupidity.* Human stupidity, as distinguished from mere igno-rance, is a vast topic, which deserves more careful and systematic treatment than can be given here. Though some people are chronically stupid in that they tend to do stupid things with alarming frequency, most people are not. (There is probably a Darwinian explanation for this.) However, nearly every-one does stupid things sometimes and thus nearly everyone is intermittently stupid. Two forms of stupidity are especially important in the present context. First, there are instances of defective cause and effect reasoning.[21] For ex-ample, unquestioned credulity, when powerful interests are at stake, indicates a failure to reason well about what causes people to say the things they do. Stupidity is an implicitly normative notion and not all defects in causal rea-soning count as instances of stupidity. It is hard to give a general character-ization of this defect, except to say that they are failures of common sense. There are certain things one should know about how the natural and social world works, and failure to know these things, or to make the correct infer-ences based on the latter, counts as being stupid. Another form of stupidity is related to what Jon Elster calls 'myopia.'[22] Roughly, this form of stupid behavior occurs when insubstantial short-term interests outweigh substantial long-term interests. A variation on this is a persistent, if intermittent, con-fusion between mere wants and genuine needs, that is, preconditions for the good life.

Intermittent stupidity is obviously something a person prefers less of to more of.[23] Some interesting work has been done on the psychological mechanisms that are involved in cognitively culpable thought.[24] Although it is not clear that it can be proved outright that intermittent stupidity is a permanent feature of the human condition, there may be a pragmatic contradiction involved in denying it. Besides, there is impressive straightforward enumerative inductive evidence that intermittent stupidity is truly indigenous to the species. If so, it is a primary evil.

*Mortality.* The fact of our mortality is obvious and consequential. It is perhaps misleading, or at least peculiar, to call this fact a primary evil, since to say that a person prefers marginally less of it to more of it is just a roundabout way of saying that time is scarce. Perhaps it is the scarcity of time that is the primary evil. One of the truly astonishing achievements of capitalism is how it effectively creates time. A wide variety of consumer goods serve to lessen the time people must spend to maintain themselves physically. This aspect of consumer society is often overlooked, but it is tremendously significant. To be sure, there are some losses when, for example, cooking the evening meal takes only thirty minutes instead of all day, but the gains are undeniable and palpable. In his biography of Lyndon Johnson, Robert Caro vividly describes the daily chores of the women of the Texas hill country in the Thirties in such a way that the contemporary reader gains a new appreciation of how modern conveniences have created extra time—time in which people can do something other than the awful drudgery that used to be required to meet basic needs.

The open-ended character of most people's conception of the good life makes the scarcity of time something a person would prefer less of to more of, all else equal. This scarcity is also, quite obviously, a permanent feature of the human condition.

*Biological Frailty.* This feature of the human condition requires little comment. Susceptibility to injury and disease is a permanent feature of the human condition that everyone would like marginally less of rather than more of. We can make ourselves less subject to the whims of nature and accident by various preventative measures, but total immunity to the vagaries of nature is unachievable. Biological frailty is, therefore, a primary evil.

The primary evils have a number of things in common; for one thing, they all either are, or imply, some kind of scarcity. Humean scarcity is a kind of general instrumental scarcity, limited benevolence and conflicting conceptions of the good life entail a scarcity of cooperation, the relative insecurity of possessions is a scarcity of security, profound ignorance implies a scarcity of useful knowledge, intermittent stupidity implies a scarcity of good judgment, mortality makes time scarce, and biological frailty implies a scarcity of assurance of physical well-being. These are permanent scarcities human beings and human societies have to face. Let us call them 'the permanent scarcities'. What makes them so important is that any one of them or any subset of them can prevent a person from achieving the good life. This is *why* people prefer marginally less of them to more of them.

The above discussion suggests a generalized version of the arguments of the second section of this chapter for the unrealizability of Marxian abundance based on the facts of the permanent scarcities. The crucial claim in that argument is the following, which I call:

*The Sad Truth*: Some subset (proper or not) of the permanent scarcities will always insure the failure of some MPCs for someone or other.

Notice that the Sad Truth does not say that the permanent scarcities will guarantee failure of some MPC or other for everyone. That would entail the obviously false proposition that no one will ever lead the good life. Though everyone must fence with the primary evils and the permanent scarcities, not everyone will lose. But some will. And that's the Sad Truth.

In light of the above discussion, the Sad Truth should seem intuitively and obviously true, but something has to be said about its epistemological status. The Sad Truth combines a prediction and the reasons for making it. What justifies the prediction can perhaps best be appreciated by an analogy with what justifies predictions in the physical sciences. When it is predicted that Halley's comet will appear at a certain time, the relevant laws and a statement of the initial conditions do not, strictly speaking, entail the prediction. For the prediction to be a deductive consequence, auxiliary assumptions to the effect that nothing else interferes must be added. Alternatively, the prediction can be thought of as only highly probable, relative to the laws and the statement of initial conditions, since it is always possible that something else will interfere.

When what is to be justified is the predicted failure of some MPCs for someone or other, that is, the failure of Marxian abundance, for as far into the future as one cares to look, the premises of the argument consist of statements to the effect that there are these primary evils, which entail the existence of the permanent scarcities. The move from these premises to the conclusion that someone or other will always fail to achieve the MPCs is inductive. Alternatively, there is a comparable tacit clause to the effect that nothing else interferes. The discussion of the second section of this chapter indicates in more concrete terms what level of systematic interference would be required to thwart each and every MPC failure. All this means that the predicted failure is, strictly speaking, only highly probable relative to the statements that there are these primary evils and the attendant scarcities.

This argument relies on an unarticulated conception of, or an intuitive judgment about, what counts as adequate justification for a prediction. However, there is nothing problematic about this reliance. This sort of thing is common practice in the sciences generally, and justified scientific (or, more generally, empirical) belief does not require a philosophical theory of what counts as adequate justification for a prediction. Of course, this means that I may be wrong that this is a reasonable prediction, but that is an unavoidable risk that accompanies any broadly empirical argument, and that is exactly the kind of argument being advanced in this section.

All that said, it is perhaps a slight overstatement to say that the failure of some MPC for someone is always insured by the permanent scarcities and that Marxian abundance cannot possibly be realized. But the probability that Marxian abundance will never be realized is so high that a touch of overstatement is at least understandable.

## Final Doubts and Reservations

Distribution in $PC_2$ society is supposed to be "according to need," which means that Marxian abundance would have been achieved in that society. But the latter means that the MPCs are met for all, and that will never happen. This means that the distribution principle for $PC_2$ society could not be instantiated, which entails that the economic system of that society could not exist. This completes my argument for the unrealizability of the second phase of post-capitalist society.

The only legitimate concern, as far as I can ascertain, is the ascription of this distributive principle to Marx himself. I have addressed this topic on a number of occasions in this chapter and the last, but one final run at it will prove useful, since it goes to the core of what this project is about.

The subject of distribution in the second phase of PC society is something Marx himself had little to say about, and a thorough search of the primary literature has turned up little beyond the discussion in the *Critique of the Gotha Program* that is useful. However, to the extent that the paucity of textual material on distribution in $PC_2$ society is evidence that Marx did not give much thought to the question, there is some doubt that the attribution of Marxian abundance, as I have defined it, to Marx's vision of PC society is accurate or fair. Given the complexity of that concept and the evidence that Marx did not give the matter much thought, is it really appropriate to say that this is what he had in mind?

The easy answer to this question is that the purpose of this project is reconstruction and not mere interpretation. The demands of the Alternative Institutions requirement call for a distributive principle for $PC_2$ society, and I have argued that no other candidate will do. This answer is adequate as far as it goes, but I want to say more than just this.

That distribution of use-values should be in accordance with need is, I believe, a common thread running through nearly all socialist thought over the past two hundred years. It is clearly expressed in the relevant passage from the *Critique of the Gotha Program*. There is also some suggestion of it in a passage in *The German Ideology*: "The communists . . . only strive to achieve an organisation of production and intercourse which will make possible the normal satisfaction of all needs, i.e., a satisfaction which is limited only by the needs themselves" (*GI*, MECW, vol. 5, p. 256). Given Marx's commitment to distribution according to need, the only question is how this is to be understood. Two suggestions were rejected out of hand as obviously too strong: That every use-value desire a person might have is met and that every use-value need a person has is met in a timely manner. The former entails the absence of Humean scarcity and the latter presupposes no human error in the production and distribution of use-values that meet needs. The next part of my case was to argue on that anything weaker than meeting the MPCs for all jeopardizes the disappearance of the State.

There is, I believe, another reason for thinking that the best way to understand distribution according to need is in terms of meeting the MPCs for all that has more of a basis in what Marx actually says. In a crucial passage in *Capital* III he contrasts the realms of freedom and necessity in $PC_2$ society as follows:

> In fact, the realm of freedom actually begins only where labour which is determined by necessity and mundane considerations ceases; this in the very nature of things it lies beyond the sphere of actual material production.... Beyond [the realm of necessity] begins that development of human energy which is an end in itself, the true realm of freedom, which, however, can blossom forth only with this realm of necessity as its basis. The shortening of the working-day is its basic prerequisite. [*Capital* III, p. 820]

This passage clearly indicates that Marx envisioned the subsidance of the economic sphere in $PC_2$ society. It will have done its job each day by noon, or even 10 A.M., so to speak. What is puzzling about this passage, in contrast to the discussion in the *Critique of the Gotha Program*, however, is his prediction of the shortening of the working day. In the latter, he tells us that one of the preconditions for the transition to $PC_2$ society is that labor has become "life's chief want."

What renders these two consistent, and what ultimately provides the key to understanding distribution according to need, is the idea that, in $PC_2$ society most "labor," conceived of as the purposive externalization of human capacities, takes place outside of the economic system, that is, outside of an organized structure of social roles. Labor becomes a form of leisure. Social roles in the economic system impose constraints on the exercise of our talents and abilities because of the externally imposed goal of meeting particular use-value needs. This is why in the Famous Passage in *The German Ideology* he has communist man engaged in solitary labor like hunting, fishing, raising cattle, and doing critical criticism, and in his criticism of Fourier's conception of labor, he uses the example of composing music.

The function of the economic system, together with the productive forces, is to provide the material basis for all this. Marx's belief in the subsidance of the economic sphere in $PC_2$ society makes sense only if he assumes that the economic sphere does everything it can do. The provision of the material preconditions for the good life for everyone by the forces of production and the economic system does exactly that. If one thinks of the economic system as a mere provider of use-values (food, clothing, shelter, etc.), this looks in principle possible. It is the grubby realm of necessity beyond which lies the realm of freedom. But in fact the economic system of any society can do and always does more, and not just in the way of meeting mere wants. This takes us to the primary evils.

Historically, the primary evils have interfered with people's pursuit of the good life as a result of productive force failure, institutional deformity, or both. But, in the Marxian vision, that is supposed to come to an end when the productive forces are sufficiently developed and the social institutions are straightened out, that is, when abundance has been achieved. At that point, though such evils do not go away (after all, people still die, get sick, etc.), they no longer prevent people from living the good life. The MPCs will be in place for everyone. This is the distributive contrast between $PC_2$ society and all hitherto existing societies. Seems like something worth starting a revolution for.

But I have argued that the economic system and the productive forces can always do more with respect to the material preconditions for the good life. This is because the primary evils will always interfere with people's pursuit of the good life. It has been my contention that it will never be possible to meet the MPCs for everyone because of the very nature of these primary evils. Marx's vision of $PC_2$ society effectively assumes a completely satisfactory solution to a whole range of problems that can never be completely solved: the achievement of the MPCs for all. In that vision, there is no room for trade-offs, there is no awareness of inherent limitations. In fact, however, the economic sphere will never subside because the primary evils will always prevent some people from living the good life. In the final analysis, this is the deepest source of social conflict facing all human societies.

# 9

## The Unrealizability
## of the First Phase
## of Post-Capitalist Society

### A Strategy for an Unrealizability Argument for the First Phase
### of Post-Capitalist Society

One of the main problems in intelligently discussing $PC_2$ society stems from the fact that there seems to be no way in principle to say when it might arrive, at least not if Marx is to be interpreted charitably. The time it takes to reach the threshold to $PC_2$ society is indeterminate and potentially significant. Since there seems to be no way of telling how long capitalism will last, it would seem that $PC_1$ society presents a similar problem; in fact, however, this isn't so.

However long capitalist society lasts, one can be sure it will not change very much, both in terms of how its major institutions function and in terms of the systematic social evils that afflict that society. Marx's analysis of the workings of the economic system of capitalist society is intended to be completely general; it is supposed to tell us how any capitalist system functions and for as long as it is a capitalist system. In addition, Marx believes, as well he should, that exploitation and alienation will continue to afflict capitalist society as long as it exists. He also predicts that the general conditions of life under capitalism will continue to be so dismal that it will become completely obvious to the workers that it is in their interests to get rid of it. The only question is how long it will take for them to see this and to recognize that they have the power to do what needs to be done. Finally, however long the transition from capitalism to $PC_1$ society takes, Marx tells us that $PC_1$ society will be "in every respect, economically, morally, and intellectually, still stamped with the birth marks of the old society from whose womb it emerges" (*CGP*, p. 16). This sobering observation should give pause to those who believe that the act of proletarian revolution will unrecognizably transform

240

the workers. All of these points taken together indicate that it is possible to know much more about what $PC_1$ society would be like, at least at the outset.

If a substantial continuity between capitalism and $PC_1$ society can be assumed, this permits a much less abstract or a priori discussion of $PC_1$ society than was necessary for $PC_2$ society. In addition, the arguments of the last chapter about the unrealizability of Marxian abundance allow us to assume that the primary evils will be around and will prevent the satisfaction of all the MPCs for everyone. The economic sphere will not subside. Finally, the arguments of Chapter 6 establish the nature of the economic system of $PC_1$ society and tell us something about the social functions the state will serve in that society.

All of the above provides some material to work with in arguing for the unrealizability of Marx's vision of $PC_1$ society, but how will it be used? The complete answer to this question will emerge over the rest of this chapter. For now, I shall indicate in a general way what the main argument will be and say something about what is required to make it work. As noted in the last chapter, unrealizability arguments are inconsistency arguments of some sort. To show that Marx's vision of $PC_1$ society is unrealizable, I shall show that there is an inconsistency in Marx's claim that $PC_1$ society will have the social institutions he envisioned, especially a centrally planned economy, and yet will eliminate systematic exploitation and alienation. As was the case with the argument of Chapter 8, the argument will make use of "outside facts," indeed a more substantial use of such facts than was found in the former argument.

The grim experience of the peoples dominated directly or indirectly by the Soviet Union over the past three-quarters of a century would seem to provide ample empirical data on which to build this case. Gross waste, inefficiency, and general impoverishment (as well as occasional famine) have characterized twentieth century societies with centrally planned economies. However, straight extrapolations from the Russian (or Cuban or Bulgarian) experience are in fact quite risky. History is shaped by a confluence of trends and circumstances, interactions among institutions (some of which are quite peculiar historically), as well as the effects of outstanding individuals—world-historical individuals, as Hegel would say. The manifest failure of the economic system in, for example, the Soviet Union and Cuba could perhaps be accounted for by factors external to central planning itself.

Indeed, the most obvious way to discount the failures of existing centrally planned economies from a Marxist perspective is to call attention to the fact that all socialist revolutions in this century took place in countries that had not really gone through a bourgeois revolution and whose forces of production were insufficiently developed. It could be claimed that a centrally planned economy can succeed if circumstances and conditions are right—and, so far, they just have not been right.

These considerations make it evident that a quick appeal to the failures of contemporary Communist societies is insufficient; if these failures are to be of any use against Marx's vision of $PC_1$ society, the evidence must be handled carefully and augmented with other considerations. Indeed, since $PC_1$ society has never existed anywhere, any empirical argument about this society will have to proceed at a much higher level of abstraction than one customarily encounters in discussions of this sort. On the other hand, the constraints imposed by this project rule out the use of models of the sort commonly employed by contemporary economists because the topic of discussion is genuine social institutions; this requires a degree of realism not generally found in these models and discussions of them.

Bearing these concerns in mind, the most obvious way to show that the economic and political systems of $PC_1$ society are incompatible with the end of alienation and exploitation is to show that these institutions will in fact produce or be responsible for these social ills. And, this is the strategy I shall pursue. The kind of arguments to be deployed will exactly parallel the critical explanations of the defects of capitalist society advanced in the first part of this book. To see what is involved in this, it would be helpful to recall some material from Chapters 2, 4, and 5 about alienation and exploitation under capitalism. Chapter 2 tells a depressing story about a wide variety of undesirable features of capitalist society that can be brought under the heading of alienation. More exactly, there were four subheadings under which various forms of alienation were grouped: (1) alienation from the products of labor; (2) alienation from the activity of laboring; (3) alienation from species being; (4) alienation from others.

What would Marx say, or what is he committed to saying, about these various forms of alienation as they pertain to PC society? Presumably, he would say that all forms of alienation would eventually be eliminated, but 'eventually' could take in $PC_2$, as well as $PC_1$, society. On the other hand, even if he would predict the end of all forms of alienation in $PC_1$ society, he would not have to say that they would all be gone from day one after the revolution. What is the minimum Marx would have to say? It is hard to be very precise in this regard, but this much seems obvious: He would say that some forms of alienation would disappear immediately, namely, those directly tied to structural features of capitalism such as the unnatural character of production for exchange. Others, such as the frustration of the need for truly human labor, might be subject to more protracted amelioration and eventual elimination, though Marx would undoubtedly predict some immediate and palpable improvement on this score. The main idea is that alternative social institutions of $PC_1$ society must not systematically or persistently reproduce substantial alienation in any of the categories identified above.

For an unrealizability argument to succeed on this score, it would have to be shown that this is false. One potential difficulty is that it is unclear what would count as substantial alienation in any of these categories. This makes for a certain indeterminacy in the level of alienation that could be found in $PC_1$ society without falsifying the claim that this society would have virtually eliminated alienation. In theory, the dispute between Marx and his critics might founder on this problem.

In practice, this may not be a serious problem, however. The dispute between Marx and his critics about alienation can be phrased simply, and sharply, as follows: Would the alternative institutions of $PC_1$ society produce substantial alienation or not? This question could have a definite though perhaps complicated answer, and that could be so even if what 'substantial' means must be left at an intuitive level. To sum up, for the unrealizability of this aspect of $PC_1$ society to be proved, it would be both necessary and sufficient to show that the institutions of $PC_1$ society would produce substantial alienation in at least some of the basic categories identified by Marx in his critique of capitalism.

The topic of exploitation, as it might pertain to $PC_1$ society, carries with it some substantial complications, since there are a number of different forms of it. Marx charges that the workers are systematically surplus value exploited by the capitalists. Chapter 3 shows that his argument for this claim is defective because of its reliance on the Labor Theory of Value. Marx's conception of surplus value exploitation can be turned against him, as it were. In the second part of the fifth section of this chapter, I shall argue that a substantial number of the workers would be surplus value exploited in $PC_1$ society. This means that this society would be plagued by one form of systematic exploitation.

Chapter 4 argues that all the various charges of systematic parasite exploitation against capitalism, save one, either fail or cannot be part of a radical critique of capitalist society. The exception is what I called 'generic parasite exploitation'. This form of exploitation is the forced, harmful utilization of one person by another for the benefit of the latter. The theory of alienation was rung in to detail the various harms that would be suffered by the workers, and people in general, under capitalism as a result of this form of exploitation. Some of the material in the sixth section and at the end of the fifth section provides some reason to believe that generic parasite exploitation will exist in $PC_1$ society, though I shall not argue for that in any detail.

The other basic kind of exploitation, discussed in Chapter 5, was property relations exploitation. John Roemer charges that the workers are capitalistically exploited under capitalism because the following three conditions hold:

i. If the workers were to withdraw from society with their per capita share of society's alienable assets and their labor and skills, they would be better off in terms of their income-leisure package.
ii. If the nonworkers were to withdraw under the same conditions, they would be worse off.
iii. If the workers were to withdraw with only their own endowments, the nonworkers would be worse off.

Two problems were discovered with Roemer's account. First, he did not establish that the workers possess valuable skills and traits (willingness to take risks, low time preference, and entrepreneurial ability) to about the same extent as nonworkers; this is necessary to show that the workers would be better off by pulling out. This empirical issue was left unresolved. A more serious problem for Roemer is that the alternatives are specified too abstractly. In other words, it is necessary to take the withdrawal metaphor more seriously. In particular, an alternative institutional framework has to be specified, and it must be argued that the workers would be better off under such a system. The discussion of Chapter 6 indicates that this specification, at least as it pertains to the economic system, must include an account of the dominant relations of production, a method of production coordination and a distribution principle.

For a Marxist to prove that the workers are capitalistically exploited under capitalism, he would have to argue that the workers would be better off under the institutions of $PC_1$ or $PC_2$ society. However, since the economic system of $PC_2$ society is unrealizable (because the distributive principle of that society cannot be realized), the Marxist would have to show that the workers would be better off under the institutions of $PC_1$ society. Because these institutions were not specified in Chapter 5, this issue was left unresolved. We are now in a position to resolve it. In the fourth section of this chapter, I shall argue that the Marxist cannot sustain a charge of capitalistic exploitation against capitalism because the institutions of $PC_1$ society will not make the workers better off in terms of their income-leisure package than they would be under capitalism. In this way, a crucial piece of unfinished business from Part I of this book will be taken care of.

It is important to understand that even if the Marxist cannot sustain a charge of capitalistic exploitation, it may still be the case that the workers are capitalistically exploited under capitalism for the simple reason that there may be realizable alternative institutional arrangements other than those of $PC_1$ society, for example, market socialism, under which the workers would be better off than they are under capitalism. So, it may turn out that the workers are in fact capitalistically exploited under capitalism, but if the main claim to be established in the fourth section of this chapter is true, it would take a non-Marxian argument to substantiate this. Indeed, it would seem next to impossible to prove that the workers under capitalism are not capitalistically exploited *tout court* because proving this would require a pairwise comparison of capitalism and all other feasible, or realizable, alternative institutional arrangements. I have no idea how one would go about constructing a list of all realizable alternatives, though I do have some thoughts on how one would proceed once these alternatives are in hand. (These thoughts are implicit in the second and third sections of this chapter.)

One form of property relations exploitation not discussed by Roemer is what I shall call 'marx exploitation'. This form of exploitation is a kind of mirror image of capitalistic exploitation. The intuitive idea behind marx exploitation can be explained as follows: Call a marx exploited coalition 'boat people'. What makes boat people exploited is that they would do better in terms of their income-leisure package under capitalism, starting only with their labor and skills, than they would under the economic system of $PC_1$ society. A marx exploiting coalition, on the other hand, differentially benefits from the institutions of $PC_1$ society in the sense that they would be worse off by withdrawing into capitalism. On the other hand, there may be some people whose position would remain unchanged as a result of withdrawing, so the marx exploited coalition and the marx exploiting coalition are not collectively exhaustive. The following makes all of this more precise:

A coalition $W$ is marx exploited if and only if:
  i. If the members of $W$ were to withdraw from $PC_1$ society with only their labor and skills into a capitalist society, they would be better off in terms of their income-leisure package than they would be by staying in $PC_1$ society.
  ii. If some subset of the complement of $W$ were to withdraw from $PC_1$ society with only their labor and skills into a capitalist society, they would be worse off in terms of their income-leisure package. Call this 'the marx exploiting coalition'.

In the first part of the fifth section of this chapter, it will be shown that the workers would be marx exploited in $PC_1$ society.

To sum up, the arguments of this chapter will attempt to establish the following:

1. The workers would not be better off in terms of their income-leisure package under the institutions of $PC_1$ society than they would be under capitalism. That is, the Marxist cannot sustain the charge that the workers are capitalistically exploited under capitalism.
2. The workers would be marx exploited under the economic system of $PC_1$ society.
3. A substantial number of workers would suffer surplus value exploitation in $PC_1$ society.
4. The social institutions of $PC_1$ society would produce substantial alienation in the four major categories of alienation discussed in Chapter 2.

Number 1 addresses unfinished business from Chapter 5, namely, whether or not a charge of capitalistic exploitation against capitalism can be part of Marx's radical critique of capitalist society. Numbers 2 and 3, on the other hand, show that there will be systemic exploitation in $PC_1$ society, and (4) shows that there will be substantial systemic alienation in that society. Given that Marx's vision of $PC_1$ society consists of the relevant institutions together with the absence of systemic exploitation and alienation, the latter three propositions prove that this vision is unrealizable. Of course, any one of these would be sufficient for this, but part of my purpose is to show that $PC_1$ society would face at least the same problems that confront capitalism. These problems may be more severe, and it may be that $PC_1$ society would face distinctive problems of its own, but I shall not try to establish any of that in this book; it is very difficult to prove things about societies that do not exist.

There is a slight but noteworthy difference between what the argument of Chapter 8 shows and what the argument of this chapter shows. The argument of Chapter 8 shows that one of the *institutions* of $PC_2$ society—the economic system—cannot be realized, because the distribution principle is unrealizable. The argument of this chapter is willing to grant the realizability of the institutions of $PC_1$ society, including the economic system. What is historically impossible, according to the present chapter, is the existence of the economic system together with the absence of exploitation and alienation. Both arguments, however, show that the respective *visions* of PC society are unrealizable. This of course shows that Marx cannot meet (and not just has not met) the Alternative Institutions requirement for a successful radical critique of capitalist society, which in turn implies that Marx's radical critique must be judged a failure.

There is an important parallel between what is required of the radical critic's opponent and the burdens the radical critic himself must shoulder. The Critical Explanations requirement for a successful critique of the existing order demands that the radical critic explain systematic social evils by appeal to essential features of the offending social institutions. This had to be done at a level of abstraction that transcends the peculiarities of particular capitalist societies, and indeed this was precisely the sort of argument deployed in Chapters 2 through 5 of this book. The radical critic's opponent faces the same challenge; that is, he must argue that the defects of the radical's social vision are rooted in the basic alternative social institutions.

The arguments of this chapter attempt to do just that. Central to all of them is an account of how the economic system of $PC_1$ society functions. This account will emerge in the next two sections in the course of a discussion of an important debate in the history economic thought. The focus of this debate is the claim by the Austrian economist Ludwig von Mises that rational economic calculation in a centrally planned economy is impossible. Though the discussion of the next section shows that Mises has overstated his case to some extent, his argument, together with some additional insights from F. A. Hayek, can be construed to show that the method of production coordination of the economic system of $PC_1$ society (i.e., central planning) is vastly inferior to the method of production coordination of capitalism (the market), no matter what the attending circumstances and conditions, including worker control of the means of production. This will provide the basis for the arguments for (1), (2), and (3) above. The understanding of how a centrally planned economy operates furnished by this discussion will also lay the groundwork for the arguments pertaining to (4).

## The Coordination of Production in a Centrally Planned Economy[1]

The argument of this section has, by contemporary standards, a long and checkered history. It was first stated in 1919 when the economist Ludwig von

Mises published a paper[2] in which he argued that rational economic calculation in a centrally planned economy is impossible. Since $PC_1$ society has a centrally planned economy, it follows that rational economic calculation in $PC_1$ society is impossible. Exactly what this claim amounts to will take some unraveling, but first some historical background would be in order.

Mises's paper inaugurated what came to be known as the Socialist Calculation Debate.[3] His challenge was taken seriously by some economists with socialist sympathies. The textbook account of the debate has it that these economists, notably Oskar Lange, in his "On the Economic Theory of Socialism," and Fred M. Taylor, in his "The Guidance of Production in a Socialist State"[4] decisively refuted Mises. It is also commonly held that an argument similar to Mises's, advanced by F. A. Hayek, was also defeated by the Lange-Taylor argument. A host of other figures took sides in the debate, but the consensus is that Mises and Hayek were decisively refuted. Despite this convergence of opinion (or conventional wisdom), we know that debates of this sort are not settled by a show of hands.

Great and long-lasting debates almost always involve some misunderstandings on both sides about exactly what is at issue. This one is an outstanding case in point. More that his proponents are willing to admit, Mises himself was responsible for much of the misinterpretation of his position by the neoclassical economists who opposed him. He often stated that he proved the impossibility of socialism, and he seems to have regarded his proof as a priori. On the face of it, neither of these claims seems plausible. However, there is a core argument that has not yet been touched by Mises's neoclassical critics. In what follows I reconstruct this argument and explain why it is basically sound. Its relevance to arguments for the unrealizability of $PC_1$ society will be systematically explored in the remaining sections of this chapter.

A first step toward understanding Mises's challenge can be taken by way of a brief review of some material in Chapter 6. Any society must have some method of coordinating production. This is not a serious problem for relatively simple societies in which the production of consumer goods involves relatively few steps and is often undertaken by the consumers themselves. Since the rise of capitalism, however, consumer goods have rarely been produced directly from original factors of production (labor and natural resources). Often many stages of production are required, and many nonspecific factors go into a variety of production lines. On the face of it, this is an allocation problem— how limited resources (i.e., specific and nonspecific factors of production, including labor power) are to be allocated to various production units to meet the ultimate goal of the economic system, namely, the production of use-values. In short, the activities of various production units, however the latter are defined, must be coordinated. Poor coordination means waste or inefficiency, and while no method of production coordination is perfectly efficient, some are better than others. One way of conceiving of Mises's charge against any centrally planned economy is that the method by which production is coordinated in such an economy must, by its very nature, be profoundly inefficient. In a way, the claim is essentially comparative in nature; the standard of comparison, however, is not a Pareto-Optimal allocation of resources (where no one can be made better off without someone else being made worse off); instead, the standard of comparison is what is achieved by the operation of the market under capitalism.

This means that to understand the nature of the challenge Mises poses for a centrally planned economy, it is essential to understand his vision of how capitalism solves the problem of production coordination. Under capitalism, the consumers transmit their needs and wants to the those who control the means of production by their acts of buying and abstentions from buying. The retailer, the wholesaler, the manufacturer, the supplier of capital goods, and the owners of the original factors of production all get information from the marketplace in terms of prices and costs.

The existence of pure or entrepreneurial profit is an indication that the existing productive structure is not meeting the needs and wants of the consumers as satisfactorily as it could. In consequence, capital flows into those lines where demand is not being adequately met; factor prices are bid up and product prices are driven down; in consequence, the system becomes better coordinated relative to consumer wants and needs as expressed by their dollar "votes." Entrepreneurial profit is not to be confused with the capitalists' return on investment (*qua* capitalist).[5] The fact that most entrepreneurs are also capitalists can obscure this fact. Entrepreneurial profit is the residual left over after all the productive factors (including means of production) have been paid for at their market price. Let us call a decision 'entrepreneurial' if it is a decision to initiate, maintain, expand, contract, or terminate a line of production (however one defines the latter). Functionally defined, the entrepreneur is *not* an owner of a factor of production. As a production organizer—and not as a manager—he sets the parameters, if not the details, on what and how much to produce, that is, the product line and the production quota. Price information is clearly central to this process, though the distinctively entrepreneurial act consists of noticing the right combination of information.[6]

Critics of capitalism have long challenged basic elements or consequences of this story: The demand for consumer goods is manipulated, some say controlled, by the owners of the means of production or their minions; the distribution of income significantly shapes what gets produced and how much it costs; monopolies distort allocations in favor of certain firms; and so on. In addition, recently, neo-Marxist economists have charged that the distribution of income can in part be explained by the ruling classes' use of economic and political power.[7] But even if these points are well-taken, they are consistent with the central claim in the above story: Resource allocation, and thus the coordination of production in a capitalist economy, is achieved by entrepreneurial decisions based on prices and motivated by the desire for profits. (Note that, by contrast, there is no theory of how prices are formed based on the use of economic and political power.) At most, some of the above complaints cite factors that can disturb or hold up the operation of the dominant tendency toward coordination that operates in a competitive capitalist economy. However, without an account of the operation of this tendency, there is no causal explanation for "reproduction" in a capitalist economy.

To be sure, capitalism operates according to no overall plan. Entrepreneurs are as myopic (and only as myopic) as their personal plans require. This is true even if there is some better way production plans can be coordinated. But, according to Mises, if markets and market prices are abolished, there will be no adequate data by which to make rational entrepreneurial (i.e., allocational) decisions for the production of consumer goods and producer goods. Those making allocational decisions will be inevitably and forever "in the dark." We have yet to detail exactly why he thinks that this must be the case. Before articulating that, let us look more carefully at the nature of the problem that must be solved by any moderately complex economy, whether or not it is a capitalist market economy or a centrally planned economy.

The basic problem of production coordination is pressing because Humean scarcity will never disappear. There will always be competing demands on limited resources. This means that some projects must be forgone if others are to be seen through—and there are always useful projects that can be done. These forgone projects represent costs—opportunity costs. If these opportunity costs are to be calculable and if competing demands on limited resources are to be rationally adjudicated, resources, including labor power, must be rendered commensurable as to their scarcity value. Calculation, as opposed to mere guesswork, requires that this value be expressible in units of something. In short, productive resources must have prices.[8] Without prices, there is no way to represent scarcity value in a calibrated way. That means there is no way to adjudicate conflicting demands on limited resources. This entails that allocational decisions in the absence of prices cannot be rational, at least not in a complex economic system.

This poses an obvious and immediate problem for central planning as it has traditionally been understood. As was pointed out in Chapter 6, central planning proceeds by what Peter Rutland has called the Method of Material Balances.[9] The plan assigns production units a certain quantity of inputs and calls for a certain quantity of outputs. Both inputs and targeted outputs are expressed in physical terms, not value terms. This is appropriate in a system of production for use, since the latter requires the conscious linking of means to ends from raw materials through producer goods to use-values (i.e., consumer goods). If rational economic calculation requires prices that reflect relative scarcity values, then Mises's claim that rational economic calculation is impossible in a centrally planned economy seems sustained simply because central planning uses the Method of Material Balances and makes no use of prices.

However, it seems at least possible for central planning to use prices as expressions of relative scarcity so as to calibrate the value of inputs and outputs.[10] But, they cannot be *market* prices, that is, prices set by individual autonomous firms, where the latter's purpose is to get exchange value in the market; in this case, the system is one of commodity production, that is, a market economy. Instead, prices must be set by some non-market mechanism. And, the Method of Material Balances must be the primary method by which resources are actually allocated (and thus production coordinated) if central planning is to be a genuinely centralized system in which decisions to produce use-values ultimately guides production in the manner indicated above, that is, if it is to be a system of production for use.

What role, then, would there be for prices in such a system? The non-market prices of a centrally planned economy could serve to inform the planning process on the question of relative costs, even if such prices do not determine what and how much gets produced. For example, the planners may not want to shut down a large factory that is operating at a loss, but they would want to know that it is operating at a loss, and they would want to have some idea of the extent of those losses. Pricing the factors of production is a necessary condition for knowing either of these.

Defenders of central planning (which includes Marx) might agree that the problem of relative scarcities cannot be ignored and must be solved by pricing the factors of production. However, they might maintain that this problem really has already been solved. After all, post-capitalist society does not come onto the scene *ex nihilo*; it replaces capitalism. The coordination achieved by market prices under capitalism can, with minor adjustments, be maintained by the new social order. Unfortunately this will result in a method of calculation that will not be rational. The reason for this fact is, in a word, *change*.

If the central planners are to allocate resources rationally to meet wants and needs, they must take into account five broad categories of change:[11]

1. *Change in the physical world.* Soil becomes exhausted, minerals and petroleum are consumed, weather and climate change. Many environmentally conscious socialists envision a world almost totally dependent on renewable resources.[12] It is pertinent to wonder whether this presupposes a technological deus ex machina that rivals that of the most fanatical advocates of growth. However, even if it does not and a decent standard of living can be achieved by an almost complete reliance on renewable resources, the problems involved in dealing with changes in the physical world remain formidable. Mutually adjusting changes that bring an ecosystem into balance can dramatically alter those features of the natural world humans happen to find valuable. An acre of fertile land can be kept at a constant level of fertility almost indefinitely, provided that one is willing to devote sufficient resources to keeping it that way. For the renewal of these resources to be rational, changing opportunity costs must be calculated; thus any historically given prices must be changed over time.

2. *Changes in population.* The absolute number and demographic profile

of a population changes over time. The rough outlines of some of these changes can be predicted, but many of the ramifications cannot.

3. *Changes in means of production.* Machinery and tools wear out and have to be replaced. However, it is not rational to do this if the social wealth that is invested could be better used to produce different means of production (either to produce the same products in a different way or to produce different products). In addition, new means of production must be produced if social welfare is to be increased or even redistributed.

4. *Technological change.* New techniques of production will be discovered and decisions must be made about whether and how they are to be used. Not all technological change counts as an improvement vis-à-vis the task of meeting the goals of production. Suppose that a change in a chemical plant can produce greater output with the same raw material input. Is the plant more efficient? Not necessarily. If the process requires more labor or more highly skilled labor (that could be better used elsewhere), it might not be. It also might be less efficient if the machinery wears out more quickly. Technical change, like all other change, ramifies in ways that require new decisions.

5. *Changes in consumer preference.* Marx reminds us that human needs are historically determined. It is an impoverished vision of post-capitalist society that assumes that human needs in the future will differ in only minor ways from what they are under capitalism.

Change is inherently disruptive to whatever coordination of production has been achieved in the economic system of any society, but it is an economically significant fact of life. In the words of G. L. S. Shackle, we live in a kaleidic world.[13] Continual and continuous change require that entrepreneurial decisions be made all the time. Schumpeter partially recognized the importance of entrepreneurial decisions under capitalism in his discussion of the "creative destruction" of the innovator-entrepreneur.[14] The vision that has been sketched here conceives of entrepreneurial activity in a capitalist economy as much more pervasive and fundamental than Schumpeter imagined. Entrepreneurial activity is not limited to the application of major technological innovations. Entrepreneurship consists of marshaling resources to meet changing conditions; and conditions are changing all the time. Even the decision to continue to produce a given product in the same quantity is speculative and entrepreneurial in a kaleidic world. In a world of no change (in the five broad categories of change outline above), there would be no entrepreneurial activity.[15]

What this discussion of change shows is that the price structure of a centrally planned economy must change to reflect changing economic realities, if the productive apparatus is to be tolerably coordinated to meet human wants and needs. This means that simply keeping the price structure of the previously existing society is irrational. To put it another way, rational economic calculation requires changes in the valuations of the various factors of production.[16] The nature of the challenge Mises poses for the pricing mechanism for a central planning should now be fairly clear. What is not yet clear is why he believes that it cannot be met.

To get a handle on this, it is again necessary to proceed by way of comparison with capitalist market economies. More specifically, it is necessary to explain how market prices—and, in particular how changes in market prices—coordinate and re-coordinate production for a capitalist market system in a changing world. It is at this juncture that Hayek's contribution to the debate takes center stage.[17] Hayek's central insight is that market prices are fraught with epistemological significance. The price at which someone is willing to buy or sell tends to reflect his plans for the money or the object he would get in exchange and his knowledge of alternatives. The prevailing market price (if there is just one price) amalgamates the knowledge and plans of (some of) the participants. Changes in prices reflect changes in those plans or beliefs about alternatives.

Let us consider an example. Suppose that a new use for tin is discovered or an existing sources of tin dries up unexpectedly. All the users of tin should know that it should be used more sparingly. A rise in the price of tin indicates this. The other users of tin do not know why tin should be used more sparingly, but for their purposes, they don't have to know. The rise in price simply says, "Use less tin." The height of the rise gives them some indication of how much less should be used. The actual drop-off in demand cannot be predicted very well, since how much less will be demanded at the new price will depend on the plans of the users and the knowledge of alternatives open to them.

A further consequence of the rise in price is that the production of substitutes will increase, as will (eventually) the production of substitutes for substitutes, and so on. Of course, what counts as a substitute will vary from one individual or group to another, depending on the uses for which they have plans. Some uses of tin will no longer be economical, and production of those items will cease. Exploration for new sources of supply may increase. In short, the structure of production will undergo a number of changes which may be more or less significant. That is, production has been re-coordinated to meet changes in the underlying economic realities (here a drop in supply or an increase in demand for tin). No one person or committee had to make all of these decisions. Prices transmit the necessary information from the owners of the tin mines to those who wanted and needed to know. The price system in a market economy is a vast and complicated information network in which producers (and consumers) communicate information by their acts of buying and abstentions from buying and by bidding prices for producer and consumer goods up and down.

Entrepreneurial profit (loss) comes from being among the first (last) to notice, either through luck or foresight, that people's plans are not as well-coordinated as they might be relative to the underlying economic realities. By the actions they take in response to changing realities, they signify to others to follow suit. For example, if a speculator learns of an ecological disaster in the cranberry bogs of eastern Massachusetts (where most of the world's cranberries are grown), he immediately begins buying up all the cranberries on hand that he can. Competition tends to wipe out entrepreneurial gains as time passes and more speculators enter the field.[18] What causes the price to go up is the fact that some owner of cranberries sees a profit opportunity and raises his prices to buyers. Other owners soon follow suit. It is an obvious but important truth that people, and not "the market," raise the price. The Mises-Hayek insight is that *all* profit is the result of essentially speculative estimates about changing economic realities. The desire for profits makes the knowledge provided by previous or existing prices effective knowledge, since those affected have a strong incentive to act on the perceived discrepancy between actual prices and prospective prices. In a market system, present prices tend to reflect present and past economic realities; price changes represent changed estimates about future realities. Since the number of different possible combinations of productive factors is enormous, successful entrepreneurship is usually a matter of reshuffling a number of different resources to meet changing realities.

The entrepreneur does not upset equilibrium as Schumpeter believed; on the contrary, it is the quest for entrepreneurial profit that moves the system toward equilibrium, insofar as entrepreneurs are successful. Equilibrium exists if and only if everyone's plans are perfectly coordinated.[19] To put it another way, the potential for entrepreneurial profit will exist as long as the plans of individuals and firms are not perfectly coordinated. As long as some factor of production (specific or nonspecific) is or could be used in some undertaking somewhere to generate entrepreneurial profit, coordination will not be perfect, and the system will not be in equilibrium. It is of considerable importance to see why this will always be the case. Part of the answer should already be obvious. Changes in the underlying economic realities disrupt coordination among various branches of production. In fact, however, the occurrence of change is not a sufficient condition for the possibility of entrepreneurial profit. After all, if all relevant changes could be foreseen by those people and groups that are affected, those changes would already be reflected in factor prices and no profit opportunities would exist. A necessary condition for entrepreneurial profit is ignorance on the part of some market participants about these underlying realities.

One might wonder why ignorance is so widespread in a market economy. Actually, it is not; the knowledge exists, but it is socially scattered. Everyone possesses or comes to possess economically significant knowledge of a very particular nature. For example, the farmer knows that his tractor cannot make it one more season and that he will need to buy a new one before next year; the tin mine operator learns that his deposits are not as extensive as he had thought; the peach orchard owners know of a local killing frost; a Marxist pamphleteer knows he will need the services of a printer to get out his latest warning about the plans of the multinational corporations. The "economic problem" is to transmit that knowledge to others who need to know about it—and it must be transmitted in a way that is effective for action. According to Hayek, the economic problem is *not*, as standard neoclassical economics maintains, a question of how to allocate scarce resources among competing ends, where both the resources and the ends are given. Rather, Hayek sees the problem as one of how dispersed and fragmented knowledge not given to any one mind can be socially mobilized. Though market systems are never perfectly efficient, in the sense that everyone's plans are perfectly coordinated, these systems possess a mechanism that allows and motivates individuals and firms to make the system more efficient, that is, better coordinated, when they discover uncoordinated plans arising from the ignorance of some people about present or future realities. Given the empirical fact that some people have the knack for becoming alert to price discrepancies (i.e., uncoordinated knowledge and plans), the market provides a systematic way to wipe out social ignorance.

An adequate understanding of how market pricing in a capitalist economic system solves the problem of production coordination helps to make clear what is required of a centrally planned economy. We have seen that rational economic calculation in a centrally planned economy requires the (non-market) pricing of factors of production. Although it is not irrational to have a production unit in a centrally planned economy producing at a loss, it would be irrational for those in control to make no effort to find out the extent of those losses. As was shown above, only by pricing factors of production can competing demands on scarce resources can be systematically adjudicated. Mises never explicitly considered a pricing mechanism for central planning because in the late nineteenth and early twentieth century, advocates of central planning envisioned the abolition of money prices.[20] However, his argument can be easily extended to cover central planning with a pricing mechanism.

Mises's basic insight is that the abolition of market prices creates insurmountable knowledge problems for the planners. *Given* a complete set of consumer preferences, *given* a certain level of technological development, *given* various quantities and qualities of natural resources, means of production, and labor power, marginal productivity theory yields a Pareto-optimal allocation of resources. Mises's most famous opponent in the "great debate," Oskar Lange, took Mises to be denying that a Pareto-optimal allocation is possible for such a system.[21] However, the problem for Mises has always been that the above "givens" are not given to any one mind or small group of minds. With the assumption of perfect knowledge, standard in neoclassical economics, the latter clearly begs the question. Mises was always willing to grant the feasibility of central planning if God were the chief planner. (This is not wholly tongue-in-cheek; some Europeans, at least since the time of Hegel, have been known to regard the State as possessing some of the attributes of the Deity.)

The basic problem is that changes in the underlying economic realities are revealed directly to individuals scattered throughout the economy and not to the planners. In extending Mises's argument, the crucial claim to be established is that any non-market pricing mechanism in a (complex) centrally planned economy must be completely inadequate to the task of tending to reflect accurately these economically significant changes.

The argument for this starts from the observation that the ultimately significant changes discussed above are of a very specific nature—the mechanical integrity of a particular machine, the fertility this year of this plot of land, Farmer Brown's need for a particular constellation of farm machinery at various times, the unexpected discovery of new coal deposits of a certain grade at a certain location, and so forth. These facts are not under the control of the planners and thus are not directly known to them. Even whether or not previous plans have been successfully carried out is not directly known to the planners. How, then, will they find out about these changes?

Presumably, reports will be filed by those at the periphery, that is, those closest to the changes. However, those at the periphery cannot transmit to the center everything they know (or, more accurately, everything they think they know) that might be relevant. The relative value-significance of the information that is revealed at the level of the individual or the firm will always be somewhat unclear in a complicated system in which the ultimate goal of creating use-values guides all production, even production that is at greatest remove from the consumer. In consequence, those who gather the information must be given guidelines (how are these to be determined?), and their knowledge must be summarized in statistical form to be manageable. Furthermore, as information flows up to individuals or committees with wider and wider areas of responsibility, those who are responsible must evaluate what information will be most significant for the ultimate decision makers. The problem with statistics as a basis for economic decision making was clearly recognized by Hayek nearly forty years ago:

> The statistics which such a central authority would have to use would have to be arrived at precisely by abstracting from minor differences between things, by lumping together as resources of one kind, items which differ as regards location, quality, and other particulars, in a way which may be very significant for the specific decision. It follows from this that central planning based on statistical information by its very nature cannot take direct account of these circumstances of time and place.[22]

Those who view aggregates (e.g., price levels, GNP figures, various indexes) as the basic economic realities often miss this point; much economically significant knowledge will always remain beyond the comprehension of any one mind or even a small group of minds. Given that the ultimate determinants of relative scarcity value are continually changing and highly particular facts, the farther removed the price setters are from the periphery, the more blind they must be as economically significant information gets screened out. As the above quotation from Hayek implies, pricing decisions will reflect misleading and inappropriate aggregations. At the center of a large, centrally planned economy, those who set the prices will always be "in the dark."

There is also a serious incentive problem facing a centrally planned economy. Traditionally, critics of existing centrally planned economies have argued that people just won't put forth the required effort for the greater good. This has been a genuine difficulty for existing centrally planned economies, and a variety of schemes have been devised to overcome it. However, this is not the problem I am calling attention to here. Rather, the problem is how to design institutional mechanisms that will give those at the periphery the incentive to report the relevant information accurately. If success for a production unit is defined in terms of fulfilling the stated goals of the plan—and in a centrally planned economy it is hard to see how else it could be defined—it would seem that there will be strong incentives for the workers or their representatives to underestimate plant capacity, to hoard means of production and labor power, and to overstate actual production. (Nor need all of this be intentional.)

Indeed, production coordination in a centrally planned economy may well have all the characteristics of a public good, in Mancur Olson's technical sense of the term. The benefits are non-excludable in that they are provided to many non-contributors (e.g., the public at large), providing this good is usually a cost for a given individual or production unit, and the contribution one person or production unit can make will usually be negligible. As Olson has argued,[23] even an altruist will find it rational not to contribute when these general conditions are met.

Similar problems do not face capitalist societies to the extent that the competitive process operates, because the tendency toward better coordination is driven by the quest for pure profits. In other words, pure profits are there for the taking by especially alert entrepreneurs, but that holds only until competitors enter the field to bid up factor prices and drive down product prices; as a result of all these actions, the structure of production is recoordinated to reflect the changes in the underlying economic realities.

If coordination in a centrally planned economy is in fact a public good, it is likely that the institutional structures that emerge will have to rely on coercion (the classic solution to a public goods problem) to secure the good in question. It is at least suggestive that the problems sketched above—and the coercive solution—have characterized existing centrally planned economies such as that of the Soviet Union.

It might be thought that pricing decisions for factors of production could be left to the discretion of individual firms as a way of giving the planners access to new information about the state of the economy. They could let their planning be guided, though not completely determined, by this information. In other words, the problems described above arise only if those making production or allocational decisions also have to set producer prices. Let us assume, then, that individuals and production units are given the authority to set prices for the factors of production they produce.

The difficulty with this solution becomes apparent when one asks what is to guide those who make these pricing decisions. Certainly they cannot be aiming at maximizing exchange value for the firm; by definition, they are not producing for exchange. Nor can they employ the neoclassical economists' equilibrium rule—'Let price = marginal cost'—for two reasons: First, this rule is appropriate only if costs are equilibrium costs, which as Hayek has shown, is never achieved in the real world. More importantly, the actual (as opposed to hypothetical) marginal cost of a product depends on which constellation of inputs a producer chooses and what kind of a deal can be arranged with suppliers. If individual production units are allowed to decide these questions for themselves, then the plan is no longer effecting production coordination by determining what is going to be produced and in what quantities. In short, central planning has been abandoned. This is production for the market, whatever the legal fictions are.

Firms in a capitalist market economy do not know what the best price to charge is either. They have to guess, but competition (including competitively determined cost prices) provides the feedback mechanism to let them know how they are doing. By contrast, in a centrally planned economy that leaves pricing decisions to the firm, there is no direct feedback mechanism to inform those who make pricing decisions about how well they are doing. They could try to make their prices bring supply and demand into balance—where what is supplied and what is demanded is already determined by the plan. Then, however, prices are not providing the planners with new information about underlying economic realities; instead, they are needlessly duplicating information already contained in the plan.

If prices are to reflect relative scarcities and yet if prices do not determine what actually gets produced in what quantities, it is hard to see how success can be distinguished from failure. The planners are supposed to be getting information from the price-setters and neither the former nor the latter has any independent way (in the absence of markets) to determine whether their estimates are even roughly accurate. Paper losses may indicate faulty pricing, poor planning, or failures of execution; there seems to be no way to know which. Plant managers (who, presumably, would set prices) can be and are evaluated, but success for them is defined in terms of fulfilling the target set by the plan. If firms have price-setting autonomy and ordinary financial criteria determine production, then the system has reverted to one of production for the market.

Successful human action requires knowledge and will. By separating production decisions (will) from pricing decisions (knowledge), this version of central planning lapses into incoherence. A system in which the planners make pricing decisions is like a market system in one important respect: Those who make the pricing decisions are the same as those who make the production decisions. Unfortunately, the former kind of system is overwhelmed by the epistemological problems outlined earlier.

Thusfar the argument has proceeded on the tacit assumption that the planners know what consumer goods are to be produced and in what quantities. The Mises-Hayek argument is powerful enough to grant this wildly implausible assumption. But, in the real world, central planning must face this problem. Undoubtedly it will begin with historical data, though advocates of central planning invariably envisions a resetting of priorities in this area. The problems facing central planning in the production of consumer goods are basically the same as those encountered with producer goods.

In another context, Oscar Lange has suggested[24] that the planners could be guided by shortages and surpluses in setting consumer prices. If there is a shortage of a product, its price is to be raised; if there is a surplus, it is to be lowered. There are a number of obvious difficulties with this approach. First, it makes prices reflect what happened in the past instead of a speculative guess about future conditions. It assumes that consumers' buying habits are constant, which in general is not true. In addition, it specifies no mechanism for innovation or valid product differentiation. The relevant information must be gathered through other means; the problems detailed above for producer goods will apply, *mutatis mutandis*, to the consumer goods sector.

There is a suggestive parallel to these problems in ethical theory; the so-called paradox of happiness is that happiness is best achieved not by aiming at it directly but by aiming at things that cause it or accompany it. In a similar vein, producing things that meet the wants and needs of the consumers is best achieved not by planning directly for the production of use-values. Instead, the more efficient procedure is for producers to produce for exchange value.

Given the severe knowledge and motivational problems facing a centrally planned economy, how will it actually operate? The most likely scenario is that the structure of production will ossify. That is, production units will simply continue to produce what they have produced in the past. This is maladaptive in a world in which the ultimate determinants of scarcity value are in a state of constant flux, but custom and tradition are powerful forces in human affairs. Besides, what else can they do? The communication network they have in place to transmit the relevant knowledge is about as sophisticated as the use of smoke signals in comparison to the fiber optic communications network that is provided by the market.

Mises's claim about the inadequacy of central planning as a method of production coordination is an empirical one. His argument does not amount to an a priori proof, though he sometimes seems to have thought so. Perhaps the reason for this is that the empirical assumptions the argument makes are extraordinarily weak. Given a moderately complex economy, given that economically significant change happens continuously and given that knowledge of these changes is widely scattered, the constraints imposed by a system of central planning create significant barriers to the accumulation and effective dissemination of this knowledge for the purposes of coordinating production. If rational economic calculation requires a pricing mechanism that effectively accumulates and disseminates knowledge about the constantly changing underlying economic realities (the ultimate determinants of scarcity value), then rational economic calculation in a centrally planned economy is indeed impossible.[25]

It might be wondered whether Mises's argument could be generalized to conclude that any market economy, whether or not it is capitalist, will have a production coordination mechanism superior to any non-market mechanism. This question might be of some interest to market socialists. At this time, it is unclear to me whether or not a generalized version of this argument will work. In the comparative account of coordination mechanisms detailed in this section, I have tacitly assumed at a number of places capitalist relations of production. Other market systems have different configurations of ownership rights, so it is unclear whether or not the argument could be extended to show that any market system will have a better coordinating mechanism than any centrally planned economy.

## Some Objections

The exact significance of the Mises-Hayek argument for Marx's vision of $PC_1$ society is not immediately evident. It does demonstrate that central planning has important inherent irrationalities ("contradictions"). As the above discussion shows, this method of production coordination is vastly inferior to the method of production coordination that characterizes a capitalist economic system. If the problem is serious enough, it becomes hard to explain why the associated producers would want to retain it, especially since $PC_1$ society is not a class society. This obviously raises a question about whether or not $PC_1$ society could persist as a stable social form. In light of this question, it would perhaps be useful to examine some strategies defenders of central planning, for example, Marxists, might adopt[26] to defeat the argument or at least to take the sting out of its conclusion.

It might be objected that the epistemological problems caused by the elimination of market prices could be satisfactorily solved by the extensive use of computers. Because computers are able to process quickly huge quantities of information, the hope is that the pricing mechanism in a centrally planned economy could be automated to the extent that the epistemological problems facing central planning would be significantly ameliorated.

There are three difficulties with this objection. First, the quantity of information that would have to be processed for computers to make a real difference is simply staggering. In the Soviet Union, for example, it has been estimated that the planning apparatus currently can process only 3 billion out of 120 to 170 billion bits of information.[27] Given that unforeseen but economically significant change is happening continuously, it is hard to see how computers could do what needs to be done in a timely manner. More importantly, the information generated only provides a picture of what has happened in the past. A simple projection of this data into the future would be as irrational as driving a car by watching the rearview mirror. If the data include estimates from the periphery about impending changes in the relevant information, then inconsistent estimates have to be resolved and unforeseen changes must be ignored. It is hard to see how a systematic solution to these problems can emerge without decentralizing actual decision making.

Another problem concerns how information is categorized or aggregated for the purposes of planning and pricing. As the quotation from Hayek in the preceding section suggests, aggregation must be very fine-grained to be useful. It is not enough for the planners to know how much steel is at their disposal; they need to know what grades are available, what its condition is, what its location is, and so on, and so forth. Ideally, two things would be of the same kind if and only if they are substitutes for one another. The problem is that things are substitutes for one another only relative to some purpose. For example, two grades of coal are substitutes for one another for the purpose of power generation in a low-pollution environment but are not substitutes for one another for power generation in a high-pollution environment.

The real problem here is not that this (ideal) level of aggregation is unachievable, though that is probably true; after all, no system is perfect. Rather, the problem is one of determining how fine-grained the distinctions should be. To assume that some sort of cost-benefit analysis can resolve this difficulty is to assume that there is a reliable pricing mechanism already in place; that, of course, is just what is at issue. Indeed, an entrepreneur in a market system can rationally decide to remain ignorant of certain information if the costs of gathering and processing information are too high relative to prospective benefits. This rational ignorance presupposes that the cost prices of information tend to be relatively accurate reflections of the scarcity value of acquiring that information. Since posted prices under central planning would not tend to be relatively accurate, central planning has a kind of second-order irrationality in that the planners cannot have any clear idea about when it is rational to seek or not to seek more information.

Finally, some economists have argued that successful entrepreneurship in a market economy requires a "sense of the situation" that cannot be articulated.[28] That is, it requires tacit knowledge that cannot be expressed in propositional form. If computers are to simulate successful entrepreneurship through the pricing mechanism, the prospects for computerizing this process are not promising.[29]

A second approach to mitigating the problem of production coordination in a centrally planned economy would be to simplify the structure of production. If the pathways from original factors of production to consumer goods were fewer and shorter, central planning might be more adequate to the task. This objection is hard to evaluate because of its vagueness. However, there is some reason to think that it won't do.

Two factors primarily determine the level of complexity in an economic system:[30] (1) the variety of consumer goods that the system ultimately produces; (2) the number of alternative pathways from the original factors of production to the consumer goods. The latter is largely determined by technology and the current state of scientific knowledge. As technology develops, the number of alternatives grows. For example, technological advances can make what were once relatively specific factors of production relatively unspecific; finally, technology brings new factors of production into existence. This obviously makes production coordination more complicated and difficult.

Marxists clearly do not envision regression in the level of technological development. What about consumer goods? Maybe there would be less product differentiation than under capitalism, but it is by no means clear there would be appreciably fewer kinds of consumer products. As previously oppressed peoples' individuality and creative potential are unleashed (or so the story goes), the challenges to the economic system will increase, not decrease.

The objections discussed above attempt to argue that the epistemological problems posed by the need for production coordination could be manageable under central planning. However, even if these objections were well-taken, they do not address the parallel motivational problem. Recall that this is *not* the problem of how to get people to work hard under a different distributional system, though that is in fact a problem in existing centrally planned economies. Rather, it is how to motivate those at the periphery to report accurately changes in economically significant information, for example, the extent to which past production targets have been met, the quantity and condition of labor power and means of production on hand. As pointed out in the last section, the public goods nature of the problem of production coordination in a centrally planned economy makes a facile appeal to changes in consciousness suspect, to say the least. This is a motivational problem for the managers, since they are the ones responsible to the planning apparatus, and because success and failure for them are defined primarily in terms of fulfilling the output target set by the plan. This problem arises irrespective of whether or not the production units are self-managed.

The general motivational problem in any economic system is how to link rewards and punishments to success and failure. As I have suggested in a number of places in this book, a necessary condition for understanding how an economic system functions is to understand the incentive structure it creates. Positive and negative incentives need not be conceived of in purely monetary terms;[31] nevertheless, however the "goods" and "bads" are specified, any economic system must link them to success and failure in some calibrated way if that system is to achieve its ends.

In my view, discussions of incentives in economic systems have focused too much on positive incentives to produce the behaviors associated with efficiency and too little on negative incentives to avoid behaviors associated with inefficiencies. Even if the appropriate positive incentives are in place, existing inefficiencies may persist if only from custom or habit. What is needed are feared consequences together with a reliable mechanism for visiting those consequences on all and only the offending parties.

To the extent that a capitalist economy achieves this happy result, it does so in two ways: First, bosses have a powerful incentive to get rid of workers who impede the firm's ability to play its part in the ensemble of production units that constitutes a market economy. The system gives them arbitrary power (less so than in the old days) to discharge workers; the fact that this power is sometimes abused or exercised unwisely should not disguise the fact that capitalism creates incentives to get rid of the unproductive worker, either by firing him or by changing him into a productive worker. It is unclear how or whether $PC_1$ society could institute something functionally similar in a manner consistent with its other features. It is suggestive that existing centrally planned economies have not solved this problem. (Actually, Stalin solved it after a fashion, but not very efficiently.)

The second way that capitalism visits feared consequences on those who do not act efficiently is perhaps more important. Impersonal market forces strip unsuccessful entrepreneurs of their control over the means of production when they judge incorrectly what products to produce, what prices to charge, or what mix of inputs to choose. This not only hurts their personal financial situation, but it also takes away their control of social wealth.

This in large measure explains why many individual capitalists are not defenders of the free enterprise system, at least insofar as it applies to them; it explains why they run to the state with all manner of explanations for why they should be protected from competitors. They're all for free competition if it is fair competition; competition has to be protected, but it can only be done by insuring the survival of existing competitors, and so on. The fantastic stories lobbyists tell state officials these days would have made Marx chuckle— if he had had much of a sense of humor.

Matters are significantly worse in a centrally planned economy, however. Those who manage production (the workers themselves or appointed managers) do not make decisions about what to produce and in what quantities. That is done by the planners. The predicted result is that blame for failures will be difficult to assess. Is it a failure of planning or a failure of execution? Each side has its excuse, and there would seem to be no impersonal forces that would decide the issue. The problem with "personal forces" is that they can be bought off or co-opted in a variety of ways—ways in which market forces cannot be bought off. There is a finality to the verdict of the market that is absent in a centrally planned economy.

As an aside, the above suggests that incentive structures themselves can be judged efficient or inefficient, depending on the kinds of behavior they encourage or discourage. Indeed, it may be that efficiency assessments of economic systems are simply efficiency assessments of their incentive structures. It seems that two necessary conditions for an efficient incentive structure are: (1) Those who are responsible for the success and failure of coordination can be clearly identified; and (2) Calibrated rewards go with success, and calibrated punishments go with failures. Capitalist systems do not perfectly achieve this, but they clearly come closer to it than centrally planned economies do.

A third motivational problem facing a centrally planned economy concerns innovation. New and better ways of doing things nearly always create problems for someone in a complex, interconnected economic system, since innovation usually disrupts existing ways of doing things. Market forces compel competitors to change or go under. However, in a centrally planned economy innovation will tend to be stifled. Those who benefit from it (usually consumers) are not well-organized, are unaware of impending developments, and will benefit only modestly from most innovations. By contrast, those who will be adversely affected will be affected more significantly and will possess the knowledge and the means to block innovation.[32] As far as innovation is concerned, centrally planned economies will tend to be permanently sclerotic.

To sum up, the motivational and epistemological problems inherent in a centrally planned economy cannot be overcome, avoided, or significantly ameliorated within the framework that defines such a system.

A more promising strategy for dealing with the Mises-Hayek argument would be to grant its conclusion about the gross inefficiency of central planning, but argue that this problem would not be so serious that the associated producers would reject the economic system of $PC_1$ society.[33] Suppose that there is a socialist revolution; the workers seize control of the means of production and proceed to "organize production consciously on a planned basis." As priorities are reordered, dis-coordination will probably increase, but (contrary to what Mises says) it is not completely obvious that utter chaos would result. Eventually, the economy might stabilize and plug along at a lower level of productive efficiency. Even economic growth might occur over the long-term, so progressive immiseration is not a worry.

Although this is abstractly possible,[34] a pressing problem with this scenario becomes evident if one asks what the payoff is. Not the elimination of exploitation and alienation. At long last, this brings us to the relevance of the Mises-Hayek position for the promised arguments for the unrealizability of $PC_1$ society as Marx envisioned it. The discussion of the last two sections about how production is coordinated in any centrally planned economy lays the foundations for justifying the key claims about exploitation and alienation identified in the first section of this chapter.

### Are the Workers Capitalistically Exploited in Capitalist Society?

Recall from the first section of this chapter that the strategy for proving Marx's vision of $PC_1$ society is unrealizable is to show that the institutions of that society will be responsible for significant exploitation and alienation. This chapter also seeks to answer the question of whether or not the Marxist can sustain the charge that the workers are capitalistically exploited under capitalism. The purpose of this section is to do the latter; the former will be taken up in the remaining sections of this chapter.

For the Marxist to sustain the claim that the workers are capitalistically exploited in capitalist society, it is necessary to show that the workers would be better off, in terms of their income-leisure package, in $PC_1$ society than they would be under capitalism. Consequently, if it can be shown that the workers would not be better off in $PC_1$ society than they would be under capitalism, then this charge of capitalistic exploitation cannot be part of Marx's radical critique of capitalist society. The purpose of this section is to establish just this.

Two preliminary clarifications: First, who are the workers? Marx identifies the workers in capitalist society as those who must sell their labor power to the owners of the means of production to survive. This excludes not only capitalists, but also peasants and state functionaries. The term 'worker' cannot be defined this way, however, since obviously there would be workers in $PC_1$ society, and yet labor power is not a commodity in that society. If a trans-historical definition of 'worker' is needed that accords with Marx's usage, the following seems about right: A worker is anyone whose labor (i) directly contributes to the production of use-values, and (ii) is part of an indefinitely larger nexus of social production. Number (i) rules out most state functionaries as workers, and (ii) rules out peasants who are largely self-sufficient. There is some vagueness in both of these conditions, but the basic idea is clear enough and in accordance with how Marx thought about it.

One other preliminary point: Following Roemer, the discussions of various forms of property relations exploitation in this section and the next abstract from the costs of the transition, so to speak. The losses many people sustain in a revolutionary transformation are often quite dramatic. Indeed, their income-leisure packages may drop to zero if they are lined up against the wall and shot! The justification for this abstraction is that what sustains any charge of property relations exploitation is the historical possibility or real-izability of better alternatives. Transition costs are relevant only for the question of whether and when revolutionary action should be undertaken to bring it about.[35]

These preliminary matters taken care of, let us turn to the main point to be demonstrated, namely, that the workers would not be better off in $PC_1$ society than they would be in capitalist society. The argument for this starts with the observation that there is a profound difference between workers' income-leisure packages in existing capitalist societies and in existing societies with a centrally planned economy. Moreover, most of this difference can be explained by the different methods by which production is coordinated in the respective economic systems, namely, markets and central planning. Because $PC_1$ society would have a centrally planned economy, it is reasonable to believe that the workers would not be better off in terms of their income-leisure packages in $PC_1$ society than they would be in capitalist society.

That is the argument; although it is brief and to the point, it is not as straightforward as it looks. Each of the premises requires considerable discussion, as does the inference to the conclusion. Let us begin with two analytical questions: (a) 'What counts as an existing capitalist society?' and (b) 'What counts as an existing society with a centrally planned economy?'. In answering (a), it is important to understand what is not included as a capitalist society. First, the only arguably market socialist economy (viz., Yugoslavia) is ruled out, since the predominance of capitalist relations of production is a necessary condition for a capitalist society. Second, and perhaps more importantly, there are other market economies in which capitalist relations of production do not predominate, for example, what can be found in many Third World nation-states. This latter point requires a brief argument.

There are two reasons why these economic systems should not be counted as capitalist. In most of them there are just too many peasants and not enough proletarians. This means that the predominant relations of production do not relate individuals qua capitalists and proletarians. (There may be no predominant relations of production.) Perhaps a more important reason that these systems should not be called capitalist is that property rights in the means of production are highly unstable because such rights are subject to arbitrary confiscation by the state.[36] Nothing approaching full liberal ownership rights in the means of production can be found in these societies. To put this in terms of the *rechtsfrei* characterization of capitalist relations of production advocated in Chapter 6, the powers associated with full liberal ownership are more tenuous, uncertain, or even nonexistent in these societies. For these two reasons, many non-socialist Third World economic systems are not capitalist, even though they have a market economy.

Which countries, then, have capitalist economic systems? Because of the growth of the international market, it is becoming obsolete to speak of nation-states, especially in the First World, as economies, though it is possible to indicate which regions of the world are capitalist, for example, North America, the European Economic Community, and the Pacific Rim in Asia. There can be disputes at the margin about who should be included and excluded, but the above discussion suggests a criterion: Are (most) liberal ownership rights in the means of production both privately held and secure or stable? If the answer to this question is yes, the society in question is capitalist; otherwise not. The boundary may remain somewhat indefinite, but the distinction is clear enough for present purposes.

On the other side of the comparison alluded to above are all centrally planned economies. Recall that an economy is centrally planned if and only if the predominant method of production coordination is the Method of Material Balances. The Soviet Union and all of Eastern Europe, with the exception of Yugoslavia, are the prime examples; Cuba, the People's Republic of China, and Vietnam should also be added to the list, though the situation is changing in China as of this writing. These economies are centrally planned because most production is carried out under a system in which output targets are assigned to production units by a central authority.

Let us now consider the key empirical issue. It is exceedingly difficult to get exact statistical evidence about workers' income, as the term 'worker' was defined above. Moreover, there are a variety of formidable problems of a more general nature in making income comparisons. To mention just a few, various currencies (e.g., the ruble) are not convertible; it is unclear whether it is individual income or household income that should be counted; demographic profiles require corrections for age distributions, and so forth, and so forth. It is even more difficult to compare the amount of leisure time enjoyed by workers in the contrasting societies. This is especially problematic because there are underground economies in both systems in which people moonlight, something that also effects income estimates. In existing centrally planned economies, this encompasses huge numbers of people. However, all of these subtleties can be ignored for present purposes. The point at issue is a very gross comparative judgment, and the basic income differentials are so significant, so consistent, and so incontrovertible that there can be no doubt that workers in existing capitalist societies have a better income-leisure package than do workers in existing societies with a centrally planned economy.[37] What is more controversial is what this comparison shows.

First, some things it does not show: The overall level of well-being, whatever that could mean, may not be higher in existing capitalist societies than in existing societies with a centrally planned economy, and workers in the latter societies may not prefer to "pull out," all things considered (love of mother Russia and all that); more importantly, there may be other advantages associated with living in these societies. In addition, there could be unemployed persons in capitalist societies who are not better off than they would be under the alternative. But none of this matters because none of it is relevant to claim under consideration; all that counts are workers' income-leisure packages, and on that score, there can be no doubt that in currently existing centrally planned economies, the workers are not better off than they are under capitalism.

But what does this tell us about the comparison between capitalist society and $PC_1$ society? Recall that what Roemer's definition of capitalistic exploitation (as amended) requires us to compare are not actually existing economies; nor does it require us to compare two different groups of workers. Instead, it requires a comparison of the same group of workers conceived of under two different kinds of economic systems. Furthermore, one of the terms of the above comparison is the set of existing centrally planned economies, and none of the latter is characterized by genuine worker control of the means of production. That is, none of these societies counts as a version of $PC_1$ society. Indeed, no versions of that society exist anywhere in the world today. What, then, is the point or relevance of dragging in socialism's twentieth century embarrassments?

The problem can be summarized as follows: The definition of capitalistic exploitation requires us to compare how workers would fare under two different kinds of economic systems. The empirical evidence cited above relates particular economic systems, and one of the terms of the comparison does not consist of instances of $PC_1$ society. How, then, can this comparison tell us anything about how ltthe workers would fare in $PC_1$ society as compared to capitalism?

The solution to this problem is to be found in what *explains* the empirical evidence. If (most of) the observed differences between workers' income-leisure packages in capitalist and centrally planned economies can be explained at a level of abstraction that transcends both historical peculiarities of the existing economic systems being compared and the fact that the workers do not control the means of production in any existing centrally planned economy, then there is good reason to believe that the workers would not be better off in $PC_1$ society than they would be in capitalist society.

The Mises-Hayek comparative evaluation of the coordinating mechanisms of capitalism and a centrally planned economy does just this. By way of an abstract comparison between the mechanisms by which production is coordinated—and the goods delivered, so to speak—in a capitalist market system versus a centrally planned economy, it explains the above-cited differences between workers' situations at a sufficiently high level of abstraction to transcend the historical peculiarities of both sides of the empirical comparison and the lack of worker control of the means of production in existing centrally planned economies. The discussion of the preceding two sections was not in terms of particular economic systems but, instead, *kinds* of economic systems: On one side, any capitalist economic system; on the other side, any centrally planned economy. The vital contrast between capitalism and $PC_1$ society is the contrast between their respective coordinating mechanisms. A brief review of that discussion might be useful at this point.

Recall that the competitive process in a capitalist market system sets in motion a powerful tendency for production to become better coordinated (that is, more efficient), and it does so in a way that is sensitive to changes in the underlying economic realities. By contrast, epistemological and motivational problems that are endemic to central planning will result in poor production coordination; the latter will manifest itself in widespread shortages, supply bottlenecks, and the overproduction of inappropriate and unusable producer and consumer goods.[38] Since worker control of the means of production is basically self-management of the production unit, it has nothing to do with how production is coordinated among those units. This means that whether or not a centrally planned economy is worker-controlled makes no difference to the point at issue. It also does not matter if the workers democratically control the planning apparatus, since the motivational and epistemological problems caused by constant change in the ultimate determinants of scarcity value will persist.

In addition, there will be systemic resistance to innovation in any centrally planned economy for reasons outlined in the last section. To use Marx's terminology, all this entails chronic and severe fettering of the forces of production, no matter who controls them. To be sure, capitalism occasionally has its own problems of "fettering" in, for example, depressions and recessions, but Marx's predictions notwithstanding, these crises have not been increasingly severe, and they have been episodic. Since its inception, the dominant trend in existing capitalist societies has been in the direction of a higher standard of living (better income-leisure packages) for the workers; technological development played a big part in this, but of course that is not exogenous to capitalism. By contrast, this development will not do much good if an appropriate mechanism of production coordination is not in place, and, it might be added, if there is no capitalist economic system and world market to provide indirect price information, as well as technology that can be bought or stolen.

It might be objected that unions and various social welfare measures have played the decisive role in improving the income-leisure packages for workers under capitalism. However, even if that is true (and the comparative account of coordinating mechanisms in the preceding two sections suggests that it is not), the argument of this section could be easily amended to take this into account. The specification of alternative institutional arrangements could be broadened to include unions and even social welfare bureaucracies; this is not ad hoc if one wants to be sensitive to realizability considerations. Mancur Olson once said that capitalism without trade unions is about as conceivable as capitalism with a negative rate of interest. The same might be said of the social welfare organs of the state.

That point to one side, the existence of the observed discrepancies between the income-leisure packages of workers in centrally planned economies and capitalist systems on the one hand, and the explanation for these discrepancies in terms of an account of how the respective methods of production coordination actually operate on the other hand, together make it reasonable to believe that the workers would not be better off in $PC_1$ society, were it ever to exist, as compared to how they would fare under capitalism. The premises of this argument do not logically entail this, but they do make it highly likely, and that is the most that can be asked of an empirical argument.

One final caveat: The above does not show that the workers are not capitalistically exploited under capitalism *tout court*, since there may be institutional arrangements other than $PC_1$ society in which the workers would be better off than they are under capitalism (e.g., market socialism). But the charge of capitalistic exploitation cannot be sustained from a Marxist perspective because, as the above discussion shows, the workers would not be better off in $PC_1$ society.

### Exploitation in the First Phase of Post-Capitalist Society

#### Marx Exploitation in $PC_1$ Society

The purpose of this section is to argue that there will be pervasive exploitation of the workers in $PC_1$ society. The first part of this section concerns a form of property relations exploitation identified in the opening section of this chapter, what I called 'marx exploitation'. The second part discusses surplus value exploitation.

Recall that the following two conditions are necessary and sufficient for a coalition W to be marx exploited in $PC_1$ society:

i. If the members of $W$ were to withdraw from $PC_1$ society with only their labor and skills into a capitalist society, they would be better off in terms of their income-leisure package than they would be by staying in $PC_1$ society.

ii. If some subset of the complement of $W$ were to withdraw from $PC_1$ society with only their labor and skills into a capitalist society, they would be worse off (in terms of their income-leisure package). This is the marx exploiting coalition.

What is to be proved in this subsection is that the workers in $PC_1$ society would be marx exploited.[39] What about the first condition? Would the workers be better off in a capitalist society than they would be in a $PC_1$ society in terms of their income-leisure package? An affirmative answer comes from the same argument that was given in the last section to show that the workers would not be better off by moving from capitalism to $PC_1$ society. The point at issue here is slightly stronger than what was at stake in the previous section, but the difference is insubstantial. That argument is sufficiently powerful to establish both that the first condition for capitalistic exploitation does not hold and that the first condition for marx exploitation does hold.

Turning now to the second condition for marx exploitation of the workers in $PC_1$ society, who is in the exploiting coalition? The most obvious candidates would be the higher-ups in the planning apparatus and the state. Trotsky's dictum, 'Those who have something to distribute seldom forget themselves', is apropos here. In existing centrally planned economies, those in the top echelons of the planning bureaucracy and the state clearly are the differential beneficiaries of the system. Their income-leisure packages would not improve if they were to withdraw into capitalist society with their labor and skills.

Two objections might be made to this. First, it might be argued that these individuals are workers, since as planners and politicians they are part of the economic system. Regarding the politicians, given the politicization of the economy in $PC_1$ society, they might well meet the two conditions for being a worker identified at the beginning of the last section. The planners have an even better claim to the status of worker. If both are workers and would not benefit by pulling out, then condition (i) would not be met and the workers would not be a marx exploited coalition. But this problem could be easily handled by "redrawing" coalition boundaries; that is, instead of saying that the workers are marx exploited, we could say that a (very large) subset of workers are marx exploited. This makes for some indeterminacy in what is being shown, but the main result is essentially unchanged. In these matters, it is very important not to let definitional questions obscure the substantive issues.

A related objection that is in a way potentially more troublesome is the following: Suppose there is no exploiting coalition; that is suppose everyone would be better off, or at least as well off, by pulling up stakes. Since condition (ii) is supposed to capture the idea that one cannot have an exploited coalition without an exploiting coalition, the absence of the latter would entail that no coalition is marx exploited. For example, one cannot simply assume that the planners and politicians of $PC_1$ society would be worse off by withdrawing into capitalism. Maybe they too would be better off, or at least as well off, as they would be under capitalism.

Suppose this is true, and thus there is no marx exploitation in $PC_1$ society. The problem now is that there is no good reason to believe that $PC_1$ society would persist as a stable social form. After all, no one is differentially benefiting from it! We get the same result if it is only the very young, the very old, and those unable to work who meet condition (ii).[40] Some uncertainty or even mere timidity may initially prevent the transformation of $PC_1$ society into something else, but not for long. (This something else need not be capitalism—it could be market socialism.)

The conclusion to be drawn from all this is that the workers, or at least a large subset of them, are marx exploited in $PC_1$ society, *or* it is likely that $PC_1$ society cannot persist as a stable social form. Concerns about who is in the exploiting coalition to one side, the root idea of any form of property relations exploitation is that exploited groups are exploited because there are realizable alternative economic systems in which they would be better off than they are in the system under which they live. The existing system, or whatever is responsible for it, weighs like a dead hand on the exploited. What makes the workers marx exploited in $PC_1$ society is the historical possibility of better alternative institutional arrangements—capitalism, according to the above argument.

The above discussion is no mere academic exercise. Something like marx exploitation is of obvious relevance to the peoples of Eastern Europe and those within the Soviet empire proper. The East Germans, the Czechs, the Poles, the Bulgarians, and yes, the Hungarians and the Great Russians suffer a form of property relations exploitation because the extensive use of the market mechanism is prohibited. Some members of those groups (and we know who they are) might not want it any other way, but that does not change the fact that the workers in those nations suffer a form of property relations exploitation in virtue of the system under which they live.

In a way, it is somewhat unfortunate that the Mises-Hayek account of the failings of production coordination in a centrally planned economy was drawn into a critique of Marx on two fairly narrow points—the inability to sustain a charge of capitalistic exploitation against capitalism (from a Marxist perspective) and the argument to sustain a charge of marx exploitation against $PC_1$ society. This discussion has a much larger significance for Marx's thought: It shows that the dream of getting economic life under "society's" conscious control is really a nightmare. The constant changes in the ultimate determinants of scarcity value (what I have called the underlying economic realities), coupled with the "social scattering" of the knowledge of those changes, make it impossible to achieve the simultaneous mastery of nature and human production envisioned by both Marxian and post-Marxian socialist thought. This dream is most clearly manifested in Engels' *Anti-Dühring*,[41] but it is also implicit in Marx's commitment to central planning as that commitment was elaborated in Chapter 6. Moreover, it suggests that if Marx's vision of $PC_1$ society includes an ever-increasing level of material wealth, a progressively better handling of the primary evils in the language of Chapter 8, that vision is probably unrealizable because part of it consists of an economic system with a centrally planned economy.[42] Finally, a faith in economic planning of some sort remains part of the ideology of the Left, whether Marxist or non-Marxist. A full examination of the implications of the Mises-Hayek argument for noncomprehensive planning lies beyond the scope of the present study, but there can be little doubt that important substantive issues in social and political philosophy are at stake.[43]

### Surplus Value Exploitation in PC₁ Society

Let us turn to surplus value exploitation in $PC_1$ society. Recall that someone is surplus value exploited in an economic system if and only if that person is doing forced, unpaid, surplus labor the product of which is not under his or her control.[44] To show that there is significant surplus value exploitation in $PC_1$ society, it would be sufficient to show that a substantial number of workers will meet all of these conditions.

To establish this, it is essential to recall how income is distributed in $PC_1$ society. In Chapter 7 the distributive principle for all workers was identified as distribution according to labor contribution. In the *Critique of the Gotha Program* Marx says:

> Accordingly, the individual producer receives back from society—after the deductions have been made—exactly what he gives to it. What he has given to it is his individual quantum of labor. [*CGP*, p.16]

But what is this contribution, what is this individual quantum of labor? In discussing the "defects" of the distribution principle of $PC_1$ society, Marx makes it clear that labor contribution is to be understood in a straight quantity-of-labor way:

> But one man is superior to another physically or mentally and so supplies more labour in the same time, or can labour for a longer time; and labour, to serve as a measure, must be defined by its duration or intensity, otherwise it ceases to be a standard of measurement. [*CGP*, p. 17]

Clearly, the Labor Theory of Value still gripped Marx's mind even at this late date in his career (1875). Problems with that theory to one side, it is clear that all, or at least nearly all, of the workers will be doing unpaid labor. To see why, consider the deductions referred to in the first quotation. Marx tells us that these funds go to cover maintenance costs for those unable to work, to provide insurance against accidents and so forth, to fund general administration, to pay for "public goods" such as health care and education, and to cover the replacement of used up means of production and for the expansion of production (*CGP*, p. 15).

To show that the workers are doing unpaid surplus labor, it is sufficient to show that they are making value contributions above and beyond what is needed to sustain themselves for which they are not compensated. The fact of these deductions and how they are distributed together imply that this is what in fact happens. Workers do benefit from most of these expenditures but not to the extent of the value they contribute. Some of it goes to those unable to work, some of it goes to future generations, and on Marx's account of what constitutes a worker, some of it goes to nonworkers involved in general administration. So, all workers are doing some unpaid labor. In addition, workers with especially scarce skills will be making more of a value contribution than those who lack such skills, yet the former will be receiving the same income (and, presumably, the same public benefits) as the latter, assuming both work about the same number of hours. This is the case even if one factors in outlays for training or education, since natural talent and time (in the form of delayed compensation) also contribute value. For these reasons they will be doing more unpaid surplus labor than the unskilled. However, doing unpaid surplus labor, in whatever quantity, is not sufficient for surplus value exploitation; it also has to be forced, and the product of that (surplus) labor must be not under the control of the worker.

What has to be shown is that both of these conditions hold in $PC_1$ society for a substantial number of workers. Let us proceed by way of an analogy with capitalism. What is supposed to make all labor forced under capitalism is that the worker has no alternative but to perform it. He may not be forced to produce this labor for any particular capitalist, but he must do it for some capitalist or other. The big difference in $PC_1$ society, of course, is that the workers control the means of production. But does this make a difference?

As was shown in Chapter 6, control of the means of production is basically self-management of the productive unit; control of the means of production also involves income rights, but in a non-exchange economy these rights can only be held collectively by all the workers. Self-management means that the workers decide among themselves how work is to be structured. However, since labor power is a producer good, and since the deployment of producer goods is determined by the plan in a centrally planned economy, the worker will have to go (or stay) where he is assigned by the plan. There may be some flexibility in that workers can arrange mutually beneficial job exchanges, but this flexibility must be quite limited if the system is at all taut. (Various skills and highly specialized knowledge are not interchangeable.) Moreover, the distribution system guarantees, in Lenin's words, that "He who does not work, neither shall he eat," which can be amended to read, 'He who does not do surplus unpaid work, neither shall he eat'. It would seem that the worker is forced to labor in about the same sense that the proletarian in capitalist society is forced to labor, though once on the job he has some say about how it will be done.

The only response available to the Marxist is to claim that, since the workers collectively control the planning apparatus, they are not really forced to labor. It might be said that freedom (in the Hegelian sense) is the recognition of necessity, or whoever wills the end wills the means. In this way, the workers are not really forced to labor. A parallel point can be made about control of the surplus social product. In a centrally planned economy, this has to be determined by the plan, so if the workers are to control the surplus social product, it must be because they control the planning apparatus.

I say 'planning apparatus' because, obviously enough, in a complex economic system there is no way for the workers to decide on the hundreds of thousands, or millions, of microlevel decisions that aim at coordinating production. Given that fact, what would democratic control of the planning apparatus amount to? In theory, it would seem that this could come down to two things. The workers could democratically decide: (i) some of the productive priorities the planners are to pursue; and (ii) who will constitute the top echelons of the planning apparatus. Those who place the list of options for (i) before the voters might have to include a price tag in terms of a projected quantity of surplus labor extraction, though the actual quantity or rate of surplus extraction for any plan period will depend on how well coordinated production turns out to be.[45]

It is obvious that decisions about productive priorities could only be at the grossest level, and it is likely that the voters will collectively express inconsistent preferences, just as in contemporary democracies people vote for lower taxes and higher social spending. But let us put that problem to one side. The real difficulty here is more fundamental.

Imagine the situation of workers whose votes did not carry the day on either (i) or (ii). These workers will be just as shut out as workers under capitalism have been, though their views may prevail when the next five-year plan or whatever comes up for a vote. But in the meantime, they are not directly or indirectly controlling the social product and from the point of view of their priorities, their surplus labor is forced. It might be said that these workers had a "say," even if their views did not prevail, but having a "say" is not the same thing as exercising control, direct or indirect. At most, it makes one feel better about the fact that one does *not* have control.

It might be objected that this assumes significant conflicts among the workers on these questions, and there is no basis for assuming a lack of consensus on these questions. However, there are in fact compelling reasons to believe that there will be significant disagreements about the disbursement of the surplus social product in $PC_1$ society. That society, like all other societies, has to deal with the primary evils. Recall that one of these is profound social ignorance. What this means in a large community is that people do not know much about the specific goals and beliefs of most other people. This virtually guarantees that there will be significant disagreements about the how much surplus social wealth the society should try to produce and how it should distributed to meet the needs of those unable to work, to provide for future generations, to fund general administration, and to provide for all the other things that surplus wealth is supposed fund; these disagreements can arise from different beliefs about what ends are worth trying to achieve or different estimates about the means needed to meet any particular end or both. In a world of profound social ignorance, different beliefs or estimates on both scores are virtually inevitable. This assumes, of course, another primary evil, namely, Humean scarcity.

These differences will be aggravated by a number of factors. The absence of market pricing makes it much less easy to assess the extent to which various social goals can be achieved, and of course it seriously limits what can be achieved. This problem is even more severe if there are no market prices for consumer goods. Two other factors that will exacerbate these disagreements are limited benevolence and conflicting conceptions of the good life; these two primary evils are variable across individuals. That is, some people's benevolence is more limited than others and some people's conceptions of the good life are more conflictive than others.

What all this means is that all those workers who disagree on the distribution of the fruits of surplus social labor will be effectively forced to contribute to all those projects they believe are unworthy or already overfunded. To make this point on a more intuitive level, in current capitalist societies, there are many people who are profoundly dissatisfied with society's productive priorities as they have emerged through the countless interactions of millions of people. If society's productive priorities are set by some combination of a planning bureaucracy and a popular vote (though the latter could be done only at the grossest level), there will still be many people dissatisfied with whatever priorities emerge. Those people may be less dissatisfied than contemporary malcontents, as they might be called, though that is by no means obvious. Be that as it may, the general point is that people will still be forced to contribute surplus labor, and they will not control some elements of the surplus social product.[46]

Who would be the exploiters in this regime? Obviously, it would be those who differentially benefit from the unpaid labor of others and those who differentially benefit from particular disbursements of the surplus social product (e.g., the highway construction cooperatives). Notice that the same individuals may be both exploited and exploiters. Just as welfare state capitalism creates the conditions for a mutual exploitation society by its game of musical chairs with wealth created by others, in like manner, $PC_1$ society would shuffle the deck of workers' surplus value contribution. One main difference, of course, is that welfare state capitalism plays with a much bigger pot.

All Marxists with pretensions to be taken seriously have admitted that there would be disagreements and conflicts in post-capitalist society. What they have failed to notice is that this implies that surplus value exploitation will be widespread. Nearly all the workers will be doing some forced, unpaid, surplus labor the product of which they do not control. It is open to Marxists to say that the democratic process is a much better way of deciding these surplus distributional questions, indeed that there are intrinsic benefits to proceeding in this manner. It might be argued (actually, admitted) that some surplus value exploitation is unavoidable and necessary, but that on a societywide scale, the benefits outweigh the costs.

All of this can be granted, however, because what the radical Marx promises for $PC_1$ society is the elimination of surplus value exploitation, not its amelioration, or to be more accurate, its repackaging. After all, in light of the discussion of the capitalist contribution to production in Chapter 4 and the discussion of the entrepreneur's contribution in the second section of this chapter, it is not at all clear that there would be less surplus value exploitation in PC$_1$ than one finds in capitalism. Be that as it may, it is nonetheless clear that extensive surplus value exploitation will be a feature of PC$_1$ society.

## Alienation in the First Phase of Post-Capitalist Society

Chapter 2 reconstructs Marx's account of alienation in capitalist society according to the Critical Explanations requirement. Recall that this requires explaining various manifestations of alienation by appeal to essential features of the capitalist economic system. Such explanations, to the extent that they are good ones, are given at a level of generality that permits us to say that any capitalist system will suffer the relevant form of alienation. If any of the various forms of alienation would afflict PC$_1$ society, this should be discoverable by the same mode of reasoning. The purpose of this section is to explain how most of the various forms of alienation found in capitalist society would also afflict PC$_1$ society. Let us proceed by considering each of the four major categories of alienation discussed in Chapter 2.

*Alienation from the Products of Labor.* Under capitalism, the workers do not own the products of their labor; they are sold on the market. It is true that under central planning the producers do not exchange their products. However, in such a system it is not plausible to maintain that the producers control the product, at least if the economic system is large and complex. Even if central planning is informed by democratic procedures, there is no question of voting on the microlevel decisions that aim at integrating production. As was pointed out in the last section, at most workers will have a say on broad policy questions, and/or they will vote for individuals who have a say on these questions. The length and complexity of production in a modern economy makes genuine control of the product by those who produce it impossible. By contrast, such control is possible in small-scale economies with simple structures of production. The replacement of the world market system with a large number of small, centrally planned economic systems would not really solve this problem, since avenues for mutually advantageous trade would soon result in the reestablishment commodity production or perhaps a reintegration of production via central planning.

A second manifestation of alienation from the products of labor under capitalism is to be found in the operation of large-scale social forces not under anyone's control. Market forces "hover over the earth like the Fates of the ancients." With the abolition of markets, these forces will no longer operate. However, it is not clear that central planning will be, or be seen to be, much of an improvement in this regard. There is the very real danger that the plan itself will be viewed by individual workers and production units as an alien force imposed on them from without. Because of the inherent irrationalities of central planning, the dictates of the plan will make even less sense than the dictates of the market. Moreover, the plan cannot be the collective design of all producers. Even if the members of the planning apparatus are demo-cratically chosen, that will not change or mitigate the inherent irrationalities characteristic of central planning.

*Alienated Labor.* Three respects in which labor is alienating under capitalism are the following: (i) The purpose of social production is not to create of use-values but to get exchange value in the market. (ii) The goal of labor for the worker is to get exchange value in the form of wages. (iii) Labor is not the creative self-expression of human capacities; instead it is a purely instrumental activity that is both mindless and physically exhausting. Let us consider each of these in turn.

Under a regime of commodity production, the fact that the goal of pro-duction is the creation of exchange value (and not use-values) is supposed to explain in part why work is experienced as meaningless. Production for use, on the other hand, is supposed to give work a point and purpose that is lacking when exchange value drives production. But is this really so in a centrally planned economy? Although the production of use-values ultimately informs production in this system, what guides production is the plan (the output target). There is no reason to think that work will be seen as more meaningful when production is driven by an output target instead of the quest for profits. Indeed, it may seem less meaningful, given the inherent irrationalities of central planning.

The contrast between production for exchange, which is really production for profit, and production for use has been a staple of socialist thought, though perhaps not always in these terms, for nearly two hundred years. But this contrast really involves a kind of utopian thinking. The idea that production in a complex industrial society could be undertaken with the production of consumer goods as the end in view for all or even most workers represents a wish to return to an idealized version of artisan labor in the Middle Ages. A certain measure of "meaninglessness" of work is inevitable in any society in which the coordination of production is a problematic affair, which means in any complex society.

Wage labor, of course, is abolished in $PC_1$ society. But what does this really amount to? In that society the distribution of consumer goods is pro-portional to the amount of labor performed. It is at best unclear how working for labor certificates differs from working for exchange value from the point of view of the worker. Certainly the incentive principle that drives each system is basically the same. The main difference between money and labor certif-icates is that the latter do not circulate and cannot be used to create capital. But working for labor certificates does not make work more meaningful.

It might be thought that since production units will be self-managed in $PC_1$ society, work would involve more creative self-expression and would be less physically and mentally damaging. But that will be the case only if the demands of the plan allow it. The basic problem is that there are potentially two positive values involved in any kind of social production: efficiency and intrinsically satisfying labor. Sometimes promoting one can promote the other (a fact that has not been lost on some capitalists) but not always. This means that there will have to be trade-offs. Given the inherent efficiency problems facing a centrally planned economy, there is no reason to believe that $PC_1$ society would have less mindless and physically exhausting labor than is to be found under capitalism; indeed, there may be more. Finally, given the basic incentive principle for $PC_1$ society, it is hard to see how or why labor would lose its instrumental character in that society and become the creative self-expression of human talents and abilities.

It can be granted for the sake of argument that $PC_1$ society will do better than capitalism has done with regard to meaningful work, although that seems unlikely. However, a radical critic promises much more than some improvement on major social ills—what is promised is their elimination, or virtual elimination. That, however, will not happen in $PC_1$ society.

*Alienation from Species Being.* Recall from Chapter 2 that man's species being, his human nature, consists in the capacity for universal production. That is, human labor has the potential to take any element or aspect of nature as its object. Furthermore, to say that man is a species being is to say that he recognizes that he has this capacity. The proletariat's alienation from their species being under capitalism consists in two facts: As individuals they have lost the capacity for universal production by being riveted to one repetitive dehumanizing task. Secondly, though collectively the proletariat produce in a universal manner under capitalism, the method of production coordination (i.e., the market) masks that fact from them.

This account of alienation from species being was presented in Chapter 2 without critical comment, but some of that would be in order here. Though there is something to be said for Marx's view, it is surely overstated. It is just false to say that the workers are riveted to one repetitive task in their working lives in capitalist society. People change jobs, acquire new skills, take on different responsibilities—and the market jostles them along in all of this. In addition, the mystification of which Marx speaks so darkly is the result of the impersonal operation of market forces. To be sure, market forces have-wrought, and continue to wreak, havoc with people's lives, as Marx correctly emphasizes. But for this complaint to have any force, there must be some alternative way to coordinate production which allows people better control of their lives.

Unfortunately, central planning doesn't do that. Indeed, in both respects, it makes the workers worse off. Labor power is a producer good whose deployment must be determined by the plan. In consequence, workers will have less freedom, not more, to deploy their labor power as they see fit. Of course, in theory they might have more freedom in the Hegelian sense (where freedom is the recognition of necessity) if central planning were a rational way to coordinate production. But it isn't, so they don't. Secondly, conscious control of the productive apparatus of the entire society is something of a fiction in a centrally planned economy. Its inability to respond to constantly changing conditions is deeply embedded in its structure. The collective capacity for universal production deteriorates—and alienation from species being increases—in such a system.

*Alienation from Others.* Marx's account of the alienation of persons from other persons under capitalism has three facets: The estrangement that characterizes the relation between the worker and the capitalist, the generalized interpersonal alienation that is reflected in the commercialization of life in capitalist society, and the alienation of the state from civil society. Marx's account of the first of these is predicated on the labor theory of value. It is only on this assumption that he can claim that the value the capitalist realizes in the form of profit comes out of the hide of the laborer and that the relationship between the capitalist and the worker is dominated by a zero-sum game. In response, it might be argued that even though the labor theory of value has to be jettisoned, intermittent conflicts of interests between workers and capitalists remain, which is itself a form of estrangement. Moreover, workers compete among themselves for the best wages and positions.

However, the inherent conflicts of interest that characterize an exchange society will not disappear with the abolition of the market. In a complex network of production, life can always be made easier for some at the expense of others. As noted above, in a centrally planned economy production units will find it in their interests to understate plant capacity, hoard means of production and labor power, overstate output, and so forth, and so forth. After all, the harm done to others (viz., the public at large) will usually be slight and the benefits to members of the production unit in question significant. The public goods nature of this problem makes it resistant to facile assumptions about changes in the workers' consciousness. Furthermore, conflicts of interests are inevitable as long as there is scarcity in the sense that not all useful projects can be pursued. These inherent conflicts of interest, coupled with the existence of alienating labor, set the stage for the occurrence of a species of generic parasite exploitation as the latter was identified in Chapter 4: Some workers will be forced to do alienating labor from which others will disproportionately benefit. Though it cannot be proved outright that there will be extensive exploitation in this sense, it seems likely that there will be some of this, especially if power differentials persist. And how could those power differentials not exist if there is a planning apparatus that controls the entire economy? The exploiters in this scenario would probably be those in the planning apparatus—and their friends and relatives.

The second form of alienation from others under capitalism arises from the inherently conflictive nature of the exchange relation. As Marx points out in his "Comments on Mill" (MECW, vol. 3, pp. 225–226), people view each other's needs and wants as levers to manipulate to satisfy their own selfish desires. For this reason there exists a kind of pervasive mutual using in capitalist society. However, the elimination of production for exchange does not entail that this mutual using will cease. In a centrally planned economy people will have an incentive to use each other for two kinds of ends: As producers, they want to get inputs to specifications in a timely manner from suppliers. Because of inherent coordination problems, this is nearly always problematic; in consequence, they have to offer individual suppliers the right incentives, which means bribes of some sort. Secondly, because of shortages, quality problems, and so forth, in the consumer goods sector, people have to cultivate others outside of official channels to get what they want and need. A vast barter system has developed in existing centrally planned economies, and who one knows is much more important than it is in capitalist societies. (The Chinese system of *quanxi* is much more elaborate and intricate than any comparable network in a capitalist society.) As a speculative point, it may be that the reciprocity implicit in any network of exchange relations (market or otherwise) is a necessary condition for the continued existence of any social institution. If that is true, Marx's complaints about mutual using look to be truly utopian.

What about the alienation of state from civil society? Marx's vision of $PC_1$ society includes the overcoming of this distinction, a kind of Hegelian synthesis between the private sphere and the public sphere. In commenting on this synthesis, Leszek Kolakowski has said,

> There is no reason to believe that the restoration of the perfect unity of the personal and communal life of every individual (i.e., the perfect, internalized identity of each person with the social totality, lack of tension between his personal aspirations and his various social loyalties) is possible, and, least of all, that it could be secured by institutional means. ... [This unity] would presuppose an unprecedented moral revolution running against the whole course of the past history of culture. To believe that a basis for such unity may be laid down first in coercive form ... and will grow subsequently into an internalized voluntary unity, amounts to believing that people who have been compelled to do something by fear are likely later to do the same thing willingly and cheerfully.[47]

The public-private distinction that Marx is so critical of in his account of the alienation of the state from civil society under capitalism is really a manifestation of the conflicting interests and beliefs of the people who make up that society. It is only by assuming that all conflict is class conflict that Marx is able to suppose that this alienation could be overcome. As the discussion of Chapter 8 makes clear, this is truly a utopian idea. Conflict is not an artifact of social classes; it is implicit in the human condition. A related problem facing $PC_1$ society is that it requires the concentration of economic and political power in the same hands. It was argued in Chapter 6 that central planning requires the state apparatus to backstop it directives. That this creates the potential for enormous abuses of power is obvious and consequential.[48]

This completes the account of alienation in $PC_1$ society. All the elements in this account could be elaborated on in greater detail, but enough is in hand for the purposes of this chapter. What can be concluded from all this? First, some things that cannot be concluded. One cannot infer, on the basis of the above, that alienation will be worse, in some all things considered sort of way, in $PC_1$ society than it is in capitalist society. The arguments proceed at too high a level of abstraction for that. Moreover, there are forms of alienation characteristic of capitalism, which were discussed in Chapter 2, that were not pinned on $PC_1$ society in this section. For all that the above shows, that society may turn out to be better than capitalism as far as alienation is concerned, though one is certainly entitled to entertain serious doubts on that score. However, and this is the important point, it has been shown that $PC_1$ society will suffer substantial alienation in all the basic categories that Marx discusses, and that is sufficient to prove the unrealizability of Marx's vision of that society. The promise of radical criticism is a set of social institutions that eliminates, or virtually eliminates, various social evils. This is what Marx implicitly promises for both exploitation and alienation, and this is what he cannot deliver. Marx's vision of $PC_1$ society, like his vision of $PC_2$ society, is unrealizable.

# 10

## The Market Socialist Alternative
## and Concluding Observations

### The Market Socialist Alternative

Nearly all the problems facing $PC_1$ society arise from the fact that it has a centrally planned economy. This has been widely suspected on more direct empirical grounds for quite some time as the failings of existing centrally planned economies have become obvious even to the most bitter critics of capitalism. The purpose of the last chapter was to demonstrate these failings at a fairly high level of abstraction. In the absence of such a demonstration, it is open to the Marxist to say that existing societies with centrally planned economies do not tell us much about what things would be like in Marx's vision of $PC_1$ society, since the latter differs in fundamental ways from existing "socialist" societies. This position cannot be decisively refuted without bringing to bear the sorts of considerations offered by Mises and Hayek.

In light of the inherent difficulties facing central planning, some socialists who have actually thought about social institutions for post-capitalist society have advocated some form of market socialism as the preferred alternative. The main idea behind market socialism is that the workers control the means of production, and yet production is for exchange on the market. Is market socialism a genuine option for a Marxist? Or, to put it another way, how much would be lost by dropping the supposition that $PC_1$ society would coordinate production via central planning? The short answer is that it would no longer be $PC_1$ society, and what remains would no longer be part of Marx's radical critique of capitalist society. The overarching theme of Chapter 2 is that the market is the chief villain in capitalist society. It is directly or indirectly responsible for widespread alienation in capitalist society, according to Marx. A necessary condition for the abolition of alienation is the abolition of the market, though as the last section of the preceding chapter shows, it is not sufficient. Moreover, Chapter 6 adduces a wide variety of textual support and other theoretical considerations to show that Marx is firmly committed to central planning as the method of production coordination for $PC_1$ society. In light of all this, it should be reasonably clear that Marxists for Market Socialism, like the group, Jews for Jesus, are more than a little confused.

However, not all those in favor of market socialism are Marxists. Indeed, the market socialist alternative deserves serious consideration in large measure because it implicitly disavows Marx's rejection of the market. Despite its acceptance of the latter, it does represent an alternative to capitalism. How does it fare as the alternative against which capitalist society is to be judged? In other words, could a radical market socialist critique of capitalist society succeed where Marx has failed? The reader will be relieved to learn that a complete answer to this question will not be attempted here. The main purpose of this section is to take some preliminary steps in the direction of an answer to that question.

To discuss a market socialist alternative, it is of course necessary to specify institutions, in the sense of 'specification' developed in Chapter 6. In principle, there could be indefinitely many versions of market socialism, individuated by how the relevant institutions are specified. I shall build on the account first presented in Chapter 6. This account comes closest to that offered by David Schweickart in his book, *Capitalism or Worker Control?*[1] By focusing on a particular specification of the market socialist alternative, this discussion runs the risk of being inapplicable to other alternatives. I see no way to avoid this problem. Any attempt to capture the "essence" of all forms of market socialism would almost certainly be too abstract to meet the Alternative Institutions requirement for a radical critique. That said, I do believe that the specification offered below is something most market socialists would feel comfortable with.

Recall from Chapter 6 the basic features of the economic system of this conception of market socialism:

    i. The firms are self-managed. That is, the workers have and exercise possession and management rights over the means of production.

    ii. All and only the workers in these firms have income rights with respect to the means of production. In short, the workers get the profits and no one works for wages. Profit in this system is the difference between sales and the costs of nonlabor inputs. These profits are subject to taxation by the state to pay for public goods, to provide for the needy, and to finance new investment.

    iii. The workers in the firm collectively have the right to sell the products of the firm for whatever they can get on the market, subject perhaps to irregular state interference.

    iv. Those who control the means of production (i.e., the workers) do not have full, liberal ownership of the latter. Full, liberal ownership includes, among other things, rights to sell, liquidate, and destroy. The workers in these firms have none of these rights.

v. All new investment (with the possible exception of investment financed out of retained earnings) is controlled by a democratic state via a planning apparatus and state-controlled banks. The state might also intervene in the economy in a variety of other ways, say by tax policy and outright subsidy.

Each of these features of market socialism has a rationale in the perceived deficiencies and strengths of contemporary capitalist and communist societies. Numbers (i) and (ii) articulate socialist relations of production: Workers control the means of production and labor power is not bought and sold, that is, it is not a commodity. Number (iii), of course, is supposed to take advantage of the superior coordinating mechanism that is the market. Number (ii), together with (iv), are designed to prevent the reemergence of a property-owning but nonlaboring class. Number (v) answers to a crucial element of nearly all versions of market socialism: "public" control of new investment. The rationale for this is twofold: First, it is intended to eliminate or mitigate the anarchy of production characteristic of capitalism. Second, it is supposed to prevent private interests from controlling the destiny of society's productive apparatus. A variety of institutional mechanisms might be proposed to bring this about; (v) specifies only one such way. The state is to be democratically controlled to prevent the gross abuses of power characteristic of contemporary communist societies. Finally, (ii), and to a lesser extent, (iv) and (v), are also intended to prevent the emergence of huge income differentials characteristic of capitalism.

Implicit in the above specification is an account of how production will be coordinated and how income will be distributed in a market socialist society: Markets coordinate production with the added wrinkle that the state has a coordinating role through its control of new investment. There is a tendency for income to be distributed among firms according to market criteria, though there is no interest income. How income would be distributed within firms is left unspecified because it is left up to the workers. Traces of a labor market would probably remain because of the need to attract workers with especially scarce skills.

Implicit in the preceding paragraph are critical explanations of various evils of capitalist (and, for that matter, communist) society. I shall not work these out in detail, but it is obvious that various forms of inequality are instrumental or intrinsic evils on the market socialist world view. As might be expected, my interest is in the Alternative Institutions requirement, which calls for not only a specification of the alternative institutions, but also for an explanation of how they would function, some reasons to believe they would be stable, and explanations for how they would avoid the defects of capitalist society.

Elsewhere I have argued in some detail that market socialism, as specified above, would not be stable.[2] I will not restate all the arguments here, but the main idea is the following: In a competitive market economy, firms that give disproportionate power to those with entrepreneurial talent will have a competitive advantage over their more collectivist rivals; from the point of view of the rest of the workers and these entrepreneurially talented individuals, it will be mutually advantageous for the latter's remuneration to be tied very closely to the firm's profitability. By way of a self-reinforcing invisible-hand process, there would be a tendency for these people to become effectively capitalists and the rest of the workers to become effectively proletarians. That is, there will be a tendency for market socialism to evolve—or degenerate— into something indistinguishable from capitalism, even if legally the workers remain the owners. If this is what will happen, then a market socialist radical critique of capitalism cannot meet the Alternative Institutions requirement, since the institutions of market socialism cannot persist as stable social forms.

In the remainder of this section, I want to pursue a line of criticism independent of the above argument. In other words, even if market socialism can persist as a stable social form, other demands of the Alternative Institutions requirement cannot be satisfied. Specifically, I shall argue that the main institutions of market socialism will result in significant alienation and/or surplus value exploitation. The question of whether or not market socialism is to be preferred to capitalism in some all things considered sort of way will not be settled here.

*Alienation in Market Socialist Society.* The most obvious place to look for alienation in market socialist society is in Marx's account of the inherently alienating character of the market. Recall that according to Marx, the market is an inherently alienating institution in three respects: First, it sets in motion large-scale social forces beyond anyone's conscious control—forces that wreak havoc with people's lives. Technological advances throw people out of work, as does a shift in demand or the activities of rival firms. Moreover, the coordination of production achieved by the market breaks down periodically, resulting in depression and recession. Second, in a regime of commodity production, the purpose of production is not the creation of use-values but instead to get exchange value; that is, labor is "unnatural." A third respect in which commodity production is inherently alienating concerns the mutually exploitative nature of interpersonal relations, which is both implicit in the exchange relationship between the capitalist and the worker and spreads to other aspects of social life in a commercial society.

I'm not sure what market socialists would say about the latter two, but a plausible response might go something like the following: The contrast between production for exchange and production for use is really a utopian notion that hard-headed socialists would do well to ignore. Work can be made more meaningful in a self-managed firm, but the complexity of industrial society precludes the Marxian ideal of artisan labor for most workers. Alienation is this respect can be ameliorated but not eliminated. A similar line might be taken on Marx's point about the "mutual using" that is characteristic of exchange (or commercial) societies. As noted in the last section of the previous chapter, it might be that something like this is characteristic of *any* society, and the most one can hope for is an amelioration of its worst effects.

It might be thought that if the market socialist makes these concessions, he has abandoned radical criticism, but this isn't really so. The radical critic can admit that there are some systematic social ills that are ineradicable and transhistorical in their incidence—provided he is willing to argue that there are some that are not. If he is not willing to do at least the latter, then he is . . . a conservative.

The point where the market socialist is most likely to object to Marx's account of alienation is the latter's charge that the market is an inherently alienating institution because it sets in motion large-scale social forces beyond anyone's control. Investment planning is supposed to harness the market in a manner analogous to the way that engineers harness a river with a dam. This planning is in the hands of the state, directly or indirectly, though it is not central planning. As Schweickart says,

> Governmental control here does give, in an important sense, control over the whole economy. That is precisely what we want. We want political control over investment to head off the anarchy of the market, to subject the growth of the economy to human direction, to eliminate the boom-bust cycles of capitalism. The appeal here is to efficiency, material well-being and autonomy.[3]

There are a number of points to be made here, the first of which is not directly relevant to alienation. The faith in the efficacy of state planning of some sort, which has been characteristic of the Left in the twentieth century, is nothing short of remarkable in the face of its uniform failure to achieve its ends throughout this century, except perhaps in limited respects in wartime. Investment planning in Yugoslavia, Western Europe, and in the developing world has repeatedly failed to deliver on its promises, as has central planning in Eastern Europe, the Soviet Union, and China. Despite these manifest failures, proponents of centralized direction of the economy (even if it is not a centrally planned economy) assume that those in control know better how to allocate scarce resources, including investment resources, than does a market responsive to consumer demand. Hayek's insights into how and why markets coordinate production explain why this cannot be the case. Since the ultimate determinants of scarcity value are highly particular and constantly changing facts, the coordination of production in a constantly changing world requires decentralized allocative decision making. To put the point more concretely, people do not need more education or more shoes or whatever. Particular individuals have particular wants and needs that are changing over time. Centralized investment planning, even if it is only on a sector-by-sector basis, faces the same sort of epistemological problem as central planning, though in a less severe form. In addition, investment planning assumes that existing means of production will always be more or less automatically replaced, since investment planning is planning for new investment. But in a constantly changing world, this is not rational. In other words, all investment is new investment in a world in which the underlying determinants of scarcity value are in a constant state of flux; the distinction between replacing existing means of production and new investment is wholly artificial. If economic policy makes a sharp distinction here, systemic irrationalities are bound to appear.

In a related vein, is it really plausible to suppose that market forces will not significantly upset people's lives in a market socialist society? Suppose a new technology is developed in another country that drives down production costs to the point where many existing firms are noncompetitive. A responsible market socialist government is not going to deny its people the benefits to be reaped by participating in the world market, which means some of its own people would suffer. On the other hand, an irresponsible government will deny those benefits, and this brings us to a much more serious problem.

Let us suppose that the state has coercive power at its disposal extensive enough to end the "anarchy" of the market—a lion tamer with a big enough whip to tame market forces, as it were. It is obvious that the state will have enough discretionary power to interfere with the verdict of the market in arbitrary and unpredictable ways. Not only does this exacerbate insecurity of possession (a primary evil) in a variety of ways, it also creates the conditions for a different form of alienation and for surplus value exploitation.

A state that has and exercises the power to intervene in the market according to its own discretion will likely come to be viewed as an alien force hovering over civil society. This seems to be the case in contemporary Yugoslavia where both the party (the League of Communists) and the state regularly interfere in arbitrary and unpredictable ways, often outside regular channels.[4] It might be objected that respect for norms of socialist legality would prevent this sort of thing from happening, but because of the nature of the problem (correcting for the "defects" of the market), the state apparatus must have wide discretionary latitude.[5] Besides, the law is not sui generis; it reflects various interests and is especially subject to change in a society in which major social institutions are conscious creations. The continual changes in the Yugoslav constitution over the past forty years make the French constitution appear to be set in concrete by comparison.

Perhaps a more serious problem with state control of the economy exercised by way of state control of new investment is that it requires an apparatus almost perfectly suited to surplus value exploitation on a massive scale. The politicization of all new investment decisions creates tremendous opportunities for the unintended redistribution of wealth. Imagine what would happen to various lobby groups after a market socialist revolution. Instead of fleeing with the former bourgeoisie, they would simply serve different clients—and increase their staffs. It would certainly be in the interests of the worker-controlled steel manufacturers, grocers, defense contractors, academicians, and so forth, and so forth, to hire these people. As many American capitalists and various other groups with their snouts in the public trough have learned, an investment in Washington pays handsome returns.

To put these considerations on a sounder theoretical footing, it is sufficient for present purposes to point out that there is an entire school of economic thought, Public Choice theory, that has been quite successful at explaining the workings of government bureaucracies on the basis of standard microeconomic assumptions about the behavior of bureaucrats and those who seek to influence them.[6] In applying the insights of Public Choice theory to the question of the allocation of new investment funds in a market socialist society, John Gray has said,

> large existing enterprises with political clout would be favoured over small and struggling ones—and certainly over enterprises projected but not in a position to lobby for capital. In these circumstances, malinvestments would be unlikely to be eliminated, but instead would be concealed by further inputs of capital. The picture derivable from theoretical considerations of the sort developed by the Public Choice School is that of a vast auction for public capital, in which successful bids would be made primarily by entrenched enterprises having political skills and connexions (a crucial point) with the ability to control the flow of information to the central allocative institutions.[7]

And, to recall Trotsky's dictum, 'Those who have something to distribute seldom forget themselves'. That there would be extensive surplus value exploitation in this process is virtually certain. Whoever is responsible for the creation of wealth in a market system, it is not those who back their trucks up to the Treasury.

It might be objected that the above account of what would happen in market socialist society assumes that people will be as greedy and self-interested as they are under capitalism, an assumption that is not only unwarranted but also unlikely to be true if revolutionary institutional change actually takes place. In particular, public-spiritedness might prevail in the political realm. The upshot is that there is inadequate justification for believing that market socialism will face the serious problems of surplus value exploitation sketched above. What can be said in response?

Two things: As Russell once said, this solution has all of the advantages of theft over honest toil. *If* we assume the revolution to have taken place, *maybe* these problems will virtually disappear. But that just pushes the question back: What reason is there to suppose that a revolution which changes consciousness in this fundamental way will actually occur? The problem of the transition looms large, as do Marx's acute observations about the impotence of moral ideals to sustain social institutions.

Second, to evaluate this response, a more careful investigation of the application of Public Choice theory to market socialism would be necessary. This cannot be undertaken here; it is worth pointing out, however, that the great attraction of market socialism, from the standpoint of feasibility or realizability, is that it can take over the assumptions and theoretical understanding of the market mechanism developed for capitalism with systematic modifications to reflect changes in ownership rights.[8] If one assumes, as advocates of market socialism do, that people's behavior in the market is motivated in certain ways and follows certain patterns, there is good reason to believe that similar motivations and patterns will exist when these people move over to the public sector. This is the essence of the Public Choice approach.[9] To put this point in terms Marx would appreciate, the problem with Capitalist Man is that you can't take him anywhere.

All of this suggests that there are good, though not logically compelling, reasons to believe that there will be significant surplus value exploitation in market socialist society and probably more than a little alienation. The prospects for a successful *radical* market socialist critique of capitalist society, at least in terms of alienation and exploitation, do not look promising.[10] It does not follow, however, that capitalist society is preferable, in some all-things-considered sort of way, to market socialist society for the following reasons: First, capitalist society suffers various forms of alienation (something its defenders rarely get around to admitting), and it may be worse than it would be in market socialist society.

Second, there is significant surplus value exploitation in capitalist society. This might seem to contradict the findings of Chapter 4, but in fact it doesn't. Chapter 4 investigated and rejected the charge that the capitalist systematically (surplus value) exploits the worker by way of capitalist relations of production. That charge foundered primarily on the facts that the capitalist contributes to production, both *qua* capitalist and *qua* entrepreneur. However, this argument leaves two avenues open for a charge of surplus value exploitation against capitalists: First, there may be occasional, irregular, or unsystematic surplus value exploitation through the economic system. Defenders of capitalism look pretty silly when they deny that economic exploitation ever occurs under capitalism. More importantly, the above discussion suggests that there can be surplus value exploitation through the state. This was not discussed in Chapter 4 because this is not the manner in which Marx believed that surplus value exploitation takes place under capitalism. But clearly this is a feature of life in contemporary capitalist society (a fact that has received no attention at all in recent discussions of exploitation in capitalist society). People are forced to contribute wealth they have created to a wide range of activities they do not endorse and indeed often do not even know about. It may be that there is more surplus value exploitation under capitalism than would be found in market socialist society, though the above discussion of investment planning in the latter certainly entitles one to be skeptical on this point.

It may be that the elimination of all systematic surplus value exploitation is not feasible in a modern complex society. To achieve its minimization would require that people's income be largely determined by their value contribution—the difference they make, on one understanding of value contribution. In this way, a certain kind of reciprocity would be assured. Though I cannot argue for it here, in my view some sort of reciprocity along this dimension is a necessary condition for a good society.

It looks like the choice between market socialism and capitalism is a choice between the lesser of two evils—to the extent that social institutions can be thought of as objects of choice, which brings us to the final point of this section. There is a common theme in discussions of market socialism by its proponents, a theme that can be found in Marx's vision of PC society: The institutions of post-capitalist society are conceived of as conscious constructions, which have no systematic unintended consequences. This stands in sharp contrast to the Smithian, and indeed Humean, conception of social institutions as rife with unintended consequences and indeed as understandable primarily in those terms. Of course, socialism is supposed to bring social forces under conscious control, but the arguments of this section and those of the last chapter suggest that the prospects for this are dim. The mistake, it seems to me, is a confusion between intended function (or purpose) and actual function. If institutions are specified in terms of intended function only, the fact that social institutions may have systematic unintended consequences is effectively masked.[11] Unmasking this requires some understanding of how social institutions, especially political institutions, actually function; at the level of abstraction that these issues are discussed, this understanding is difficult, but not impossible, to achieve. This puts considerable demands on social science, but this is something radical criticism must face.

### Perspectives on Social Criticism

The concept of radical criticism employed throughout this book provides a useful perspective from which to view the strengths and weaknesses (mostly the latter) of various forms of social criticism. It is easy enough to describe a variety of social ills and defects of contemporary society and to imagine what society would be like in the absence of these social problems. For some people, this is enough to warrant condemnation of contemporary society and a call to arms. This form of criticism can be called 'cheap criticism' in light of its obvious deficiencies as compared to radical criticism. One of these is that the radical critic has systematic explanations of various social ills, that is, explanations in terms of the fundamental social institutions of contemporary society. The cheap critic lacks such explanations and is prone to think in pictures. It is an interesting comment on the human condition that cheap critics often wield considerable political power.

Radical criticism itself can be more or less cheap. The program of some radical critics extends only to mere assertions that there are systematic institutional defects in contemporary society. These people are fond of asserting that capitalist society has some social defect or other, without offering any plausible explanation of how or why capitalist relations of production are responsible for the problem. In their enthusiasm for radical criticism, they take out huge intellectual loans from more serious radical critics, often to the point where there is no hope of repayment. They can be called 'insolvent radical critics'. Both cheap critics and insolvent radical critics warrant little serious attention from those who think carefully about social problems.

More interesting is the serious radical critic. Serious radical criticism starts, but does not end, with genuine critical explanations of the sort that Marx offers in his account of alienation in capitalist society. Though his account of exploitation is defective, it has the right general form or structure. Providing successful critical explanations is in actuality an enormously complicated and difficult task. It requires an understanding of how social institutions function and change. It would be a distortion to represent all of Marx's writings on economic theory and the theory of history as directed toward providing the foundations for critical explanations. His scant systematic attention to normative questions is obvious, and his conception of himself as a scientist is much more accurate than the picture this book suggests. However, if what one is interested in is the radical critique of capitalist society implicit in Marx's thought, then this is how the economic theory and the theory of history should be viewed. As pointed out above, serious radical criticism does not end with the provision of critical explanations. In general, the seriousness of radical criticism is proportionate to the extent to which all four of the requirements for a successful radical critique are, or at least can be, addressed. On this score, Marx counts as a serious radical critic, despite the fact that his radical critique is both incomplete and ultimately unsuccessful.

The framework of, and requirements for, a successful radical critique provides a useful vantage point for judging recent work on the broadly normative aspects of Marx's thought. Nearly all of this work on Marx has addressed the Critical Explanations requirement or the Normative Theory requirement.[12] That philosophers have found the latter of special interest is obvious enough, and those with radical sympathies have shown considerable interest in developing the former. As was the case with Marx himself, however, there has been little attention paid to alternative institutions, and this is a serious omission. Absent an account of the institutions that should replace those of capitalist society and an explanation of how these institutions will preclude the recurrence of the defects of capitalist society, these partial critiques of capitalism have lost touch with reality in an important sense. This problem is pressing in light of the fact that Marx's dual vision of PC society is unrealizable. To put this point in a slightly different way, it is fair to ask: What is the point of saying, for example, that capitalism systematically alienates people, denies positive freedom, prevents self-realization through labor, and so forth, if the critic has in view no realizable alternative institutions where these social problems are absent?

Perhaps one way this question might be addressed is to think about the implications of Marx's thought for moderate social criticism. After all, most people are not radical critics (a fact that Marx's theory of ideology attempts to explain). By definition, the moderate critic is distrustful of, or even openly hostile to, radical institutional change. In spite of this, the moderate critic can accept the radical's critical explanations and/or the latter's normative theory. That is, she can admit that contemporary society has the social ills that the radical calls attention to, and she can admit that the explanations for these ills are to be found in the basic institutions of contemporary society. It's just that she believes that these problems probably cannot be eliminated and that they can be ameliorated through marginal institutional change.

This looks to be a more manageable enterprise, since institutions that do not exist in reality do not have to first be constructed in the imagination (or, to be more accurate, in the understanding). But in fact the demands are not much less stringent than they are for a radical critique in this connection. An understanding of how existing institutions function is required, as well as knowledge of the effects of institutional change. Moreover, purposive descriptions of institutional change, that is, descriptions that cite intended functions of (altered) institutions, are inadequate. We will want to know what will really happen if certain changes are made, not what the reformer intends to happen. The demands on economic and political theory remain significant.

If successful moderate criticism requires that these tasks be executed, it suggests a rather harsh judgment on much of contemporary political philosophy. The two-decade long preoccupation with theories of justice runs a grave risk of being just so much talk. For example, John Rawls does not explain how income redistribution by the democratic state is to be effected without organized pressure groups diverting rivers of tax revenues to themselves and their constituencies. To take another example, Robert Nozick offers no explanation for why the minimal state would stay minimal. These theorists of justice ignore not only the question of how to get from here to there, which is perhaps understandable, but also the question of how the changed institutions would actually function there. But then capitalist society is known for riveting people to just one task.

By now, all readers of this book who are social critics, whether moderate or radical, are probably wondering whether or not there may be some way to repudiate the burdens of proof that have been thrust upon them. One possibility that warrants some investigation is the following: If theory and practice are truly unified, perhaps it is impossible to talk about alternative, or even modified, institutions in the absence of a particular plan or agenda for social change. It might be claimed that large-scale theorizing about alternative institutions is neither necessary nor possible.

The problem with this line of criticism is that it does not really reduce the intellectual burdens the social critic bears; it simply relocates them. The burden now falls more heavily on a theory of social change, that is, how and why institutional change takes place (the Transition requirement). It seems that social science is even worse off on this sort of question than it is on questions about how economic systems function. Revolutionary action, of course, can succeed without a theory, but what is at issue is not the practicality of revolution but its wisdom. Besides, as it applies to Marx, the problem of the transition was not completely ignored. In the last section of Chapter 6, there was some discussion of Marx's views on how trends in capitalism would lead to a centrally planned economy. More generally, the basic assumption on the question of the transition was that Marx is right! That is, it was assumed for the sake of discussion that capitalism will give way to $PC_1$ society. The reconstruction of his vision of that society was largely directed by, and implicit in, his materialist theory of history.

But maybe the above does not quite get at what is behind the idea of the unity of theory and practice and its application to the notion of radical critique. Perhaps the problem is that talk about theories in connection with either the process of social change or its results (i.e., alternative institutions) is somehow misplaced—at least in the sense of 'theory' employed in this book. This concern gains credibility if one looks at just how much the various sub-requirements of the Alternative Institutions requirement call for. For example, even the moderate critic has to have a theory of distribution of income for the existing order, and this theoretical issue is by no means settled at this time.

There are two responses that can be made to this point, both of which have something to be said for them and yet are perhaps too facile. The first comes from an appreciation of Marx's achievements. One of Marx's truly outstanding intellectual contributions is his theory of history. This is a large-scale theory of how social institutions are sustained and how and why they undergo revolutionary change. Complaints that societies are just too complicated or difficult to understand are decidedly un-Marxist and indeed un-scientific. The entire scientific enterprise is predicated on the regulative ideal of bringing conceptual order out of chaos. A high order of intelligence is required, but as Hegel might say, that is what Galileo, Kepler, Newton, Darwin, Smith, and Marx, and others are for. The quest for knowledge is not easy. Second, the project of radical critique is a cognitive enterprise. The knowledge gained may or may not be useful for the sake of action, but knowledge is the goal, and the burdens of proof should be viewed as challenges to be met, not avoided.

I say these responses are perhaps facile, and that is because this "praxis objection," as it might be called, arises from the fact that the prospect of social change is no mere abstraction for the social critic. Unlike the physicist or the astronomer, it is very difficult for the social critic to remain aloof from his or her society. Besides, it is probably neither necessary nor desirable. At last we come to the question that books on social criticism always ask: 'What is to be done?'.

This book began with an extended analogy between the role of the skeptic in challenging our claims to knowledge and the role of the social critic in challenging our pretheoretical normative beliefs about our own society. Recall that the basic skeptical gambit is to undermine whole categories of claims to knowledge: knowledge about the external world, other minds, mathematics, and so forth. If these arguments succeed, they confute the dogmatist who claims these categories for the domain of knowledge. In a similar manner, if the radical critic's critical explanations succeed, he confutes the apologists for the existing order, for example, the "vulgar economists," for Marx. (There are interesting parallels between Descartes's response to skepticism and the reconstructed Marxian response to the critical explanations of the ills of cap-italist society.)

Assuming the success of the skeptical arguments, the next question is 'What ought we to believe?'. If some pragmatists are right in maintaining that believing is a form of doing, this question can be rephrased as 'What is to be done?'. It may be that responses to the skeptic canvassed in the first chapter offer more than a parallel for responses to radical criticism: The latter may be a special case of the former.

In Chapter 1, three nondogmatic responses to skepticism were discussed, which can be identified as follows:

1. *The Academic Response.* Following the Greek philosopher Carneades, some skeptics maintain that, while certainty and hence true knowledge cannot be achieved, some beliefs are more reasonable to accept than others. Hume sums up this attitude concisely when he says in his essay, "Of Miracles,"

   > though experience be our only guide in reasoning concerning matters of fact; it must be acknowledged, that this guide is not altogether infallible, but in some cases is apt to lead us into errors. ... A wise man, therefore, proportions his belief to the evidence.[13]

   The advice is threefold: Believe nothing with absolute certainty; believe more strongly that for which there is better evidence; disbelieve proportionately those propositions against which the evidence runs.

2. *The False Pyrrhonian Response.* The false Pyrrhonian advises that we completely suspend judgment on all those matters about which there is uncertainty, which is nearly everything, if skeptical arguments succeed. As Hume points out, such a creature is not likely to be met with in experience. This is called 'the False Pyrrhonian Response' because Pyrrho himself never advocated such a position, Hume's claims to the contrary notwithstanding.

3. *The True Pyrrhonian Response.* What Pyrrho actually advocated in the face of skeptical arguments was that we hold no belief with certainty but that we let our lives be guided by appearances alone. The True Pyrrhonian response differs from the Academic Response in that the True Pyrrhonian maintains (though without much conviction) that, from an epistemic point of view, all beliefs are equally groundless, a claim the Academic denies (with mental reservations, of course!). The True Pyrrhonian obviously differs from the False Pyrrhonian in advocating that we accept some beliefs—those that accord with appearances.

How does all this apply to social criticism? General skeptical arguments to one side, social science provides very little knowledge about social reality. Even the arguments of the central sections of Chapter 9 are directed at only one aspect of a market system and how it works; other trends and tendencies were ignored. (On the other hand, the object of discussion was real world market systems, not idealized models that could not exist in reality.) In particular, no unified theory of political and economic systems, for capitalism or PC society, was offered. More generally, the imposing demands of both radical and moderate social criticism are going to be very difficult to meet in the foreseeable future. Which brings us back to the question 'What is to be done?' Once again, there are three options:

1′. *The Academic Response.* Try to initiate institutional change to the extent that proposals for institutional change are epistemically warranted. The paucity of relevant knowledge makes this a relatively conservative strategy.

2′. *The False Pyrrhonian Response.* Do not act at all to change existing institutions to alleviate perceived social ills.

3′. *The True Pyrrhonian Response.* Act according to appearances, that is, act to change institutions on the basis of what seems appropriate, all the while recognizing that one does not know much about what is best to do.

A word needs to be said about the difference between (2′) and (3′). The former is obviously the more quietistic position. It is a conservative position in a fairly strict sense of being satisfied with (acting as if one is satisfied with) whatever institutions happen to exist. It is worth pointing out that most people who call themselves 'conservatives' these days are not conservative in this sense. They want institutions to be different from the way they currently are: Social institutions should be the way they used to be. Or the way such people think these institutions used to be. This might be called 'reactionary idealism'. Number (3′), on the other hand, urges some activism, if that appears to be a good idea, provided that it is tempered with a recognition of the fallibility of human belief and action. Number (1′) is closer to (2′) or (3′), depending on the extent to which we have, and can be guided by, reasonable beliefs about social institutions and institutional change.

It is not at all obvious that reason dictates any one of these responses. Even (2′) seems rational for people who have better things to do with their time than work for social change. My own inclination is toward (1′), though in the spirit of Hume, this may only be a matter of taste or sentiment, not something dictated by reason.

There are, of course, other options, namely, dogmatic responses on both sides. Dogmatic defenders of the existing order see no room for improving social institutions, often, as Marx would point out, because they see no room for improvement in their class status. Dogmatic critics of the existing order, on the other hand, are certain of the desirability and practicality of social change. Reason does not support the dogmatists on either side, but whether we in turn should support them is a question of practical rationality, which in turn may be a matter of taste or sentiment.

# NOTES

## Chapter 1

1. This position has been defended by, among others, Jon Elster. See his *Making Sense of Marx* (Cambridge: Cambridge University Press, 1985), pp. 28–29. Other theorists, such as G. A. Cohen, have maintained that, while such elaborations are desirable, they are not necessary for genuine explanation. See his, *Karl Marx's Theory of History: A Defence* (Princeton: Princeton University Press, 1978), pp. 271, 285–289. Clearly, this issue touches on controversies surrounding methodological individualism. For a nuanced discussion of the latter, see Harold Kincaid, "Reduction, Explanation, and Individualism," *Philosophy of Science* 53 (December 1986):492–513.

2. Richard Miller tries to draw this distinction in the course of arguing that Marx offers a nonmoral critique of capitalism. See his *Analyzing Marx* (Princeton: Princeton University Press, 1984), pp. 15–35.

3. This seems a pretty banal truth today, but that may only indicate the extent to which times have changed. When Hegel's *Geist* ruled Germany, or at least German intellectuals, this was a truly revolutionary thought!

4. Implicit in my discussion of the Alternative Institutions requirement is that this is not the only way to test the adequacy of ideas about alternative institutions. This claim will be given a more elaborate defense in Part II of this book.

5. Marx to Bracke, 5 May 1875 in *Critique of the Gotha Program* (Moscow: Progress Publishers, 1971), pp. 9–10.

6. See my "Recent Work on Marx: A Critical Survey," *American Philosophical Quarterly* 24 (October 1987), Sections III–VII for a review of this literature.

7. The view that Historical Materialism contains tendency laws is implicit in much of the recent work on Marx. See, for example, Cohen, *Marx's Theory of History*, pp. 150–157 and my "Marx and Disequilibrium in Market Socialist Relations of Production," *Economics and Philosophy* 3 (Spring 1987), Section II. For this way of construing the laws of Marx's economics, see Daniel Little, *The Scientific Marx* (Minneapolis: University of Minnesota Press, 1986), pp. 24–27.

8. On the other hand, Robert Brenner has argued that this sort of mechanism could not have operated in the rise of capitalism from feudalism. See his "The

Social Basis of Economic Development," in *Analytical Marxism*, John Roemer, ed. (Cambridge: Cambridge University Press, 1986), pp. 23–53.

9. There are many different forms of ethical relativism. Some concern the meaning of moral language, others concern how moral judgments are justified, and others are about ethical principles. My concern is with the last of these. Roughly, this version of ethical relativism says that different ethical principles are correct for different societies. More on this shortly.

10. Allen Wood, "The Marxian Critique of Justice," *Philosophy and Public Affairs* 3 (Summer 1972):244–282, and Allen Wood, "Marx on Right and Justice: A Reply to Husami," *Philosophy and Public Affairs* 8 (Spring 1979):267–295. Both are reprinted in *Marx, Justice, and History*, Thomas Nagel, Marshall Cohen, and Thomas Scanlon, eds. (Princeton: Princeton University Press, 1980). See also Robert Tucker, *The Marxian Revolutionary Idea* (New York: W.W. Norton, 1969), pp. 37–53.

11. For a useful summary of this debate, see Steven Lukes, *Marxism and Morality* (Oxford: Oxford University Press, 1985), pp. 48–59.

12. Shlomo Avineri, *The Social and Political Thought of Karl Marx* (Cambridge: Cambridge University Press, 1968), p. 221. To be fair, immediately following this passage, Avineri proceeds to discuss some features of Marx's conception of post-capitalist society. The quotation is perhaps most charitably regarded as involving an unfortunate choice of words. It does, nonetheless, express a widely held view.

13. Interestingly, Avineri himself reports (ibid., p. 81) that the Soviets have altered the text here by omitting from their English translation an illuminating qualifying phrase at the end of this passage: "*also schon ideel vorhanden war*," which he translates as: "i.e., has already pre-existed ideally."

14. This might be one way of understanding the unity of theory and practice. See Cohen, *Marx's Theory of History*, p. 339.

15. The possibility for self-defeating predictions is present whenever the person making the prediction interacts with the person or group who is the object of his predictions.

16. Some of the remaining discussion in this chapter depends on a very close reading of a page or so that surrounds the following quotation. The reader might find it useful to have this part of *The German Ideology* ready to hand.

17. Cohen, *Marx's Theory of History*, pp. 132, 133.

18. This functional definition may appeal to some goal(s) or purpose(s) that the institution serves. For more on the goals or purposes of social institutions, see my "Why Profits Are Deserved," *Ethics* 97 (January 1987), Section II.

19. There may be a transcendental argument about the presuppositions of social life that could be developed on this foundation. The parallels with Kant's Transcendental Deduction of the Categories in the *Critique of Pure Reason* are suggestive.

20. Cohen, *Marx's Theory of History*, p. 133.

21. Ibid.

22. For another example of semantic overkill by Marx, consider the following passage that occurs about a page before the Famous Passage: "Division of labor and private property are, after all, identical expressions: in the one the same thing is affirmed with reference to activity as is affirmed in the other with reference to the product of the activity," (*GI*, MECW, vol. 5, p. 46). Of course, they are not identical expressions, nor are they co-intensional or even co-extensional, as Marx's own gloss makes clear; at most, their referents are closely related as a matter of empirical fact. It is very hard for analytic philosophers like Cohen and myself to make sense out of what Marx says when he was writing under the spell of Hegel.

# Chapter 2

1. G. A. Cohen, *Karl Marx's Theory of History: A Defence* (Princeton: Princeton University Press, 1979), pp. 181–193.

2. For a brief survey of recent discussions of Marx's theory of alienation, see my "Recent Work on Marx: A Critical Survey," *American Philosophical Quarterly* 24 (October 1987), Section V.

3. The fact that no one—not even the capitalists—controls the society-wide division of labor is of considerable importance. Among other things, it is a crucial element in Marx's account of crisis under capitalism.

4. See Cohen, *Marx's Theory of History*, p. 115.

5. Of course, by itself, this does not explain stability in capitalist relations of production, since a similar mystification does not characterize feudal relations of production.

6. P. C. Roberts and Matthew Stephenson effectively argue that Marx's explanation of both the possibility and necessity of crisis in the capitalist mode of production crucially depends on the fact that capitalism is a system of commodity production. See their *Marx's Theory of Exchange, Alienation, and Crisis* (Stanford, Calif.: Hoover Institution Press, 1973), Chapter 6 and Appendix. This point is of obvious significance for the Transition requirement for a radical critique.

7. The term 'laboring activity' sounds a bit pedantic and awkward, but its reference is quite clear. The more natural term is 'labor', though it is multiply ambiguous in both Marx's writings and the secondary literature. To avoid the awkwardness of 'laboring activity', I shall use the term 'labor' to refer to laboring activity or to the results of that activity as it exists in the product. The latter is what Marx refers to when he modifies the term 'labor' with adjectives such as 'embodied', 'crystallized', 'ossified', and 'congealed'. The context will make clear which sense is intended.

8. The fact that labor power is a commodity also reflects the "unnaturalness" of labor under capitalism: That labor power is *sold* reflects the fact that the natural purpose of laboring—the creation of use values—has been replaced by another goal or purpose: wages. The fact that labor power is *bought* signals that the purpose of the use of labor power (by the capitalist) is to get exchange value.

9. There is a certain "slippage" in my interpretation of this passage. Marx says that it is labor power, and not the laborer, which is conceived of by the laborer himself or herself as a mere commodity. However, Marx regarded his discovery that it is labor power, and not labor, that the capitalist buys as an important and recent contribution to the science of political economy—a discovery that required considerable scientific acumen. The distinctions among labor, labor power, and the laborer are, in this connection, subtle ones, and it is easy to understand how the worker might conflate them in his or her self-conception.

10. This statement of the Labor Theory of Value has to be qualified in various ways. Strictly speaking, it is the quantity of socially necessary labor required to make the product; this is labor of average skill and intensity. Otherwise, especially inefficient labor would make the object more valuable. This and related complications are taken up in the next chapter.

11. See Chapter 3, pp. 65–66 for a more elaborate discussion and defense of Marx's use of this terminology.

12. Allen Buchanan makes a similar point in *Marx and Justice* (Totowa, N.J.: Rowman and Allenheld, 1982), p. 43.

13. The claim that competition is a necessary condition for commodity production is defended in Chapter 6. See below, p. 158 and note 27 of that chapter.

14. On this last point, see Cohen, *Marx's Theory of History*, p. 183f.

15. See, for example, Thesis VI of *Theses on Feuerbach* (MECW, vol. 5, p. 4) and *Capital* I, p. 579n2. For a good discussion of Marx's criticisms of various conceptions of human nature, especially Feuerbach's, see Buchanan, *Marx and Justice*, pp. 19–21, 28.

16. Since the publication in this century of some of Marx's early writings, such as the *1844 Manuscripts*, hardly any commentator on Marx denies this anymore. For a brief survey of recent discussions in the secondary literature of Marx's conception of human nature, see my "Recent Work on Marx: A Critical Survey," Section VI.

17. Perhaps the most important consequence in this regard is that humans create social relations (usually not self-consciously) through which production takes place. A full tracing out of the consequences of this would take us through most of the major doctrines of Historical Materialism. Throughout this section, I want to keep the focus on the nature of individual human beings; I recognize that this is a ruthless abstraction, given Marx's emphasis on man's sociality, but there are two points that can be made in defense of this procedure: First, Marx does discuss the nature of human beings in abstraction from their social relations (notably in the *1844 Manuscripts*). Secondly, in the next section, I shall discuss in detail the social dimension of alienation.

18. Marx seemed to have thought that no animal engages in genuinely purposive behavior. There seems to be an emerging consensus in contemporary behavioral sciences that this is false, but I shall not press that point here.

19. The German word Marx uses, *entausserung*, often translates as 'loss.'

20. It is central to Marx's account of the transition to socialism that these people will create the new society. Some of Marx's critics have worried about the character of post-capitalist society based on this fact. This is a complicated issue I shall not pursue here.

21. For this explanation to be complete, it is probably necessary to appeal to another essential feature of capitalism, viz., that the workers do not control the means of production.

22. Given their degraded status as human beings, commodity production is probably the only way production can be organized. One of the main tasks of socialism will be to restore to individual proletarians the capacity for free, universal labor. Though they lack the capacity for such labor under capitalism, their biological nature probably guarantees that they have the potential to develop this capacity. Analogously, fetuses lack the capacity to reason but, in virtue of their biology, they have the potential to develop that capacity.

23. This functional explanation can be elaborated (via an invisible-hand explanation) in terms of the rational behavior of individuals in the marketplace; the standard explanation of the rise and predominance of precious metals as media of exchange proceeds in just this manner. See, for example, Robert Nozick, *Anarchy, State, and Utopia* (New York: Basic Books, 1974), p. 18.

24. Marx's concurrence with Feuerbach on this point is clearest in the Introduction of his *Contribution to the Critique of Hegel's Philosophy of Law*. See MECW, vol. 3, pp. 175–176.

25. That Marx had in effect two theories of the State has been noticed by many

commentators. One of the best discussions of this, which I follow in this chapter, can be found in Avineri, *The Social and Political Thought of Karl Marx*, pp. 22–27, 48–52. For references to additional secondary literature, see Buchanan, *Marx and Justice*, p. 185n16.

26. The context of Marx's discussion of private property almost always makes it clear that he is concerned with private property in the means of production. As far as I am aware, personal private property is never the target of Marx's attack.

27. Marx has a deeper explanation for the egoism characteristic of capitalist society, which will be discussed in the last part of this section. At most, ideology can help to explain its persistence under capitalism.

28. Karl Marx, *The Eighteenth Brumaire of Louis Bonaparte* (New York: International Publishers, 1963), p. 122. See also, Karl Marx, *The Civil War in France*, MECW, vol. 22, pp. 329–30.

29. Avineri, *The Social and Political Thought of Karl Marx*, p. 51.

30. Karl Marx, "Comments on James Mill's *Elemens D'Economie Politique*," MECW, vol. 3, p. 215. Hereafter, quotations from this work will be referenced in the text to MECW under the short title, "Comments on Mill."

31. The commentators who have appreciated most fully the significance of commodity production in Marx's theorizing are P. C. Roberts and Matthew Stephenson. See their *Marx's Theory of Exchange, Alienation and Crisis*, passim.

32. This is explicit or implicit in the writings of those who attribute to Marx an ethics of positive freedom or an ethics of self-realization. See, for example, George Brenkert, *Marx's Ethics of Freedom* (London: Routledge and Kegan Paul, 1983), Chapter 4; Steven Lukes, *Marxism and Morality* (Oxford: Oxford University Press, 1985), Chapter 5; Hilliard Aronovitch, "Marxian Morality," *Canadian Journal of Philosophy* 10 (September 1980):357–376.

33. See above, note 16.

## Chapter 3

1. I use the term 'capitalist *qua* capitalist' to refer to the capitalist in his role or function as a capitalist. This is of some importance because some capitalists labor (conceived of broadly to include materially necessary managerial labor), some innovate, some take risks, etc. But some do none of these things, and they don't get kicked out of the ruling class for it. These are the pure coupon-clippers. The fact that it is possible to be a capitalist without laboring, innovating, etc., makes it appropriate to speak of the capitalist in his role as capitalist—a controller of the means of production. For ease of exposition, unless otherwise indicated, when I speak of the capitalist, this is to be understood to mean the capitalist *qua* capitalist.

2. As does Engels, for whom this was one of the two most important contributions attributable to Marx. See his "Karl Marx," in Karl Marx and Frederick Engels, *Selected Works*, vol. 3, (Moscow: Progress Publishers, 1970), pp. 77–87. For a good discussion of the location of this problem and Marx's solution to it in the history of classical political economy, see Robert Paul Wolff, *Understanding Marx* (Princeton: Princeton University Press, 1984), Chapters II–IV.

3. As Marx repeatedly stresses, the worker is fully compensated for his labor *power*; that does not change the fact that he is not fully compensated for the labor—his labor—that gets embodied in the product.

4. This way of construing Marx's argument is suggested by Wolff. See *Understanding Marx*, pp. 109–110.

5. Eugen Böhm-Bawerk, "Unresolved Contradiction in the Marxian Economic System," Alice Macdonald, trans., reprinted in *Shorter Classics of Eugen Böhm-Bawerk* vol. I (South Holland, Ill.: Libertarian Press, 1962), pp. 202–302. This work can also be found under the title *Karl Marx and the Close of His System*, Paul M. Sweezy, ed. (London: Merlin Press, 1974).

6. Marx makes some effort to argue for this assumption at *Capital* III, p. 175. See also Wolff, *Understanding Marx*, p. 123.

7. See Böhm-Bawerk, "Unresolved Contradiction," pp. 208–211 for some interesting details.

8. It might be argued that a more "global" justification is available to Marx, to wit, that his entire economic theory best explains the range of phenomena it is supposed to explain. This potential line of justification is explored in the next section and critically evaluated in the final section of this chapter. My focus here, as I indicated at the beginning of this chapter, is on Marx's actual chain of reasoning.

9. This way of conceiving of explanation via models is articulated by Ronald Giere in his *Understanding Scientific Reasoning* (New York: Holt, Rinehart, and Winston, 1979). See especially chapters 5 and 6. The term 'model-theoretic explanation', however, is my own. It should not be confused with model theory in logic.

10. After a careful examination of the textual evidence, G. Catephores and M. Morishima conclude that Marx did not believe this. See their "Is There an Historical 'Transformation Problem'?," *The Economic Journal* 85 (June 1975):314–315.

11. See Daniel Little, *The Scientific Marx* (Minneapolis: University of Minnesota Press, 1986), pp. 80–82.

12. This assumption has been challenged at least as early as Böhm-Bawerk. Although Marx defines a commodity as a something that is produced, and thus is a product of labor, our ordinary understanding of the term 'commodity' would include unimproved natural resources (e.g., standing timber) as commodities. That this is not a mere terminological dispute is evident in the fact that these things have an exchange value. This problem, however, provides further confirmation for the view that the LV is a simplifying assumption to be evaluated in terms of its explanatory adequacy under restricted conditions (e.g., in a regime of simple commodity production where all land is free).

13. For useful discussions of the problem of heterogeneous labor, see Jon Elster, *Making Sense of Marx* (Cambridge: Cambridge University Press, 1985), pp. 130–131, and John Roemer, *A General Theory of Exploitation and Class* (Cambridge: Harvard University Press, 1982). Ironically, it was Böhm-Bawerk who first stated this problem, though not in the context of a reproduction model. See Böhm-Bawerk, "Unresolved Contradiction," pp. 270–273.

14. Sraffa's results have been further developed and extended by Ian Steedman in his *Marx After Sraffa* (London: New Left Books, 1977). For a concise appreciation of these developments, see Little, *The Scientific Marx*, pp. 82–85.

15. A group of people stranded on a desert island are trying to figure out how to get back to civilization. The economist among them says, "It's simple. First assume a boat."

16. For the definitive exposition and defense of these elements of contemporary methodology, see Part I of Milton Friedman's *Essays in Positive Economics* (Chicago:

University of Chicago Press, 1953). Some of this is reprinted as, "The Methodology of Positive Economics," in *Philosophy and Economic Theory*, Frank Hahn and Martin Hollis, eds. (Oxford: Oxford University Press, 1979), pp. 18–35.

17. See *Capital* III, p. 177. This is consistent with Catephores and Morishima's finding that Marx did not believe that there ever existed a full-blown regime (a "mode") of simple commodity production.

18. By contrast, it seems that the Identity Thesis cannot be interpreted as a tendency law. Its status is as yet unclear on this interpretation of what Marx is up to, a problem that will be remedied toward the end of this chapter. For now, it is enough to note that Marx believed it to hold for any regime of commodity production and that the *Capital* I argument for it proceeds from the LV. In the reconstruction that follows, it will be assumed to be true.

19. On the other hand, an application of a tendency law to a particular situation may have the character of an approximation. For example, if Marx's LV is right, it might be that in a particular market at a particular time it is correct to say that two bundles of commodities that exchange have approximately equal values. However, if the law is conceived of as an unrestricted universal generalization on this instance, it would be false, since it is often (usually) the case that not all else is equal. For this reason, the *ceteris paribus* clause cannot be interpreted as an approximation qualifier.

20. The heart of the argument is given at *Capital* I, pp. 64–65.

21. Böhm-Bawerk, "Unresolved Contradiction," p. 259. (Page references to *Capital* I have been added.)

22. Ibid., p. 260.

23. Theft, fraud, and fluctuations in supply and demand are covered by the fact that the explanandum is equilibrium exchange ratios.

24. Given Marx's commitment to the Identity Thesis, it is not surprising to find, outside of the *Capital* I arguments, that he does not sharply distinguish the LV ("Commodities that exchange have equal values") and the LTC ("Commodities that exchange contain equal quantities of socially necessary labor").

25. Barry Hindess, Paul Hirst, et al., *Studies in Marxist Philosophy*, vol. 4 (New York: Mepham, 1974), p. 36.

26. Böhm-Bawerk, "Unresolved Contradiction," p. 240.

27. In other words, Marx maintains that the higher price is accounted for solely by a greater quantity of used up constant capital. Böhm-Bawerk is claiming that even if the quantity of used up constant capital were the same over a given period, say six years, the relative capital-intensiveness of the production of *I* and *II* would still be reflected in exchange ratios.

28. Adam Smith, *The Wealth of Nations* ([London, 1776] New York: Random House, 1985), p. 48 (Book I, Chap. 6). For a discussion of the general significance of this model for classical economics, see Wolff, *Understanding Marx*, pp. 30ff.

29. G. A. Cohen, "The Labor Theory of Value and the Concept of Exploitation," in *Marx, Justice, and History*, Marshall Cohen, Thomas Nagel and Thomas Scanlon, eds., *A Philosophy and Public Affairs* reader (Princeton: Princeton University Press, 1980), p. 142. See ibid., pp. 143, 144 for quotations from Marx supporting Cohen's interpretation.

30. Strictly speaking, the LTV is the conjunction of the Identity Thesis and the Labor Time Corollary ("Bundles of commodities that exchange in the market in equilibrium contain the same quantity of socially necessary labor time"). But since

the term 'the value of a commodity' means 'what determines equilibrium exchange ratios of commodities,' the Identity Thesis means the same thing as the Labor Time Corollary. What is at issue in this appendix is, in effect, the question of whether or not Marx subscribed to the Identity Thesis.

31. Cohen, "The Labor Theory of Value," pp. 144.

## Chapter 4

1. John Roemer, "Property Relations vs. Surplus Value in Marxian Exploitation," *Philosophy and Public Affairs* 11 (Fall 1982):281–314.

2. The seminal article in this debate is G. A. Cohen, "The Structure of Proletarian Unfreedom," *Philosophy and Public Affairs* 12 (Winter 1983):3–33. For a thoughtful appraisal and critical evaluation of this and some related articles by Cohen, see John Gray, "Against Cohen on Proletarian Unfreedom," *Social Philosophy & Policy* 6 (Fall 1988):77–112. See also George Brenkert, "Cohen on Proletarian Unfreedom," *Philosophy and Public Affairs* 14 (Winter 1985):91–98.

3. Cheyney Ryan has explored the historical antecedents of this view in the writings of the Ricardian socialists. See his "Socialist Justice and the Right to the Labor Product," *Political Theory* 8 (November 1980):503–524.

4. Different specifications of the alternative systems yield different forms of exploitation (e.g., feudalist, capitalist, socialist). See John Roemer, *A General Theory of Exploitation and Class* (Cambridge: Harvard University Press, 1982), pp. 194–195. See also Roemer, "Property Relations vs. Surplus Value." Roemer's various conceptions of exploitation will be discussed in detail in Chapter 5.

5. The idea that exploitation can be conceived of as a kind of parasitism is first developed by Allen Buchanan. See his *Ethics, Efficiency and the Market* (Totowa, N.J.: Rowman and Allenheld, 1985), pp. 90–95. The organization of this chapter owes a great deal to Buchanan's suggestive remarks.

6. *CGP*, pp. 14–15. However, as Nancy Holmstrom points out, the collective consumption that results from the "withholding tax" under socialism is a way of giving workers something closer to the total value of their products. See Nancy Holmstrom, "Marx and Cohen on Exploitation and the Labor Theory of Value," *Inquiry* 26 (September 1983):291.

7. Nancy Holmstrom, "Exploitation," *Canadian Journal of Philosophy* 7 (June 1977):359. See also the article referred to in the preceding note.

8. I do not mean to suggest that the rate of exploitation or degree of exploitation is a function of capitalist consumption. My claim is only that a necessary condition for surplus value exploitation is that the worker is doing unpaid labor for the capitalist. Since there are other necessary conditions for exploitation, it may be that the rate or degree of exploitation is determined in some other way.

9. Holmstrom does discuss the LTV in her second article (see note 6) on exploitation in the context of her critique of G. A. Cohen. It is unclear where she stands in the end on the LTV, though she does seem to think that many of the standard objections are misconceived. See also Cohen's response to Holmstrom, G. A. Cohen, "More on Exploitation and the Labor Theory of Value," *Inquiry* 26 (September 1983):309–332.

10. Marxists routinely assume that what the worker sells to the capitalist is his labor power. As noted in the last chapter, the nature of the phenomenon of exchange between the capitalist and the worker cannot settle the issue of what the laborer sells. Cohen, in *Marx's Theory of History* (p. 43), claims that the worker

cannot sell his labor, since activities are not the sort of thing that can be owned, but he offers no argument for this. Furthermore, this would come as quite a surprise to performers. It is doubtful that Frank Sinatra sells his audiences or concert promoters his capacity to sing; it seems that what he sells them is a show. My claim is not that the worker always sells labor; rather, it's that the assumption that what the worker sells is his labor power is unwarranted, since it is driven by the demands of a defective theory of value.

11. Jon Elster, *Making Sense of Marx* (Cambridge: Cambridge University Press, 1985), p. 167. In fairness to Elster, it should be noted that he quickly points out some difficulties with this formulation.

12. Richard W. Miller, "Marx and Aristotle: A Kind of Consequentialism," in *Marx and Morality*, Kai Nielsen and Steven C. Patten, eds. Supplementary volume VII, *Canadian Journal of Philosophy*, (Guelph, Ont.: Canadian Association for Publishing in Philosophy, 1981), p. 337.

13. Jeffrey Reiman, "Exploitation, Force and the Moral Assessment of Capitalism: Thoughts on Roemer and Cohen," *Philosophy and Public Affairs* 16 (Winter 1987):4.

14. See, e.g., Arthur DiQuattro, "Value, Class, and Exploitation," *Social Theory and Practice* 10 (Spring 1984):70–71, and Gary Young, "Justice and Capitalist Production: Marx on Bourgeois Ideology," *Canadian Journal of Philosophy* 8 (September 1978):441–444. There are also suggestions of this in Marx, independent of his identification of labor time with value. See, e.g., *Capital* III, p. 819.

15. This parallels worries about how to compute the labor content of a commodity when labor is qualitatively heterogenous. If labor content is simply the absolute *quantity* of labor embodied in the product, there is no problem in principle in computing that amount. On the other hand, if the concern is to identify the *value* of the labor content, then this difficulty is much more serious. See Elster, *Making Sense of Marx*, pp. 130–131 and the references cited therein for discussions of the problem of identifying the labor content, in value terms, of the product.

16. It might be thought that the Marxist could support a charge of parasite exploitation directly from the fact of unequal labor exchange, that is, without going through the preliminary conclusion that the worker does unpaid labor for the capitalist. This possibility will be considered in the next section.

17. This is why the LTV fails even if there is some determinate mathematical relation between price and embodied labor (i.e., if the so-called Transformation Problem has a solution). If in fact there is such a relation, that does not constitute an explanation for how prices are determined. The formalism so beloved by economists is particularly ill-suited to address the 'How questions' that must be answered to explain real-world economic phenomena like price formation. A theory of price formation in a capitalist economic system is sketched in Chapter 9.

18. Other necessary conditions for exploitation might still hold (e.g., the worker is forced to work for the capitalist), and they might be part of a radical critique. Indeed, other forms of exploitation may be present. My claim here is only that the charge of surplus value exploitation (which, by the way, remains unproved) would not be part of that critique.

19. G. A. Cohen, "The Labor Theory of Value and the Concept of Exploitation," *Philosophy and Public Affairs* 8 (Summer 1979):338–360. This article is reprinted in, *Marx, Justice, and History*, Marshall Cohen, Thomas Nagel, and Thomas Scanlon, eds. (Princeton: Princeton University Press, 1980), pp. 135–158. All subsequent references will be to the latter.

20. Ibid., p. 151.

21. Ibid., p. 153. I have taken the liberty of renumbering the premises of this argument.

22. Ibid., p. 140.

23. Ibid., p. 151.

24. Ibid., p. 152.

25. This section draws heavily on my article, "Capitalists and the Ethics of Contribution," *Canadian Journal of Philosophy* 15 (March 1985):87–102. My account of entrepreneurship owes a great deal to the writings of Israel M. Kirzner. See, e.g., Israel M. Kirzner, *Competition and Entrepreneurship* (Chicago: Chicago University Press, 1973) and *Perception, Opportunity and Profit* (Chicago: University of Chicago Press, 1979). See also Harold Leibenstein's seminal article, "Allocational Efficiency vs. 'X-Efficiency,'" *American Economic Review* 56 (1966):392–415.

26. See, e.g., Paul A. Samuelson, *Economics*, 7th ed. (New York: McGraw Hill, 1967), pp. 591–594.

27. David Schweickart, *Capitalism or Worker Control?* (New York: Praeger Publishing, 1980), p. 20.

28. Ibid., p. 11.

29. Since Schweickart does not use the term 'parasite', it is not clear that he would endorse this trailer on his argument, but it is relevant for the issue at hand.

30. One answer that might come to mind is that there is some risk the money will not be repaid. That this cannot entirely explain the returns to capital is evident in the fact that it implies that there would be no returns to capital in a world of perfect information, an implication any economist would rightly reject. The point can be empirically verified by the fact that there is a positive rate of interest even on investments that are, for all practical purposes, certain. I return to this point below.

31. It is at this juncture that the results of the preceding chapter bear fruit. It should come as no surprise that the first economist to appreciate adequately the economic significance of time was Böhm-Bawerk. See Eugen Böhm-Bawerk, *Capital and Interest*, G. H. Huncke and H. F. Sennholz, eds. and trans. (South Holland, Ill.: Libertarian Press, 1959 [First published, 1902]). The first pure time preference theory of interest can be found in the writings of the early twentieth century American economist, Frank Fetter. See the following: "The 'Roundabout Process' in Interest Theory"; "The Relations Between Rent and Interest"; "Interest Theories, Old and New"; "Interest Theory and Price Movements." These and other essays by Fetter are reprinted in Frank Fetter, *Capital, Interest, and Rent*, Murray N. Rothbard, ed. (Kansas City, Kans.: Sheed, Andrews and McMeel, 1977).

32. Schweickart, *Capitalism or Worker Control?*, p. 25.

33. I have in mind objections to neoclassical general equilibrium analysis, such as the so-called "Cambridge Controversies" about capital reswitching. For a brief but lucid summary of these controversies, see Arthur DiQuattro, "Alienation and Justice in the Market," in *Marxism and the Good Society*, John P. Burke, Lawrence Crocker, and Lyman Legters, eds. (Cambridge: Cambridge University Press, 1981), p. 122. Though I shall not argue it here (since nothing hangs on it), I believe that marginal productivity theory in general, and the time preference theory of interest in particular, can be stated in a way that does not depend on aspects of general equilibrium analysis called into question by the Cambridge Controversies.

34. Quoted in Schweickart, *Capitalism or Worker Control?*, p. 16.

35. Socialists sometimes recognize the contribution of entrepreneurship by con-

ceiving of it as a form of labor. That any productive activity must be conceived of as a form of labor suggests an interesting bias in socialist thought.

36. There may be other inequalities, such as inequalities of power, but the mere fact of inequality does not imply a lack of reciprocity. In the case of inequality of power, it seems that its proper role would be to explain why there is a lack of reciprocity, assuming the latter can be established.

37. Lawrence Becker, *Reciprocity* (London: Routledge and Kegan Paul, 1986), p. 143.

38. Allen Buchanan, *Marx and Justice* (Totowa, N.J.: Rowman and Allenheld, 1982), p. 38.

39. The above paragraph locates exploitation in the character of interpersonal relations between bureaucrats and citizens. It might be argued that the state also engages in straight surplus value exploitation of the citizens in collecting taxes. The Marxist account of the state could then be rung in to blame this on capitalist relations of production. Although the modern welfare state does redistribute much of this wealth among the citizenry, it might be argued that the state is the central exchange in what might be called a 'mutual exploitation society'. This idea is worth further investigation, but it is unclear whether or not Marx would subscribe to it.

## Chapter 5

1. John Roemer, *A General Theory of Exploitation and Class* (Cambridge: Harvard University Press, 1982), pp. 194–195. The notion of dominance is undefined and does not appear in Roemer's definitions of the various species of exploitation. In an article published in the same year as the book, Roemer expresses some doubts about the utility of this concept in accounts of exploitation. See his "New Directions in the Marxian Theory of Exploitation and Class," reprinted in *Analytical Marxism*, John Roemer, ed. (Cambridge: Cambridge University Press, 1986), pp. 103–104n15.

2. Roemer considers (*A General Theory*, pp. 200–201) and rejects possible counterarguments to the effect that the lord provides services or skills to the serf, which would be missing were the serfs to pull out.

3. John Roemer, "Property Relations vs. Surplus Value in Marxian Exploitation," *Philosophy and Public Affairs* 11 (Fall 1982):285. My quotation corrects an obvious misprint in the text: The latter contains the word 'worse' in condition (1), which I have replaced by the bracketed 'better'.

4. Ibid., p. 305.

5. See ibid., pp. 305–308.

6. At least with respect to the issue of entrepreneurial ability. See ibid., p. 307. Given the obvious importance of these empirical issues to the question of whether or not the workers are exploited in some distinctive way under capitalism, it is surprising and more than a little puzzling that Roemer makes no effort to gather and present any relevant data.

7. Ibid., p. 309.

8. Ibid.

9. Roemer, *A General Theory*, p. 267.

10. The relation between the level of development of the forces of production and how well off people are is not straightforward, as Roemer and other Marxists point out. Following Roemer, however, the problem is simplified by understanding well-being in terms of an income-leisure package and by assuming a fairly close

correlation between level of development of the forces of production and the proletariat's income-leisure package.

11. Perhaps what Roemer is offering in his series of definitions is what Carnap
calls an explication. I follow Carnap's account of how explications can be criticized.
See *The Logical Foundations of Probability* (Chicago: University of Chicago Press, 1950),
Chapter I.

12. There may be other forms of unfairness and indeed other forms of exploitation under capitalism. For example, Roemer claims that there is socialistic exploitation under capitalism (as well as under socialism). What is at issue here is
whether or not there is a form of exploitation under capitalism that is unique or
distinctive.

13. Actually, there are some constraints imposed by the ethical ideals Roemer
seeks to model. However, there are no realizability constraints on the alternatives.

14. Jon Elster, *Making Sense of Marx* (Cambridge: Cambridge University Press,
1985), p. 203.

15. Roemer, *A General Theory*, pp. 212–216. In fairness to Roemer, he explicitly
recognizes that there may be a realizability problem in this connection. For Elster's
criticism of Roemer's model, see Elster, *Making Sense of Marx*, p. 203n2.

16. Much of the remainder of this section is a preview of issues that will be
addressed in a more rigorous and systematic fashion in Chapters 6 and 7, where
the Alternative Institutions requirement is discussed in more detail. Ultimately,
much more is at stake than the relatively narrow question of whether or not the
proletariat satisfy Roemer's definition of a capitalistically exploited coalition.

17. Roemer, *A General Theory*, p. 248. For further evidence that Roemer believes
that incentives are the decisive issue, see the two preceding quotations from
Roemer.

18. For an account of how this might work in an market system with an egalitarian distribution of income, see Joseph Carens, *Equality, Moral Incentives and the
Market* (Chicago: University of Chicago Press, 1981).

19. For an illuminating account of all logically possible relations of production,
see G. A. Cohen, *Karl Marx's Theory of History* (Princeton: Princeton University
Press, 1979), pp. 63–69.

20. See Lawrence Becker, *Property Rights* (London: Routledge and Kegan Paul,
1977), Chapter II, for an appreciation of some of the complexities involved in the
concept of property rights. This issue will be more thoroughly explored as it pertains
to Marx's vision of post-capitalist society in the second section of Chapter 6.

21. These problems will be discussed in detail in Chapter 9.

22. I have argued for this claim in some detail in "Marx and Disequilibrium in
Market Socialist Relations of Production," *Economics and Philosophy* 3 (Spring
1987):23–48. See also "Further Thoughts on the Degeneration of Market Socialism:
A Reply to Professor Schweickart," *Economics and Philosophy* 3 (Fall 1987):320–330.

23. These problems will be cleared up by Chapter 9 where it will finally be
possible to adjudicate the question of whether or not the workers are capitalistically
exploited under capitalism.

24. There is more to justice than justice in the distribution of wealth and income
(e.g., retributive justice, civil justice), but for the purposes of this section, the term
'justice' will refer to justice in the distribution of wealth and income. That is what
this controversy is about. I will have something to say about the other kinds of
justice in Chapter 7.

25. See Allen Wood, "The Marxian Critique of Justice," *Philosophy and Public*

*Affairs* 3 (Summer 1972):244–282, and Allen Wood, "Marx on Right and Justice: A Reply to Husami," *Philosophy and Public Affairs* 8 (Spring 1979):267–295. Both are reprinted in *Marx, Justice, and History*, Thomas Nagel, Marshall Cohen, and Thomas Scanlon, eds. (Princeton: Princeton University Press, 1980), pp. 3–41 and 106–134, respectively. See the works cited in this note and the next two for quotations supporting the contending interpretations.

26. See, for example, Donald van de Veer, "Marx's View of Justice," *Philosophy and Phenomenological Research* 33 (March 1973):366–386; Ziyad Husami, "Marx on Distributive Justice," *Philosophy and Public Affairs* 8 (Fall 1978):27–64, reprinted in *Marx, Justice, and History*, pp. 42–79; G. A. Cohen, review of Allen Wood's *Karl Marx,*" *Mind* 92 (July 1983):440–445.

27. For the former, see Gary Young, "Justice and Capitalist Production: Marx and Bourgeois Ideology," *Canadian Journal of Philosophy* 8 (September 1978):421–457, and "Doing Marx Justice," in *Marx and Morality*, Kai Nielsen and Steven C. Patten, eds. supplementary volume VII, *Canadian Journal of Philosophy* (Guelph, Ont.: Canadian Association for Publishing in Philosophy, 1981):251–268. Also see Derek H. P. Allen, "Marx and Engels on the Distributive Justice of Capitalism," in ibid., pp. 221–250, and Allen Buchanan, *Marx and Justice* (Totowa, N.J.: Rowman and Allenheld, 1982), pp. 50–60. For a rejectionist view, see Richard Miller, *Analyzing Marx* (Princeton: Princeton University Press, 1980), pp. 80f. Steven Lukes provides a useful summary of the debate. See his *Marxism and Morality* (Oxford: Clarendon Press, 1985), pp. 48–59.

28. Nancy Holmstrom, "Exploitation," *Canadian Journal of Philosophy* 7 (June 1977):359.

29. Buchanan, *Marx and Justice*, p. 59.

## Chapter 6

1. For a concise and useful discussion of nineteenth century utopian socialist thought and Marx's relationship to it, see Leszak Kolakowski, *Main Currents of Marxism*, vol. 1 (Oxford: Oxford University Press, 1981), Chapter X. See also Vincent Geoghegan, *Utopianism and Marxism* (New York: Methuen & Co., 1987).

2. Marx and Engels's attitude toward the Utopian Socialists was highly ambivalent. This was obscured in their later writings where they tended to misrepresent Utopian Socialist thought to heighten the contrast with their "scientific socialism." For a useful discussion, see Geoghegan, *Utopianism and Marxism*, Chapters I and II.

3. See Kolakowski, *Main Currents of Marxism*, vol. 1, pp. 222–227.

4. See *CCPE*, p. 21. Notice the complete generality that characterizes Marx's most famous statement of the basic principles of Historical Materialism in the "Preface."

5. For a general defense of laws in social science and an account of the function of *ceteris paribus* clauses in such laws, see Harold Kincaid, "Defending Laws in the Social Sciences," *Philosophy of the Social Sciences*, forthcoming.

6. See, e.g., *CCPE*, pp. 21–22.

7. This assumes that if $x$ explains $y$, then $y$ can be inferred from $x$. Since not all inference is deductive, this assumption is unproblematic.

8. This is probably oversimplified as an account of contemporary capitalist society in which many of those who own means of production also labor as, e.g., managers. This problem can be avoided by defining the terms of the relations of

production functionally, as Marx himself sometimes seems to have done. In other words 'capitalist' and 'proletariat' refer to social roles or to individuals *qua* occupiers of social roles. Another alternative for determining the occupants of capitalist relations of production is in terms of what individuals must do to optimize. This approach is explored by John Roemer in a number of his writings on the concept of class. See, e.g., *A General Theory of Exploitation and Class* (Cambridge: Harvard University Press, 1982), Chapters II and IV. Note that our interest in this chapter is not in social classes per se (a vast and confusing topic) but only in capitalist relations of production and the occupants of the terms of the latter.

9. See above, Chapter 2, p. 33.

10. A. M. Honoré, "Ownership," in *Oxford Essays in Jurisprudence.*

11. Lawrence Becker, *Property Rights* (London: Routledge and Kegan Paul, 1977), p. 19.

12. There are a number of potential problems with explicating worker control of the means of production in terms of rights. First, it might be that rights-claims are appropriate only in class societies, since rights are boundary markers whose purpose is to adjudicate the kinds of conflicts that will disappear in PC society. This view, as an interpretation of Marx, has been articulated and defended by Allen Buchanan in his *Marx and Justice* (Totowa, N.J.: Rowman and Allenheld, 1982), Chapter IV. I return to this point in Chapter 7. Secondly, as G. A. Cohen points out, ownership is a superstructural notion, since it is defined in terms of legal property rights. If the base and superstructure are to be logically independent and if the relations of production are part of the base, then it is necessary to explicate ownership in such a way that it does not bring in superstructural notions like legal rights. Cohen deals with this problem by pointing out that (*de jure*) rights talk can be replaced by (*de facto*) *rechtsfrei* talk about corresponding powers. See G. A. Cohen, *Karl Marx's Theory of History*, (Princeton: Princeton University Press, 1978), Chapter VIII, Section (2). In other words, if a person has the right to decide how a machine will be used, one can instead talk about his power to decide how that machine shall be used. Of course, these ways of talking are not equivalent; a person can have the right to do something without having the corresponding power, and vice versa, but replacing rights talk with talk about *de facto* powers allows us to understand how the relations of production can be kept out of the superstructure. However, in explicating the concept of control of the means of production, let us follow Cohen's lead and retain rights talk for reasons of economy of exposition, all the while bearing in mind that this sort of talk can, in principle, be replaced by *rechtsfrei* characterizations in terms of powers.

13. Frederick Engels, *A-D*, MECW, vol. 25, pp. 259, 267.

14. Suggestions of this model can be found, among other places, in David Schweickart, *Capitalism or Worker Control?* (New York: Praeger Publishers, 1980); Jaroslav Vanek, *The General Theory of Labor-Managed Economies* (Ithaca, N.Y.: Cornell University Press, 1970), pp. 1–5. See also Svetojar Pejovich, *The Market-Planned Economy of Yugoslavia* (Minneapolis: University of Minnesota Press, 1966), Chapter III.

15. See Schweickart, *Capitalism or Worker Control?*, pp. 51, 71–72 on this point.

16. Peter Rutland has called this "The Method of Material Balances." See his *The Myth of the Plan* (Lasalle, Ill.: Open Court, 1985), pp. 80–81, 115–116. Since central planning does away with markets and market prices, coordination is achieved by directly assigning inputs and outputs to production units. This point is further explored in the next section.

17. By contrast, Marx need not specify how production will be organized within

production units, beyond saying that the workers will decide that question for themselves. That much is implicit in the concept of worker control of the means of production, and it is easy to envision a variety of ways in which the workers could solve this problem.

18. As noted in Chapter 3, Marx does discuss a system of production for exchange in which the workers control the means of production, what he calls 'simple commodity production'. However, as pointed out in Chapter 3 (see note 10 of that chapter), he never maintains that this is an actual, stable historical form of social organization.

19. See note 14.

21. This way of designating the two stages of PC society seems to have been first done by Lenin. See V. I. Lenin, *State and Revolution* (Moscow: Progress Publishers, 1971), Chapter V. In this book I generally avoid the terms 'socialism' and 'communism' because they are ambiguous. Such terms could refer to a set of relations of production, a mode of distribution, an economic system, or what I have earlier called a 'social vision'.

22. See *CGP*, pp. 14–18, 26.

23. Ibid., pp. 14–16.

24. Karl Marx, *Sochineniia*, vol. XIII (Moscow, n.d.), pp. 241–242; as cited in Peter Wiles, *The Political Economy of Communism* (Cambridge: Harvard University Press, 1962), p. 358.

25. Frederick Engels, *Socialism: Utopian and Scientific* (Moscow: International Publishers, 1985), pp. 70, 75. These passages can also be found in MECW, vol. 25, pp. 268 and 642.

26. Of course, any explanation consists of more than just one statement, and there may be tacit auxiliary assumptions as well. This is why the definition of 'inherently alienating' is formulated in the rather convoluted way that it is. The main purpose here is to restate those critical explanations of alienation in capitalist society that do not appeal to other essential features of capitalist society identified by Marx (e.g., lack of worker control of the means of production). In this manner, commodity production will be shown to be inherently alienating. The implication, of course, is that any system of widespread commodity production will be shown to be inherently alienating.

27. Engels excoriates Rodbertus for failing to recognize the functional necessity of competition for market economies. See Frederick Engels, "Preface to the First German Edition" in Karl Marx, *The Poverty of Philosophy* (New York: International Publishers, 1963), pp. 16ff.

28. For an articulation and defense of this interpretive claim, See Paul Craig Roberts and Matthew Stephenson, *Marx's Theory of Exchange, Alienation, and Crisis* (Stanford, Calif.: Hoover Institution Press, 1973), Chapter VI and Appendix.

29. Although it is clear that the elimination of commodity production is a sufficient condition for eliminating the mystification it causes, it is less clear that it is a necessary condition as well. For Marx, the false beliefs induced by commodity production are like mirages, which do not "go away" even if their explanation is known. Why Marx thought this is a complicated story, which cannot be adequately discussed here. For an excellent account, see Cohen, *Marx's Theory of History*, pp. 115–133; 326–338.

30. See Chapter 2, pp. 42–43.

31. A more sophisticated variant of this argument has been offered by the Hungarian economist Janos Kornai. Kornai argues that certain developments in modern production favor planning over markets, specifically, the length and com-

plexity of some production processes and economies of scale in many industries. See Janos Kornai, *Anti-Equilibrium* (Amsterdam: North Holland Publishing, 1981), pp. 339.

32. Lenin, *State and Revolution*, p. 96.

33. For a more elaborate discussion of the points in this paragraph and the next, see my "Marx and Disequilibrium in Market Socialist Relations of Production," *Economics and Philosophy* 3 (April 1987):26–42.

34. For an explanation of why noncomprehensive planning cannot achieve this result, see ibid., pp. 42–46.

## Chapter 7

1. The first part of the last section of the preceding chapter tells part of that story. However, explaining this transition is a complex matter; it must include, after all, an account of socialist revolution. Though a successful radical critique requires that the whole story be told, I shall not pursue this requirement in any detail in this book. I will, however, have something to say in the next section of this chapter about the transition from the first to the second phase of PC society.

2. On Marx's account of distribution in capitalist society, workers do not get the value of what they contribute, which is the embodied labor they give up. Instead, they get the value of what they *sell*, which is their labor power. As Marx frequently emphasizes, this distinction is mystified by the wage contract. This allows bourgeois ideologists to promulgate an "ethic of contribution" while allowing the process of surplus extraction to go on undisturbed. As Chapter 3 explains, this gap between contribution and income is the basis for Marx's charge of exploitation against capitalism.

3. For a discussion and defense of this claim, see Allen Buchanan, *Marx and Justice* (Totowa, N.J.: Rowman and Allenheld, 1982), pp. 60–69.

4. It is ironic that physical birthmarks on human beings are permanent. This discussion makes it evident that Marx did not believe that the "birthmarks" on PC society would have a comparable permanence.

5. The Famous Passage in *The German Ideology* comes readily to mind; this was discussed in detail in Chapter 1.

6. John McMurtry, *The Structure of Marx's World View* (Princeton: Princeton University Press, 1978), p. 80n.

7. Karl Marx, *Grundrisse* (New York: Random House; Penguin Books, 1973), p. 611.

8. The following ellipsis omits some peculiar metaphysical speculations about the inherently defective nature of rights, but the remainder of the quotation makes it clear that the defects are substantive.

9. Allen Wood has argued that the egalitarian overtones to Marx's description of the defects of distribution in $PC_1$ society are to be explained by the fact that Marx conceived of his audience, at least in this section, as egalitarians. See Allen Wood, "Marx and Equality," in *Analytical Marxism*, John Roemer, ed. (Cambridge: Cambridge University Press, 1986), p. 292.

10. See, e.g., *Capital* III, p. 820, and *GI*, MECW, vol. 5, p. 42.

11. See G. A. Cohen, *Marx's Theory of History* (Princeton: Princeton University Press, 1978), Chapter II, for a thorough catalogue and discussion of the forces of production.

12. The term 'incentive' might be objected to on Marx's behalf on the grounds

that the economic system of PC$_2$ society involves the *aufhebung* of all incentives. This is merely a terminological difficulty. A more neutral term might be 'motivational structure.' Some such structure must be presupposed as long as human behavior is purposive.

13. See Max Weber, *Theory of Social and Economic Organization* (New York: Free Press, 1947), p. 156.

14. This characterization of the state, as well as the one given below in Marx's other account, are called 'theories' to emphasize their explanatory import. As such, the major claims of these theories are construed to be empirical and contingent. This is occasionally at variance with Marx's use of the term 'state' ('staat') in that he sometimes conceives of the state as necessarily distinct from society (and indeed malevolent). For reasons of economy and clarity of exposition, I prefer to stick to the standard definition given in the preceding paragraph.

15. See Karl Marx, *Selected Correspondence*, 2d ed., I. Lasker, trans., and S. Ryazanskaya, ed. (Moscow: Progress Publishers, 1975), p. 318.

16. The third quotation of this chapter (*CGP*, p. 17) makes it clear that Marx regarded payment according to labor contribution as a right possessed by the workers in PC$_1$ society.

17. See Mancur Olson, *The Logic of Collective Action* (New York: Schocken Books, 1965), Chapters I and II.

18. Richard W. Miller, "Democracy and Class Dictatorship," *Social Philosophy & Policy* 3 (Spring 1986):64.

19. Actually, given the mystifying nature of ideological beliefs in capitalist and pre-capitalist societies, it may be that actual distribution need only *seem* to be in accordance with the relevant beliefs. For example, under capitalism, the wage contract makes it seem that the worker is being paid the full value of his contribution (the labor he expends) when in fact he is only paid the value of his labor power. However, the following quotation in the text suggests that Marx believes that systematic differences between appearance and reality in this connection would disappear after the revolution.

20. Jeffrie G. Murphy, " Marxism and Retribution," *Philosophy and Public Affairs* 2 (Spring 1973):217–243; reprinted in *Marx, Justice, and History*, Thomas Nagel, Marshall Cohen, and Thomas Scanlon, eds. (Princeton: Princeton University Press, 1980), p. 175.

21. This interpretation of Marx has been more fully articulated and defended by Allen Buchanan. See his *Marx and Justice*, pp. 60–62.

22. In attempting to facilitate the transition to communism, the Soviets have done very poorly in unclogging the springs of cooperative wealth; by contrast, they did an exceptional job in expediting the demise of bourgeois specialists under the leadership of Stalin.

23. See Buchanan, *Marx and Justice*, pp. 66–67.

24. Bertell Ollman, "Marx's Vision of Communism: A Reconstruction," *Critique* 8 (Summer 1977):4–41.

25. Nancy Holmstrom, "Exploitation," *Canadian Journal of Philosophy* 7 (June 1977), p. 359.

26. It is worth noting that this is separate from the question of whether or not the proletariat are in fact capitalistically exploited under capitalism. For the sake of this discussion, this has been assumed to be true, though as I argued in Chapter 5, this depends on whether or not the economic system of PC society can deliver the goods. If it cannot, that is, if what I have called 'the Monstrous Hypothesis' is

true, then the workers are not capitalistically exploited under capitalism. Moreover, if that hypothesis is true, it would seem that the workers do suffer some form of property relations exploitation in $PC_1$ society. I return to this point in Chapter 9.

27. See, for example, his description of talk about 'equal right' and 'fair distribution' as "obsolete verbal rubbish," "ideological nonsense," and most simple of all, "trash," in the *Critique of the Gotha Program* (p. 18). To be accurate, it is not at all obvious that these warm predicates are intended to describe all such notions; the context suggests that they might be intended to apply only to the ideas of the French Socialists and "other democrats." But I am not sure. In any case, I have sworn off quotation-hurling on this question.

28. Buchanan, *Marx and Justice*, p. 58.

29. See Engels, *A-D*, MECW, vol. 25, pp. 266–270.

## Chapter 8

1. See note 11 of Chapter 7.

2. Allen Wood, "Marx and Equality," in *Analytical Marxism*, John Roemer, ed. (Cambridge: Cambridge University Press, 1986), p. 296.

3. Maybe this poverty, or the failure to satisfy the MPCs this poverty entails, should be added as an independent element to Marx's radical critique of capitalist society as it was developed in Part I of this book. The Critical Explanations requirement could be easily satisfied in the manner sketched in the paragraph to which this note is appended. However, dealing with the Alternative Institutions requirement on this point would be problematic, as the main argument of this chapter will show.

4. For more or less comprehensive discussions of what normative theory might lie behind Marx's critique of capitalist society, see the following: Derek Allen, "Does Marx Have an Ethic of Self-Realization?" *Canadian Journal of Philosophy* 10 (Fall 1980):517–534; Hilliard Aronovitch, "Marxian Morality," *Canadian Journal of Philosophy* 10 (Spring 1980):357–376; George Brenkert, *Marx's Ethics of Freedom* (Boston: Routledge and Kegan Paul, 1983); George Brenkert, "Marx's Critique of Utilitarianism," in *Marx and Morality*, Kai Nielsen and Steven C. Patten, eds. Supplementary volume VII, *Canadian Journal of Philosophy* (Guelph, Ont.: Canadian Association for Publishing in Philosophy, 1981), pp. 193–220; Richard Miller, "Marx and Aristotle," ibid., pp. 323–352; Jon Elster, *Making Sense of Marx* (Cambridge: Cambridge University Press, 1985), Chapter 2.2; Alan Nasser, "Marx's Ethical Anthropology," *Philosophy and Phenomenological Research* 35 (June 1975):484–500; Thomas Wartenberg, "Species Being and Human Nature in Marx," *Human Studies* 5 (April–June 1982): 77–95.

5. If MPCs are thought of as use-values, they would have to be thought of as kinds of use-values since, for the most part, particular use-values are not material preconditions for the good life. The specification of the kind, however, would bring in all the other relevant aspects of the states of affairs as defining properties of the kind.

6. This remark is not wholly tongue in cheek. What counts as an acceptable level of risk is culturally and historically relative. See Mary Douglas and Aaron Wildavsky, *Risk and Culture* (Berkeley: University of California Press, 1982). It may be that as a society's capacity to lower risks increases, what will be tolerated as an acceptable level of risk diminishes. Recent experience seems to bear this out in that

a wide range of risks that historically had been deemed perfectly acceptable are no longer thought to be acceptable.

7. Robert Nozick, *Anarchy, State, and Utopia* (New York: Basic Books, 1974), p. 240. The present argument owes much to Nozick's illuminating discussion of the nature of self-esteem.

8. Unlike some others who have given thought to these matters (see the next note), I am not sure that high self-esteem is a necessary condition for the good life for every person. (It is even more doubtful that it is a material precondition for the good life for everyone.) People who are described as "laid back" and are also modest may not require high self-esteem to live the good life, unless the notion of high self-esteem is stretched to the point where a twinge of pride in one's modesty makes for high self-esteem. I suspect many of these people are to be found living the good life on Caribbean islands. By contrast, the avoidance of low self-esteem is more plausibly thought to be a necessary condition for the good life. To have neither high nor low self-esteem would require, in Nozick's terminology, that one score around the middle on many socially significant dimensions.

9. Recent work by Jon Elster on self-realization expands on these points. See his *Making Sense of Marx*, pp. 521–527. See also his "Self-Realization in Work and Politics: The Marxist Conception of the Good Life," *Social Philosophy and Policy* 3 (Spring 1986):101, 106.

10. These problems are identical with or connected to some of today's outstanding problems in the philosophy of economics, notably the problems involved in the normative assessment of economic systems. For a road map to these problems, see Allen Buchanan's *Ethics, Efficiency, and the Market* (Totowa, N.J.: Rowman and Allenheld, 1986), Chapter 3.

11. Aside from complaining about the existing order, radicals are fond of talking about two things: ideals and revolutionary tactics. The question they seem chronically unable to answer is 'How is all this supposed to work?'. Reformers, by contrast, are chronically unable to answer the question 'How are you going to pay for all this?'.

12. In the third section of Chapter 7 and the first section of this chapter, I claimed that the only alternative to Marxian abundance for which there is any textual evidence is an egalitarian distributive principle, which, theoretically, would allow for something less than the satisfaction of all the MPCs for everyone. However, the following argument applies to any distributive principle that falls short of Marxian abundance, whether or not it is egalitarian; another reconstruction or interpretation of Marx's distributive principle for $PC_2$ society would, therefore, succumb to this argument if the latter succeeds.

13. See John Rawls, *A Theory of Justice* (Cambridge: Harvard University Press, 1970), pp. 90–95.

14. As will become clear shortly, moderate scarcity in Hume's sense does not entail the absence of Marxian abundance.

15. See Steven Lukes, *Marxism and Morality* (Oxford: Clarendon Press, 1985), pp. 32–33; and Allen Buchanan, *Marx and Justice* (Totowa, N.J.: Rowman and Allenheld, 1982), p. 167.

16. David Hume, *A Treatise of Human Nature*, L. A. Selby-Bigge, ed. (Oxford: Oxford University Press, 1973), pp. 494–495. Hereafter, all references to Hume will be made in the text to the Selby-Bigge edition of the *Treatise*.

17. Notice that this account requires that the selfish person make interpersonal

comparisons of utilities. This may not be epistemically legitimate, but that would not prevent people from doing it.

18. All forms of status are necessarily scarce. However, the quest for status is usually less absolutistic than the text suggests. The economist Robert Frank has noticed that high local status in any respect is usually exchangeable in the sense that most people will trade relative ranking along some socially significant dimension for other goods at the margin. This fertile insight permits explanations of a wide range of phenomena in contemporary societies. See Robert Frank, *Choosing the Right Pond* (New York: Oxford University Press, 1985). Frank's insights are compatible with my claim that conceptions of the good life are at least sometimes at stake in the quest for status.

19. See Frank Parkin, *Marxism and Class Theory* (New York: Columbia University Press, 1985) for a good discussion of the explanatory failure of Marx's account of social conflict.

20. For discussions of Marx's conception of positive freedom, see the following: George Brenkert, *Marx's Ethics of Freedom* (Boston: Routledge and Kegan Paul, 1983), pp. 88–89, 213–214, 224–225, and passim; John Plamenatz, *Karl Marx's Philosophy of Man* (Oxford: Clarendon Press, 1975), pp. 170f; Steven Lukes, *Marxism and Morality*, Chapter 5; Steven Lukes, "Marxism and Dirty Hands," *Social Philosophy & Policy* 3 (Spring 1986):220f; Hilliard Aronovitch, "Marxian Morality," *Canadian Journal of Philosophy* 10 (Spring 1980):371f; Jon Elster, "Self-Realization in Work and Politics: The Marxist Conception of the Good Life," *Social Philosophy & Policy* 3 (Spring 1986):101–102. A full discussion of the Normative Theory requirement for a radical critique would pursue this aspect of Marx's thought systematically. See note 4 above.

21. Implicit in Hume's discussion of the causes of belief in Part (iii) of Book I of the *Treatise* are many interesting suggestions about how to think about human stupidity. See especially Sections 8–10 and 13.

22. See Jon Elster, "Self-Realization in Work and Politics," p. 107.

23. But compare what Descartes says at the beginning of the *Discourse on Method*: "Good sense is of all things in the world most equally distributed, for everybody thinks himself so abundantly provided with it that even those most difficult to please in all other matters do not commonly desire more of it than they already possess." Rene Descartes, *Discourse on Method* in *The Philosophical Writings of Descartes*, trans. Elizabeth S. Haldane and G. R. T. Ross (Cambridge: Cambridge University Press, 1969), p. 81.

24. For a good synthesis and survey of some of this work, see Richard E. Nisbett and Lee Ross, *Human Inference: Strategies and Shortcomings of Social Judgment*, Century Psychology Series (New York: Prentice Hall, 1980).

## Chapter 9

1. This section and parts of the third and sixth sections of this chapter are adapted from my article, "Marx, Central Planning, and Utopian Socialism," *Social Philosophy & Policy* 6 (Spring 1989):160–199.

2. Ludwig von Mises, "Economic Calculation in the Socialist Commonwealth," reprinted in *Collectivist Economic Planning*, F. A. Hayek, ed. (London: George Routledge and Sons, 1935), pp. 87–130.

3. For a useful analysis and summary of the debate, see Don Lavoie, *Rivalry or Central Planning?* (Cambridge: Cambridge University Press, 1986).

4. Both are reprinted in *On the Economic Theory of Socialism*, B. E. Lippincott, ed. (1938; reprint, New York: McGraw Hill, 1964).

5. For a discussion of the contribution of the capitalist *qua* capitalist, see the fourth section of Chapter 4.

6. On all this, see Israel M. Kirzner, *Competition and Entrepreneurship* (Chicago: University of Chicago Press, 1973), pp. 154–181.

7. This approach was inspired by the work of Piero Sraffa. See Piero Sraffa, *Production of Commodities by Means of Commodities* (Cambridge: Cambridge University Press, 1963). See also Ian Steedman, *Marx After Sraffa* (London: New Left Books, 1977). For the purposes of this chapter, the explanations characteristic of this approach are not helpful because they do not attempt to explain how prices are actually formed. That is, no effort is made to identify causal mechanisms.

8. The question of whether consumer goods should be priced, and if so, how won't be considered here.

9. Peter Rutland, *The Myth of the Plan* (LaSalle, Ill.: Open Court, 1985), pp. 114–117. See also Chapter 6, note 16 of this book and the material to which that note is appended.

10. Sometimes neo-Austrian defenders of Mises insist on the "impossibility" of a centrally planned economy by defining the latter in such a way that it can have no pricing mechanism. This is sheer obscurantism. More than anyone else, these economists should be sensitive to the demand to discuss real-world economic systems or institutions. As I use the term, an economic system is centrally planned if and only if it uses the Method of Material Balances to allocate resources. This serves to cast the issue as one of how well central planning meets the goal or purpose of any economic system, which is to serve the wants and needs of consumers by the production of use-values. Moreover, as indicated in Chapter 6, a centrally planned economy may include some production for exchange; all that is required is that the Method of Material Balances is the predominant method of allocating resources.

11. See Ludwig von Mises, *Socialism: An Economic and Sociological Analysis* (New Haven: Yale University Press, 1951), pp. 196–208.

12. See, e.g., David Schweickart, *Capitalism or Worker Control?* (New York: Praeger Publishing, 1980), pp. 219–220.

13. "It will be a kaleidic society, interspersing its moments or intervals of order, assurance, and beauty with sudden disintegration and a cascade into a new pattern. . . . It invites the analyst to consider the society as consisting of a skein of *potentiae* and to ask himself not what *will* be its course, but what the course is capable of being in case of the ascendancy of this or that ambition entertained by this or that interest." G. L. S. Shackle, *Epistemics and Economics* (Cambridge: Cambridge University Press, 1972), p. 76.

14. Joseph Schumpeter, *Capitalism, Socialism, and Democracy* (New York: Harper and Row, 1942), pp. 81–86.

15. This suggests that reproduction models, which both Marxist and neoclassical economists are so fond of, are largely irrelevant to explaining how real world economic systems actually function.

16. This assumes that no one can foresee all changes or their effects, i.e., that no one is omniscient in this connection. Of course, to the extent that the effects of change can be foreseen, they will be reflected in current factor prices. More on this below.

17. F. A. Hayek, "The Use of Knowledge in Society," *American Economic Review*

35 (September 1945):517–530; reprinted in F. A. Hayek, *Individualism and Economic Order* (London: Routledge and Kegan Paul, 1949), pp. 77–91.

18. Mises and Hayek do not presuppose perfect competition in the neoclassical sense. Under "perfect competition" there are no entrepreneurial profits at all. Everyone is a "price-taker" and all factors receive their marginal value. Paradoxically, under perfect competition, no one competes! That is, there is no rivalrous competition, as Kirzner calls it, which involves activities such as price cutting, product differentiation, etc., etc.

19. See F. A. Hayek, "Economics and Knowledge," *Economics*, vol. 4 (February 1937); reprinted in *Individualism and Economic Order*, pp. 33–56.

20. See David Ramsey Steele, "The Failure of Bolshevism and its Aftermath," *Journal of Libertarian Studies* 5 (Winter 1981):99–104, for a discussion of this point.

21. Lange, "On the Economic Theory of Socialism," pp. 59–61 (see note 4).

22. F. A. Hayek, "The Use of Knowledge in Society," p. 524.

23. Mancur Olson, *The Logic of Collective Action* (New York: Schocken Books, 1965), Chapters I and II.

24. Oskar Lange, "On the Economic Theory of Socialism," pp. 86ff.

25. For other discussions of the Socialist Calculation Debate and the inherent problems of a centrally planned economy, see the works referred to in notes 2, 3, and 9 above. Also, see Trygve J. B. Hoff, *Economic Calculation in the Socialist Society* (London: Armamento, 1949). For a less abstract, yet highly illuminating discussion of the Soviet experience, see Alec Nove, *The Economics of Feasible Socialism* (Cambridge: Cambridge University Press, 1983), Part 2.

26. I say 'might adopt' because, as far as I am able to determine, few Marxists are even aware of this problem in the manner it is posed by Mises and Hayek; among those who are, none has dealt with it at a theoretical level without retreating to some form of market socialism. Of course, nearly everyone recognizes that existing centrally planned economies have serious efficiency problems, but a correct diagnosis of this problem eludes anyone who does not see the issues in roughly the way that Mises and Hayek did. For an example of one socialist who has seen these problems in their proper light (largely independently of Mises and Hayek), see the reference to Alec Nove in the preceding note.

27. Rutland, *The Myth of the Plan*, p. 192.

28. See Don Lavoie, *National Economic Planning: What is Left?* (Washington, D.C.: The Cato Institute, 1985), pp. 76–87.

29. There have been some efforts to program computers to simulate abilities of this sort in expert systems research. Results thusfar have been disappointing. For some principled objections to this research program, see Hubert Dreyfus, *What Computers Can't Do*, rev. ed. (New York: Harper and Row, 1979).

30. A third factor is the absolute size of the population served by the economy in question. How large this population is depends in part on how "economies" are individuated. As international trade has developed in recent years, the idea of national economies is becoming increasingly irrelevant. At this time it is unclear to me how much absolute population size is responsible for complexity in the structure of production. Pre-capitalist economic systems were comprised of relatively small autarchic units; their small size was surely a factor in making a system of production for use feasible.

31. For a discussion of how a market system could function without huge income differentials, see Joseph Carens, *Equality, Moral Incentives and the Market* (Chicago: University of Chicago Press, 1981).

32. For more details on the problem of innovation in a centrally planned economy, see Rutland, *The Myth of the Plan*, pp. 146–149.

33. Indeed, there is an instructive parallel here with the development of socialist thought since Marx. Originally, socialists, including Marx and Engels, argued that the superiority of socialism lay in its huge potential for creating wealth. When actually existing socialist regimes failed in this mission, socialism was, and indeed continues to be, advocated on "noneconomic" grounds.

34. It might be thought that existing centrally planned economies provide a model for this scenario. However, they have access to prices on the world market to use as a rough guide to scarcity values, and they are limited participants in the world market. If worldwide socialist revolution resulted in one centrally planned world economy or a number of relatively autarchic centrally planned economies, the planners would be significantly worse off than existing central planners are, since there would be no market prices anywhere to serve as guides to scarcity values.

35. Actually, it may be that much more must be abstracted from, or held constant, in making these comparisons, which could lead to misgivings about Roemer's entire project. The basic problem is that this discussion involves some highly abstract cross-system comparisons. I shall ignore misgivings about such comparisons on the grounds that there does seem to be a form of exploitation in which the comparison of alternative systems is the central intuition. Roemer's general theory of property relations exploitation is the most completely articulated version of this intuition. Moreover, I hope that the kind of argument offered in this section and the next will address some of the concerns that the reader might have (and that I share) about these highly abstract cross-system comparisons.

36. Confiscation need not take the form of total expropriation. It involves what Richard Epstein calls a "taking," which is, roughly, any coerced transfer of a valuable right or power. See Richard Epstein, *Takings* (Cambridge: Harvard University Press, 1985), Chapter IV. The obvious impact that this species of insecurity of possession can and does have on the operation of a market economy, as it was outlined in the preceding two sections of this chapter, makes it clear that the exclusion of the economic systems of some Third World countries on this basis does not involve special pleading.

37. Peter Rutland's article, "Capitalism and Socialism: How Can They Be Compared?" *Social Philosophy & Policy* 6 (Autumn 1989):197–227, contains a useful discussion of the relevant empirical evidence. For a recent illuminating comparison of Soviet and Western living standards in terms of how many hours someone must work to purchase various common consumer goods, see V. Kuvarin, "Dolya tseny v zarplate," *Argumenty i fakty* 28 (1988):4ff.

38. For detailed empirical confirmation of this general claim, see Rutland, *The Myth of the Plan*, pp. 131–133; 146–149.

39. See note 35 above.

40. It might be thought this definition of marx exploitation has to be rejected if it could turn out that these groups are exploiters. However, as Jon Elster has pointed out [*Making Sense of Marx* (Cambridge: Cambridge University Press, 1985), p. 203], no form of property relations exploitation is causal because the relevant counterfactuals neither entail nor are entailed by a causal claim. The relevance of this point in the present context is that if the very young, etc., are in the exploiting coalition, it does not follow that they are *causing* the exploitation of the workers. All that follows, if indeed these groups are in an exploiting coalition, is that they

are differentially benefiting from the economic system of $PC_1$ society. To put the point in general terms, if a coalition is property relations exploited (in any of its exemplifications), it does not follow that there is anyone causing the exploitation.

41. See Frederick Engels, *A-D*, MECW, vol. 25, pp. 264–271.

42. The reader might wonder why the discussion of central planning was not brought to bear against Marx's vision of $PC_2$ society. After all, it too is supposed to have a centrally planned economy. The reason this was not done was because, in the absence of the discussion of the primary evils, it is not at all obvious that the forces of production, etc., could not develop to the extent that the economic sphere would "subside." In that never-never land, central planning might be adequate to the task of coordinating production.

43. For a discussion of this point at a less abstract level, see Lavoie, *National Economic Planning: What is Left?*

44. As in Chapter 4, I follow Nancy Holmstrom's account of surplus value exploitation, which seems to capture best Marx's intentions. See Nancy Holmstrom, "Exploitation," *Canadian Journal of Philosophy* 7 (June 1977):359.

45. Note that this shows that the rate of economic growth could not be subject to a vote, since that rate depends on how well production is coordinated. For reasons discussed earlier in this chapter, central planning cannot do a very good job on this, though it remains indeterminate (for us and for them) how poorly they will do. What the voters will vote on is a wish list with a projected price tag. There are obviously horrendous problems in all of this, but I overlook them with the intention of getting at the central issues for surplus value exploitation in $PC_1$ society.

46. It might be objected that, by staying in $PC_1$ society, they have implicitly agreed to abide by the results of the vote or the decisions of the planners. This contractarian objection is surely far from anything that Marx would endorse. More to the point, "constitutional" consent is perfectly consistent with being forced to contribute surplus value on specific projects, and the latter is what is at stake here.

47. Leszek Kolakowski, "The Myth of Human Self-Identity: Unity of Civil and Political Society in Socialist Thought," in *The Socialist Idea: A Reappraisal*, Leszek Kolakowski and Stuart Hampshire, eds. (New York: Basic Books, 1974), pp. 32–33.

48. This has been a concern of critics of Marxism since the Bolshevik revolution. For the most up to date discussion of this problem, see Allen Buchanan, "The Marxist Conceptual Framework and the Origins of Totalitarian State Socialism," *Social Philosophy & Policy* 3 (Spring 1986):127–145.

## Chapter 10

1. David Schweickart, *Capitalism or Worker Control?* (New York: Praeger Publishing, 1980), pp. 48–55. Schweickart's discussion is especially useful since he is willing to discuss actual institutions in a way that answers to some of the concerns raised in this book. It is primarily for this reason that I have devoted so much attention to his work. (See the next note.) In what follows I do not mean to suggest that Schweickart conceives of himself as offering a radical critique of capitalist society, at least as the latter has been defined here.

2. N. Scott Arnold, "Marx and Disequilibrium in Market Socialist Relations of Production," *Economics and Philosophy* 3 (April 1987):23–47. This article occasioned a two-round exchange between the author and Professor Schweickart and a separate

exchange with Louis Putterman. These exchanges can be found in the October 1987 and April 1988 issues, respectively, of *Economics and Philosophy*.

3. Schweickart, *Capitalism or Worker Control?*, p. 143.

4. See, for example, D. D. Milenkovitch, *Plan and Market in Yugoslav Economic Thought* (New Haven: Yale University Press, 1971), p. 115.

5. This point was made by Hayek over forty years ago. See F. A. Hayek, *The Road to Serfdom* (Chicago: Phoenix Books, University of Chicago Press, 1944), especially Chapter IV, "Planning and the Rule of Law."

6. A vast literature on this has developed over the past quarter of a century inspired by the seminal work of James Buchanan and Gordon Tullock. The classic statement of Public Choice principles is their book, *The Calculus of Consent* (Ann Arbor: University of Michigan Press, 1962). For a recent statement, see James M. Buchanan and Geoffrey Brennan, *The Reason of Rules: Constitutional Political Economy* (Cambridge: Cambridge University Press, 1985), especially Chapter IV.

7. John Gray, "Contractarian Method, Private Property, and the Market Economy," in *Liberalisms: Essays in Political Philosophy* (London: Routledge and Kegan Paul), forthcoming.

8. This is the intent of Jaroslav Vanek's *The General Theory of Labor-Managed Economies* (Ithaca, N.Y.: Cornell University Press, 1970).

9. In all of his theoretical writings James Buchanan has gone to great lengths to make the least restrictive and most plausible assumptions possible. See, for example, Buchanan and Brennan, *The Reason of Rules*, p. 65.

10. I have not discussed the egalitarian aspect of a radical market socialist critique of capitalism. This would require critical explanations of various objectionable forms of inequality under capitalism, a normative theory explaining why these inequalities are objectionable, and an explanation of how the institutions of market socialism would avoid them. An investigation of these elements of a radical market socialist critique goes far beyond what can be undertaken here.

11. Schweickart's discussion of state control of new investment is always in terms of intended functions with no account of the mechanisms by which any of this operates. (See, for example, *Capitalism or Worker Control?*, pp. 73–74; 80; 143–144; 152.) The failure to provide such mechanisms obscures the very real possibility that the actual function of the institution may systematically diverge from its intended function (purpose).

12. This literature is surveyed in my "Recent Work on Marx: A Critical Survey," *American Philosophical Quarterly* 24 (October 1987):277–293.

13. David Hume, "Of Miracles," in *Enquiry Concerning Human Understanding*, L. A. Selby-Bigge, ed. Rev. 3d ed. with notes by P. H. Nidditch (Oxford: Oxford University Press, 1975), p. 110.

# BIBLIOGRAPHY

## Books

Avineri, Shlomo. *The Social and Political Thought of Karl Marx*. Cambridge: Cambridge University Press, 1968.
Becker, Lawrence. *Property Rights*. London: Routledge and Kegan Paul, 1977.
———. *Reciprocity*. London: Routledge and Kegan Paul, 1986.
Böhm-Bawerk, Eugen. *Karl Marx and the Close of His System*. 1896. Edited by Paul M. Sweezy. London: Merlin Press, 1974.
———. *Capital and Interest*. 3 vols. 1902. Edited and translated by G. H. Huncke and H. F. Sennholz. South Holland, Ill.: Libertarian Press, 1959.
Brenkert, George. *Marx's Ethics of Freedom*. London: Routledge and Kegan Paul, 1983.
Buchanan, Allen. *Marx and Justice*. Totowa, N.J.: Rowman and Allenheld, 1982.
———. *Ethics, Efficiency and the Market*. Totowa, N.J.: Rowman and Allenheld, 1985.
Buchanan, James M. and Gordon Tullock. *The Calculus of Consent*. Ann Arbor: University of Michigan Press, 1962.
Buchanan, James M. and Geoffrey Brennan. *The Reason of Rules: Constitutional Political Economy*. Cambridge: Cambridge University Press, 1985.
Burke, John P., Lawrence Crocker, and Lyman Legters, eds. *Marxism and the Good Society*. Cambridge: Cambridge University Press, 1981.
Carens, Joseph. *Equality, Moral Incentives and the Market*. Chicago: University of Chicago Press, 1981.
Carnap, Rudolph. *The Logical Foundations of Probability*. Chicago: University of Chicago Press, 1950.
Cohen, G. A. *Karl Marx's Theory of History: A Defence*. Princeton: Princeton University Press, 1978.
Cohen, Marshall, Thomas Scanlon, and Ernest Nagel, eds. *Marx, Justice, and History*. A Philosophy and Public Affairs Reader. Princeton: Princeton University Press, 1980.
Descartes, René. *Discourse on Method*. In *The Philosophical Writings of Descartes*, trans-

lated by Elizabeth S. Haldane and G. R. T. Ross. Cambridge: Cambridge University Press, 1969.

Douglas, Mary and Aaron Wildavsky. *Risk and Culture*. Berkeley: University of California Press, 1982.

Dreyfus, Hubert. *What Computers Can't Do*. Rev. ed. New York: Harper and Row, 1979.

Elster, Jon. *Making Sense of Marx*. Cambridge: Cambridge University Press, 1985.

Engels, Frederick. *Socialism: Utopian and Scientific*. Moscow: International Publishers, 1985.

Epstein, Richard. *Takings*. Cambridge: Harvard University Press, 1985.

Fetter, Frank. *Capital, Interest, and Rent*. Edited by Murray N. Rothbard. Kansas City, Kans.: Sheed, Andrews and McMeel, 1977.

Frank, Robert. *Choosing the Right Pond*. New York: Oxford University Press, 1985.

Friedman, Milton. *Essays in Positive Economics*. Chicago: University of Chicago Press, 1953.

Giere, Ronald. *Understanding Scientific Reasoning*. New York: Holt, Rinehart, and Winston, 1979.

Geoghegan, Vincent. *Utopianism and Marxism*. New York: Methuen & Co., 1987.

Gray, John. *Liberalisms: Essays in Political Philosophy*. London: Routledge and Kegan Paul, forthcoming.

Hayek, F. A., ed. *Collectivist Economic Planning*. London: George Routledge and Sons, 1935.

————. *The Road to Serfdom*. Chicago: Phoenix Books, University of Chicago Press, 1944.

————. *Individualism and Economic Order*. London: Routledge and Kegan Paul, 1949.

Hindess, Barry and Paul Hirst, et al. *Studies in Marxist Philosophy*. Vol. 4. New York: Mepham, 1974.

Hoff, Trygve J. B. *Economic Calculation in the Socialist Society*. London: Armamento, 1949.

Hume, David. *A Treatise of Human Nature*. Edited by L. A. Selby-Bigge. Oxford: Oxford University Press, 1973.

Kirzner, Israel M. *Competition and Entrepreneurship*. Chicago: University of Chicago Press, 1973.

————. *Perception, Opportunity and Profit*. Chicago: University of Chicago Press, 1979.

Kolakowski, Lezsek. *Main Currents of Marxism*. 3 vols. Oxford: Oxford University Press, 1981.

Kornai, Janos. *Anti-Equilibrium*. Amsterdam: North Holland Publishing, 1981.

Lavoie, Don. *National Economic Planning: What is Left?* Washington, D.C.: The Cato Institute, 1985.

————. *Rivalry or Central Planning?* Cambridge: Cambridge University Press, 1986.

Lenin, V. I. *State and Revolution*. Moscow: Progress Publishers, 1971.

Little, Daniel. *The Scientific Marx*. Minneapolis: University of Minnesota Press, 1986.

Lukes, Steven. *Marxism and Morality*. Oxford: Clarendon Press, 1985.

Marx, Karl. *Capital*. Vol. I. Moscow: Progress Publishers, 1977.

————. *Capital*. Vol. II. New York: International Publishers, 1967.

————. *Capital*. Vol. III. New York: International Publishers, 1967.

————. *Critique of the Gotha Program*. Moscow: Progress Publishers, 1971.

————. *The Eighteenth Brumaire of Louis Bonaparte*. New York: International Publishers, 1963.

————. *Grundrisse*. New York: Random House, Penguin Books, 1973.

————. *Selected Correspondence*. 2d ed. Translated by I. Lasker, and edited by S. Ryazanskaya. Moscow: Progress Publishers, 1975.

Marx, Karl and Frederick Engels, *Collected Works*. 50 volumes, projected. Moscow: Progress Publishers; New York: International Publishers; London: Lawrence and Wishart. Various dates.

McMurtry, John. *The Structure of Marx's World View*. Princeton: Princeton University Press, 1978.

Milenkovitch, D. D. *Plan and Market in Yugoslav Economic Thought*. New Haven: Yale University Press, 1971.

Miller, Richard. *Analyzing Marx*. Princeton: Princeton University Press, 1984.

von Mises, Ludwig. *Socialism: An Economic and Sociological Analysis*. New Haven: Yale University Press, 1951.

Nisbett, Richard E. and Lee Ross. *Human Inference: Strategies and Shortcomings of Social Judgment*. Century Psychology Series. New York: Prentice Hall, 1980.

Nove, Alec. *The Economics of Feasible Socialism*. Cambridge: Cambridge University Press, 1983.

Nozick, Robert. *Anarchy, State, and Utopia*. New York: Basic Books, 1974.

Olson, Mancur. *The Logic of Collective Action*. New York: Schocken Books, 1965.

Parkin, Frank. *Marxism and Class Theory*. New York: Columbia University Press, 1985.

Pejovich, Svetojar. *The Market-Planned Economy of Yugoslavia*. Minneapolis: University of Minnesota Press, 1966.

Plamenatz, John. *Karl Marx's Philosophy of Man*. Oxford: Clarendon Press, 1975.

Rawls, John. *A Theory of Justice*. Cambridge: Harvard University Press, 1970.

Roberts, Paul Craig and Matthew Stephenson. *Marx's Theory of Exchange, Alienation, and Crisis*. Stanford, Calif.: Hoover Institution Press, 1973.

Roemer, John. *A General Theory of Exploitation and Class*. Cambridge: Harvard University Press, 1982.

————, ed. *Analytical Marxism*. Cambridge: Cambridge University Press, 1986.

Rutland, Peter. *The Myth of the Plan*. Lasalle, Ill.: Open Court, 1985.

Samuelson, Paul A. *Economics*. 7th ed. New York: McGraw Hill, 1967.

Schumpeter, Joseph. *Capitalism, Socialism, and Democracy*. New York: Harper and Row, 1942.

Schweickart, David. *Capitalism or Worker Control?* New York: Praeger Publishing, 1980.

Shackle, G. L. S. *Epistemics and Economics*. Cambridge: Cambridge University Press, 1972.

Smith, Adam. *The Wealth of Nations*. 1776. Reprint. New York: Random House, 1985.

Sraffa, Piero. *The Production of Commodities by Means of Commodities*. Cambridge: Cambridge University Press, 1963.

Steedman, Ian. *Marx After Sraffa*. London: New Left Books, 1977.

Tucker, Robert. *The Marxian Revolutionary Idea*. New York: W.W. Norton, 1969.

Vanek, Jaroslav. *The General Theory of Labor-Managed Economies*. Ithaca, N.Y.: Cornell University Press, 1970.

Weber, Max. *Theory of Social and Economic Organization*. New York: Free Press, 1947.

Wiles, Peter. *The Political Economy of Communism*. Cambridge: Harvard University Press, 1962.

Wolff, Robert Paul. *Understanding Marx*. Princeton: Princeton University Press, 1984.

## Articles and Correspondence

Allen, Derek H. P. "Does Marx Have an Ethic of Self-Realization?" *Canadian Journal of Philosophy* 10 (Fall 1980):517–534.

———. "Marx and Engels on the Distributive Justice of Capitalism." In *Marx and Morality*, edited by Kai Nielsen and Steven C. Patten, 221–250. Supplementary volume VII, *Canadian Journal of Philosophy*. Guelph, Ont.: Canadian Association for Publishing in Philosophy, 1981.

Arnold, N. Scott. "Capitalists and the Ethics of Contribution." *Canadian Journal of Philosophy* 15 (March 1985):87–102.

———. "Why Profits Are Deserved." *Ethics* 97 (January 1987):387–402.

———. "Marx and Disequilibrium in Market Socialist Relations of Production." *Economics and Philosophy* 3 (April 1987):23–48.

———. "Final Reply to Professor Schweickart." *Economics and Philosophy* 3 (October 1987):335–338.

———. "Further Thoughts on the Degeneration of Market Socialism." *Economics and Philosophy* 3 (October 1987):320–330.

———. "Recent Work on Marx: A Critical Survey." *American Philosophical Quarterly* 24 (October 1987):277–293.

———. "Reply to Professor Putterman." *Economics and Philosophy* 4 (April 1988): 338–340.

———. "Marx, Central Planning, and Utopian Socialism." *Social Philosophy & Policy* 6 (Spring 1989):160–199.

Aronovitch, Hilliard. "Marxian Morality." *Canadian Journal of Philosophy* 10 (September 1980):357–376.

Böhm-Bawerk, Eugen. "Unresolved Contradiction in the Marxian Economic System." 1896. In *Shorter Classics of Eugen Böhm-Bawerk*. Vol. 1. Translated by Alice Macdonald. South Holland, Ill.: Libertarian Press, 1962.

Brenkert, George. "Marx's Critique of Utilitarianism." In *Marx and Morality*, edited by Kai Nielsen and Steven C. Patten, 193–220. Supplementary volume VII, *Canadian Journal of Philosophy*. Guelph, Ont.: Canadian Association for Publishing in Philosophy, 1981.

———. "Cohen on Proletarian Unfreedom." *Philosophy and Public Affairs* 14 (Winter 1985):91–98.

Brenner, Robert. "The Social Basis of Economic Development." In *Analytical Marxism*, 23–53. See Roemer 1986.

Buchanan, Allen. "The Marxist Conceptual Framework and the Origins of Totalitarian State Socialism." *Social Philosophy & Policy* 3 (Spring 1986):127–145.

Catephores, G. and M. Morishima. "Is There an Historical 'Transformation Problem'?" *The Economic Journal* 85 (June 1975):309–335.

Cohen, G. A. "The Labor Theory of Value and the Concept of Exploitation." *Philosophy and Public Affairs* 8 (Summer 1979):338–360. Also in *Marx, Justice, and History*, 135–158. See M. Cohen, et al. 1980.

———. "The Structure of Proletarian Unfreedom." *Philosophy and Public Affairs* 12 (Winter 1983):3–33.

———. Review of *Karl Marx*, by Allen Wood. *Mind* 92 (July 1983):440–445.

———. "More on Exploitation and the Labor Theory of Value." *Inquiry* 26 (September 1983):309–332.

DiQuattro, Arthur. "Alienation and Justice in the Market." In *Marxism and the Good Society*, 121–156. See Burke, et al. 1981.

———. "Value, Class, and Exploitation." *Social Theory and Practice* 10 (Spring 1984):55–80.

Elster, Jon. "Self-Realization in Work and Politics: The Marxist Conception of the Good Life." *Social Philosophy & Policy* 3 (Spring 1986):97–126.

Engels, Frederick. "Preface to the First German Edition." In Karl Marx, *The Poverty of Philosophy*, pp. 7–24. New York: International Publishers, 1963.

———. "Karl Marx." In Karl Marx and Frederick Engels. *Selected Works*. Vol. 3, 77–87. Moscow: Progress Publishers, 1970.

Friedman, Milton. "The Methodology of Positive Economics." In *Philosophy and Economic Theory*, edited by Frank Hahn and Martin Hollis, pp. 18–35. Oxford: Oxford University Press, 1979.

Gray, John. "Against Cohen on Proletarian Unfreedom," *Social Philosophy & Policy* 6 (Fall 1988):77–112.

———. "Contractarian Method, Private Property, and the Market Economy." In *Liberalisms: Essays in Political Philosophy*. London: Routledge and Kegan Paul, forthcoming.

Hayek, F. A. "Economics and Knowledge." *Economics* 4 (February 1937). Reprinted in *Individualism and Economic Order*, edited by F. A. Hayek, pp. 33–56. London: Routledge and Kegan Paul, 1949.

———. "The Use of Knowledge in Society." *American Economic Review* 35 (September 1945):519–530. Reprinted in *Individualism and Economic Order*, edited by F. A. Hayek, pp. 77–91. London: Routledge and Kegan Paul, 1949.

Holmstrom, Nancy. "Exploitation." *Canadian Journal of Philosophy* 7 (June 1977): 353–369.

———. "Marx and Cohen on Exploitation and the Labor Theory of Value." *Inquiry* 26 (September 1983):287–308.

Honoré, A. M. "Ownership." In *Oxford Essays in Jurisprudence*.

Hume, David. "Of Miracles." In *Enquiry Concerning Human Understanding*, edited by L. A. Selby-Bigge, rev. 3d ed. with notes by P. H. Nidditch. Oxford: Oxford University Press, 1975.

Husami, Ziyad. "Marx on Distributive Justice." *Philosophy and Public Affairs* 8 (Fall 1978):27–64. Also in *Marx, Justice, and History*, 42–79. See M. Cohen, et al. 1980.

Kincaid, Harold. "Reduction, Explanation, and Individualism." *Philosophy of Science* 53 (December 1986):492–513.

———. "Defending Laws in the Social Sciences." *Philosophy of the Social Sciences*, forthcoming.

Kolakowski, Leszek. "The Myth of Human Self-Identity: Unity of Civil and Political Society in Socialist Thought." In *The Socialist Idea: A Reappraisal*, edited by Leszek Kolakowski and Stuart Hampshire, 18–35. New York: Basic Books, 1974.

Kuvarin, V. "Dolya tseny v zarplate." *Argumenty i fakty* 28 (1988).

Lange, Oskar. "On the Economic Theory of Socialism." In *On the Economic Theory of Socialism*, edited by B. E. Lippincott, 55–130. Minneapolis: University of Minnesota Press, 1938.

Leibenstein, Harold. "Allocational Efficiency vs. 'X-Efficiency'." *American Economic Review* 56 (1966):392–415.

Lukes, Steven. "Marxism and Dirty Hands." *Social Philosophy & Policy* 3 (Spring 1986):204–223.

Marx to Bracke, 5 May 1875. In Marx, Karl. *Critique of the Gotha Program.* Moscow: Progress Publishers, 1971.

Miller, Richard. "Marx and Aristotle: A Kind of Consequentialism." In *Marx and Morality,* edited by Kai Nielsen and Steven C. Patten, 323–352. Supplementary volume VII, *Canadian Journal of Philosophy.* Guelph, Ont.: Canadian Association for Publishing in Philosophy, 1981.

———. "Democracy and Class Dictatorship." *Social Philosophy & Policy* 3 (Spring 1986):59–76.

von Mises, Ludwig. "Economic Calculation in the Socialist Commonwealth." Reprinted in *Collectivist Economic Planning,* edited by F. A. Hayek, 87–130. London: Routledge and Kegan Paul, 1935.

Murphy, Jeffrie G. "Marxism and Retribution." *Philosophy and Public Affairs* 2 (Spring 1973):217–243. Also in *Marx, Justice, and History,* 158–184. See M. Cohen, et al. 1980.

Nasser, Alan. "Marx's Ethical Anthropology." *Philosophy and Phenomenological Research* 35 (June 1975):484–500.

Ollman, Bertell. "Marx's Vision of Communism: A Reconstruction." *Critique* 8 (Summer 1977):4–41.

Reiman, Jeffrey. "Exploitation, Force and the Moral Assessment of Capitalism: Thoughts on Roemer and Cohen." *Philosophy and Public Affairs* 16 (Winter 1987):3–41.

Roemer, John. "Property Relations vs. Surplus Value in Marxian Exploitation." *Philosophy and Public Affairs* 11 (Fall 1982):281–314.

———. "New Directions in the Marxian Theory of Exploitation and Class." In *Analytical Marxism,* edited by John Roemer, 81–113. Cambridge: Cambridge University Press, 1986.

Rutland, Peter. "Capitalism and Socialism: How Can They be Compared?" *Social Philosophy & Policy* 6 (Autumn 19880:197–227.

Ryan, Cheyney. "Socialist Justice and the Right to the Labor Product." *Political Theory* 8 (November 1980):503–524.

Steele, David Ramsey. "The Failure of Bolshevism and its Aftermath." *Journal of Libertarian Studies* 5 (Winter 1981):99–111.

Taylor, Fred M. "The Guidance of Production in the Socialist State." In *On the Economic Theory of Socialism,* edited by B. E. Lippincott, 55–130. Minneapolis: University of Minnesota Press, 1938.

van de Veer, Donald. "Marx's View of Justice." *Philosophy and Phenomenological Research* 33 (March 1973):366–386.

Wartenberg, Thomas. "Species Being and Human Nature in Marx." *Human Studies* 5 (April–June 1982):77–95.

Wood, Allen. "The Marxian Critique of Justice." *Philosophy and Public Affairs* 3 (Summer 1972):244–282. Also in *Marx, Justice, and History,* 3–41. See M. Cohen, et al. 1980.

———. "Marx on Right and Justice: A Reply to Husami." *Philosophy and Public Affairs* 8 (Spring 1979):267–295. Also in *Marx, Justice, and History,* 106–134. See M. Cohen, et al. 1980.

———. "Marx and Equality." In *Analytical Marxism,* 283–303. See Roemer 1986.

Young, Gary. "Justice and Capitalist Production: Marx on Bourgeois Ideology." *Canadian Journal of Philosophy* 8 (September 1978):421–457.

———. "Doing Marx Justice." In *Marx and Morality,* edited by Kai Nielsen and Steven C. Patten, 251–268. Supplementary volume VII, *Canadian Journal of Philosophy.* Guelph, Ont.: Canadian Association for Publishing in Philosophy, 1981.

# Name Index

# Subject Index